The Book of Infinity

Joel David Hamkins

The MIT Press
Cambridge, Massachusetts
London, England

The MIT Press
Massachusetts Institute of Technology
77 Massachusetts Avenue, Cambridge, MA 02139
mitpress.mit.edu

This book was set in Times Roman by the author using LaTeX and TikZ. Printed and bound in the United States of America.

Library of Congress Cataloging-in-Publication Data

Names: Hamkins, Joel David author
Title: The book of infinity / Joel David Hamkins.
Description: Cambridge, Massachusetts : The MIT Press, [2026] | Includes bibliographical references and index.
Identifiers: LCCN 2025048306 (print) | LCCN 2025048307 (ebook) | ISBN 9780262054010 paperback | ISBN 9780262054027 pdf | ISBN 9780262054034 epub
Subjects: LCSH: Infinite | LCGFT: Textbooks
Classification: LCC BD411 .H36 2026 (print) | LCC BD411 (ebook)
LC record available at https://lccn.loc.gov/2025048306
LC ebook record available at https://lccn.loc.gov/2025048307

10 9 8 7 6 5 4 3 2 1

EU Authorised Representative:
Easy Access System Europe, Mustamäe tee 50, 10621 Tallinn, Estonia |
Email: gpsr.requests@easproject.com

The Book of Infinity

Dedicated to Horatio and Hypatia, with infinite love

Contents

Preface

Come, let us explore infinity! We shall visit all my favorite paradoxes and conundrums. The ancient puzzles, confounding or intractable, will yield at times to our analysis. And what a joy it is to experience those *Aha!* moments—a flash of clarity lights the way out of the labyrinth. But alas, having escaped one maze, we shall often find ourselves immediately lost in another—a new paradox with new questions to answer. The puzzles of infinity are endless riddles nestled within one another.

When I first came from Oxford to the University of Notre Dame, I was asked by my new department what course I would most want to teach. My answer was a new course on infinity that I had long dreamed about—what fun it would be to share my ideas and puzzles with enthusiastic students, tracing the concept from ancient times to contemporary issues. I set furiously to work preparing this book, a series of vignettes on infinity, and we offered the course in spring 2023.

I was intrigued by Substack, a new way of publishing that enabled a more direct connection between author and reader, and so, as an experiment, I serialized my chapters as they were completed on my substack, Infinitely More (www.infinitelymore.xyz), just in time each week for the students. The experiment and the course were both a success, and we ran the Infinity course again in fall 2024—this will clearly be a regular offering. How pleased I am that, with new chapters and a thorough polishing, the book is now being published with the MIT Press.

In twenty-six chapters, we shall encounter all manner of infinity and paradox—Zeno's paradox, supertasks, the paradox of giants, the infinite coastline paradox; we shall roll out Aristotle's wheel and make a holiday at Hilbert's Grand Hotel; we shall climb into the ordinal numbers and submerge ourselves in the bubbling froth of the surreal numbers. Are you troubled by the paradox of the largest tweetable number? Or the infinite liar paradox? We shall prove Georg Cantor's theorems on countability and uncountability. We shall undertake mathematical thought experiments aimed at settling the continuum hypothesis and see how the axiom of choice enables the nearly perfect prediction process. And much more.

It will be clear before long that I aspire in this book to whimsy, and also to a serious, scholarly investigation. I approach paradox with a sense of wonder and curiosity, to be sure, but I carry readily at hand the rigorous analytical tools of an expert. Many chapters open with an absurd or fantastical premise, upon which I aim to build a serious philosophical or mathematical account.

Advice for the Reader—About Mathematical Prerequisites

This book is aimed at a general, intellectually curious reader—anyone who wonders about infinity. I have tried not to presume any particular mathematical or philosophical background. Nevertheless, one must recognize that any study of infinity will engage seriously with mathematical ideas. My readers have varied mathematical backgrounds, and I aim to provide something for everyone, but it is a delicate task to communicate simultaneously on many levels with diverse readers. Chapters generally focus on the key elementary ideas or questions initially and build in sophistication as the chapter progresses. I have strived to treat the deeper ideas with a soft touch that makes it possible even for beginners to gain some insight. Even when things seem unfamiliar, therefore, I encourage you to give it a go! But also, please feel free to skip over any bits that are truly too difficult. I can let you in on a secret: This is also how mathematicians read mathematics papers—one seldom understands everything on the first pass.

Advice to the Instructor

I wrote this book for my Infinity lecture course at the University of Notre Dame, an introductory course open to students of all majors and fulfilling a general university philosophy requirement, and I find the book suitable for any kind of introductory course on infinity, whether in philosophy or mathematics. Each chapter serves as the basis for one lecture, but in my own teaching, I have usually given a few chapters extra attention, such as chapters 13 and 17, and many chapters have sufficient material for multiple lectures. The chapters vary in difficulty and mathematical sophistication, but there is plenty of material for you to design your own course.

A Note on the Illustrations

The book contains two hundred illustrative color figures created by the author using TikZ in LaTeX specifically for this book, and in several instances also for my previous books *Lectures on the Philosophy of Mathematics* (2021) and *Proof and the Art of Mathematics* (2020), both available from the MIT Press. Other illustrations include the map of Europe on page 63 (released with CC3.0 license) and of the coastline of Great Britain in figure 6.1, as well as the image of the painting of the *RMS Aquitania* on page 158 and the image of the painting *Buried Treasure* by Howard Pyle on page 358, the latter of which are in the public domain and available on the Wikimedia Commons. The illustration of the stock traders of the platonic realm on page 321 and that of the blue-eyed islander on page 356 were both created by the author using DALL·E.

1 The Book of Numbers

Abstract. Shall we explore a number puzzle? All the numbers, placed into a curious order.

1.1 A Curious Order on the Natural Numbers

Consider the natural numbers up to ten in the following curious order:

8 5 4 9 1 7 6 10 3 2 0

What order is this? Is it random? No—I have placed these numbers in a very definite and logical order, using an underlying order idea with which I am sure that you, gentle reader, are familiar. Can you discover that underlying rule? Surely one often gets insight into a problem by thinking about it from different perspectives, so perhaps it may help for me to write out the numbers in a different manner, like this:

eight, five, four, nine, one, seven, six, ten, three, two, zero

OK, well ...how could that possibly help? Am I trolling? Perhaps not *every* new perspective is helpful.

Let me mention that the underlying order idea here applies just as much with larger numbers, and we can extend the list to the numbers up to one hundred, or one thousand, or one million. The new larger numbers will generally appear between and amongst the original numbers, however, rather than after them. Here are the numbers up to one hundred, for example, ordered according to the same underlying rule, with the original numbers up to ten appearing in the same order as they did before, highlighted in bold:

8 18 80 88 85 84 89 81 87 86 83 82 11 15 50 58 55 54 59 51 57 56 53 52 **5** 40 48 45 44 49 41 47 46 43 42 **4** 14 **9** 19 90 98 95 94 99 91 97 96 93 92 **1** 100 **7** 17 70 78 75 74 79 71 77 76 73 72 **6** 16 60 68 65 64 69 61 67 66 63 62 **10** 13 30 38 35 34 39 31 37 36 33 32 **3** 12 20 28 25 24 29 21 27 26 23 22 **2** **0**

What is this crazy order? Observe carefully and think about it before turning the page.

Did you find the underlying rule? Let me provide a further hint by making the same trolling suggestion of writing the numbers out in words. Does that help?

eight, eighteen, eighty, eighty-eight, eighty-five, eighty-four, eighty-nine, eighty-one, eighty-seven, eighty-six, eighty-three, eighty-two, eleven, fifteen, fifty, fifty-eight, fifty-five, fifty-four, fifty-nine, fifty-one, fifty-seven, fifty-six, fifty-three, fifty-two, **five**, forty, forty-eight, forty-five, forty-four, forty-nine, forty-one, forty-seven, forty-six, forty-three, forty-two, **four**, fourteen, **nine**, nineteen, ninety, ninety-eight, ninety-five, ninety-four, ninety-nine, ninety-one, ninety-seven, ninety-six, ninety-three, ninety-two, **one**, one hundred, **seven**, seventeen, seventy, seventy-eight, seventy-five, seventy-four, seventy-nine, seventy-one, seventy-seven, seventy-six, seventy-three, seventy-two, **six**, sixteen, sixty, sixty-eight, sixty-five, sixty-four, sixty-nine, sixty-one, sixty-seven, sixty-six, sixty-three, sixty-two, **ten**, thirteen, thirty, thirty-eight, thirty-five, thirty-four, thirty-nine, thirty-one, thirty-seven, thirty-six, thirty-three, thirty-two, **three**, twelve, twenty, twenty-eight, twenty-five, twenty-four, twenty-nine, twenty-one, twenty-seven, twenty-six, twenty-three, twenty-two, **two**, **zero**.

What is this order?

Think about it...

So I was not trolling with the hints at all, for the list is simply the number words in alphabetical order. Thus, *eight* is first, and then the other various numbers whose names start with the letter *e*, followed by the numbers having names starting with the letter *f*, and so on finally to *zero*, which is the very last number, as it is the only one starting with *z*.

We could just as easily alphabetize the numbers up to one thousand, or one million, one billion, and so on, and again the new larger numbers would appear interspersed amongst the numbers that we already have. For example, the number *seven hundred two* would appear between *seven* and *seventeen*, using the usual dictionary custom that extensions of a short word appear after it, so that *zookeeper* appears alphabetically after *zoo*.

The *Book of Numbers*

Let us consider in our imagination the *Book of Numbers*, the number dictionary consisting of all the natural numbers, placed in alphabetical order. The ever larger numbers will appear eternally between and amongst the smaller numbers already considered.

Question 1. What is the nature and order type of the set of natural numbers when placed in alphabetical order? What is the nature of the *Book of Numbers*?

Will the order be discrete, with each number having a next number and a previous number on the list? Or will it be densely ordered like the rational numbers are, with any two numbers having another between them? Will the book partition naturally into chapters?

The order type will not be the same as the usual order on the set of natural numbers $\mathbb{N}$, of course, since there will be infinitely many numbers starting with the letter *f*, as in forty, four hundred, four million, four thousand, and so on, and these will all alphabetically precede two hundred, two million, two thousand, and so on. In short, some pages in the *Book of Numbers* will have infinitely many preceding pages. Clearly, the nature of the order will depend on syntactic features of the naming convention. But for that matter, what actually is the naming convention for extremely large numbers?

1.2 Number Nomenclature

In order to put all the numbers in alphabetical order, we shall need to know their names. But does every natural number have a definite name? What are the names of the extremely large numbers? Of course, we are all familiar with the standard nomenclature for the smallish natural numbers, such as

> one, two, three, …, one hundred, one hundred one, …, fifteen thousand two hundred forty-nine, …

And we all know that the number names continue much higher, using such names as

> one thousand, one million, one billion, one trillion, …

But what happens after that? Perhaps some of you are familiar with the number googol and perhaps googolplex, which we shall discuss in more detail in chapter 8. But this would be skipping a lot of numbers between one trillion and googol. What are their names?

Question 2. Does every natural number have a determinate name?

The standard American naming system, the so-called *short scale*, has terms for naming various milestone powers of ten, as shown in the following table:

Value	Short Scale	Long Scale
10^0	one	one
10^1	ten	ten
10^2	hundred	hundred
10^3	thousand	thousand
10^4	ten thousand	ten thousand
10^5	hundred thousand	hundred thousand
10^6	million	million
10^9	billion	thousand million
10^{12}	trillion	billion
10^{15}	quadrillion	thousand billion
10^{18}	quintillion	trillion
10^{21}	sextillion	thousand trillion
10^{24}	septillion	quadrillion
10^{27}	octillion	thousand quadrillion
10^{30}	nonillion	quintillion
10^{33}	decillion	thousand quintillion
$\vdots$		
10^{57}	octodecillion	
10^{60}		decillion
$\vdots$		
10^{108}		octodecillion
$\vdots$		
10^{303}	centillion	
$\vdots$		
10^{600}		centillion

And so on—I have skipped many entries.

The short-scale naming system is based on every third power of ten 10^{3n+3}, in each case taking the Latin roots for units, tens, and hundreds place in n and appending the suffix "illion." The alternative *long-scale* system, in contrast, is based on every sixth power 10^{6n} but confusingly uses the very same names for these different values! The long scale was traditionally used in Britain and British-affiliated territories, but much less so in recent decades, as the short scale is becoming dominant in most English-speaking countries. Much of continental Europe and South America, meanwhile, uses a form of the long scale. So there is a clash in terminology.

Same Number Different Name, Same Name Different Number

How absurd and confusing that large numbers have had fundamentally different names in different countries, even in different English-speaking countries! It is frankly an intolerable mess. The systems agree on the meaning of one million, which is 10^6, but they disagree on one billion, which on the short scale is a thousand millions (10^9) but on the long scale is a million millions (10^{12})—a thousand times larger. The divergence grows with larger numbers—a trillion on the short scale is a thousand billions (10^{12}), but on the long scale, it is a million (long scale) billions (10^{18})—a million times larger. For names above one million, the same number name generally refers on the long scale to a far larger number than on the short scale. The term "milliard" refers to one billion on the short scale, but it is now rarely used, except that UK financiers do use it, sometimes shortened to a "yard," to refer to what on the short scale would be a billion pounds. A tidy sum.

Sometimes this difference in number names has caused confusion in trans-Atlantic communication, such as with international economic matters or astronomical and other scientific data. On the short scale, for example, you have about eighty-six billion neurons in your brain, whereas this same number would correctly be described on the long scale as a small fraction of a billion. How embarrassing it would be to be misunderstood about this, say, if you should happen to be discussing the matter on a first date.

Regrettably, since they both consider values of n only up to one thousand, both the short scale and long scale eventually run out of names. In this sense, neither nomenclature is fit for our purpose since they do not provide names for extremely large numbers.

Meanwhile, of course, one can extend the number-naming systems in various sensible manners. One such extension was proposed by John H. Conway and Richard Guy (2012), continuing the Latinized root and "illion" suffixes, and this system in principle provides a determinate name for every natural number, thereby providing an answer to question 2.

For very large numbers, however, the Conway-Guy system calls for iterated repetitions of the primitive Latinized roots, as in "sesexagintasescentilli-sesexagintasescentilli-sesexagintasescentillion." Worse, in my view, is that this iterative naming might be required many times within a single number name. The system is principally designed to provide names for the milestone powers of ten, but when a number has many nonzero digits, the corresponding exaggerated milestone names are needed for every block of three digits (or six on the long scale) with a nonzero digit, so even a moderate-sized large number, with several hundred nonzero digits, could have a name involving hundreds of different, absurdly iterated Latinized root names. For this reason, I find the system unwieldy. A simpler number naming system is needed.

Question 3. Is there a simple, sensible, efficient naming system that provides a canonical name for every natural number?

The Digit-Pronunciation Naming System

Let me propose a simple possible answer, what I call the *digit-pronunciation* naming system, by which one simply pronounces the decimal digits of a number in order, so 7216 is pronounced *seven-two-one-six*, and so on for any number. Thus, we obtain a naming system for numbers, and while it does not extend the standard Latinized nomenclature, nevertheless I find it to be perfectly sensible and efficient while also providing a definite, unique name for every natural number. The digit-pronunciation system is efficient mainly because it entirely avoids the need for the Latinized roots to name the milestone powers of ten, as these powers are determined simply by their position in the name, just as the decimal notation system itself is a place value system. The digit-pronunciation system is also easily modified for any base.

Let me add, furthermore, that the digit-pronunciation naming system is already in common use with very large numbers, such as reciting a phone number or reading off the number on a credit card, and it is also commonly used to help disambiguate small numbers, such as 15 and 50. So I find it to be a reasonable naming system—efficient, already in use, and providing a definite name for every number.

1.3 The *Book of Numbers*

So let us finally place the natural numbers in alphabetical order. We shall form the *Book of Numbers*, the number dictionary, listing all the natural numbers in alphabetical order according to the digit-pronunciation naming system. The ten digits themselves appear in alphabetical order like this:

eight, five, four, nine, one, seven, six, three, two, zero

Larger numbers are placed in alphabetical order by comparing the first difference in letters in the iterated digit-pronunciation names, combined with the standard proviso that extensions of a given word appear after it in the dictionary order.

Can you put the numbers 40, 87, 88, 873, and 882 in alphabetical order, using the digit-pronunciation naming system?

Think about it...

The alphabetical order would be 88, 882, 87, 873, 40, in light of the digit-pronunciation names of these five numbers, which are ordered alphabetically like this:

eight-eight, eight-eight-two, eight-seven, eight-seven-three, four-zero

Notice that the ten fundamental digit names are prefix-free—none of them is an initial segment of another. Thus, when comparing the names of two numbers, we will never be in a situation where part of one digit is combined with part of another to make the alphabetical comparison. Rather, the alphabetical order is determined completely by comparing the digits themselves according to the alphabetical order of the ten fundamental digits.

Let us begin to understand the nature of the alphabetical order on the numbers by making several elementary observations.

The Last Number

Although there is no final, largest number in the usual order on the natural numbers, nevertheless in the alphabetical order there actually is a last number. Namely, the number *zero* appears last in the alphabetical order, the final entry on the very last page of the *Book of Numbers*, because no other number starts with the letter *z*, so this number will appear as the very last entry alphabetically.

The First Number

Next, observe also that there is a first number alphabetically—the number 8 is first since it begins with the letter *e* and the only other numbers beginning with *e* also begin with 8, followed possibly by additional digits. Therefore, they will appear after the single-digit 8.

The First Chapter

For a similar reason, the next number after 8 alphabetically is 88, and then 888 and 8888, and so on. There is no number alphabetically between any of these numbers since 8 is the very first number that could be added to the end alphabetically. Thus, the beginning of the alphabetical order on the natural numbers, what might be seen as the first chapter of the *Book of Numbers*, looks like this:

$$8 \quad 88 \quad 888 \quad 8888 \quad 88888 \quad \cdots$$

Successors

Pressing this idea a bit harder, what I claim is that every number (except zero) has an alphabetical successor, which is obtained simply by adding 8 as a digit at the end of the decimal representation of the number.

For example, the next number alphabetically after 532 is 5328 because any other digit sequence alphabetically after the first number must either extend it or deviate from one of those digits. But 5328 will be below any other higher deviation or extension, so it is a successor. Similarly, 53288 and 532888 are the next few numbers, simply adding more

8s at the end. In this way, locally, there is a chapter beginning with 532 and its iterated-8 successors:

532 5328 53288 532888 ...

Every number except zero in the alphabetical order has a successor, and a next successor, and a next successor after that, making a little copy of the natural numbers proceeding forward alphabetically just from that number. Locally, one simply tacks on additional 8s to move to later numbers alphabetically.

Predecessors

What comes just before 532? Is there a predecessor number? Well, the digit immediately before 2 is 3 (remember *three* is alphabetically before *two*), so one natural candidate for the predecessor might be 533, which certainly comes before 532 alphabetically. But 533 is not the predecessor of 532, since 5338 and 53388 and even 53345634738 are all after 533 but before 532 alphabetically. Indeed, what I claim is that there is no number at all that is an immediate predecessor of 532 since it would have to start with 533..., but then tacking on another 8 would make an alphabetical increase, but still below 532. So there is no immediate predecessor.

The same argument works with any number that does not end with the digit 8. If a number does not end with 8, then for any number preceding it alphabetically, we can tack an extra 8 on the end, and this would then come between them. If by contrast a number does end with 8, then it will have a predecessor simply by removing the 8.

Thus we see that every number (except zero) is sitting in its own little chain of successors, a chapter initiated by a number not ending with the digit 8, followed by the iterated-8 successor sequence. For example, the entire chapter starting with 532 appears before the chapter starting with 5325, like this:

(532 5328 53288 532888 ...)   ...   (5325 53258 532588 5325888 ...)

Density of the Chapters

Let us now become a bit more sophisticated. I should like to ask whether there is anything between these two chapters.

Yes, indeed, there is—there is another totally different and entire iterated-8 successor sequence strictly between them. Simply consider the sequence starting with 53287, for example, which is after 532 but before 5325. And since this number does not end in 8, its entire iterated-8s sequence will also appear in the gap, like this:

(532 5328 53288 ...)   (53287 532878 5328788 ...)   (5325 53258 532588 ...)

And indeed, there are entire further chapters between any two of these, and so on infinitely. For example, the sequence beginning with 53284 is after the 532 chapter and before the 53287 chapter. What I claim is that between any two iterated successor sequence chapters, there is another such chapter. Thus, the chapters of the *Book of Numbers* are themselves *densely* ordered, which means that strictly between any two of them, we can find another.

No Last Chapter Before Zero

There is no last chapter before the number zero because to any number alphabetically before zero we can append a 7, 5, 4, or what have you, which will be later alphabetically, but since it doesn't end in 8, it starts a new chapter—the beginning of its own iterated-8s successor sequence, which will lie entirely before zero. So there is no last chapter before zero.

The Overall Order Type

The analysis that we have given so far is enough to conclude the overall order type of the *Book of Numbers*. Namely, we have divided the book into the iterated-8 successor sequence chapters, each starting with a number whose decimal representation does not end with 8, except for the very first chapter, which consists of 8 88 888 ... and so on. After that first chapter, the chapters continue in a densely ordered manner, with no final chapter before the single final number zero appears on the last page of the book.

From this information, it follows that the order type of the *Book of Numbers* is exactly

$$\mathbb{N} \cdot (1 + \mathbb{Q}) + 1.$$

I recognize that this notation may not be familiar to all readers, so let me try to explain a little about it. The symbol $\mathbb{Q}$ refers to the order of the rational numbers, which is a countable dense linear order, and $1 + \mathbb{Q}$ is such a dense linear order with an initial point. This is the order type of the chapters since we have a first chapter, but then after that first chapter, they are ordered as a dense linear order, just like the rational numbers. (It is a remarkable theorem of Georg Cantor, which I won't get into here, that all such orders built from the natural numbers are realized as a copy of the rational order like this.) So we have $1 + \mathbb{Q}$ many chapters, but each chapter consists of an iterated-8s successor sequence, which is ordered just like the natural numbers. The product $\mathbb{N} \cdot (1 + \mathbb{Q})$ refers to $1 + \mathbb{Q}$ many copies of $\mathbb{N}$, and this is exactly what the *Book of Numbers* consists of, plus the number zero by itself at the very end, and this explains the $+1$ in the order type expression. So the order type of the *Book of Numbers* is precisely $\mathbb{N} \cdot (1 + \mathbb{Q}) + 1$.

Questions for Further Thought

1.1 What comes next: O T T F F S S E ?

1.2 Carry out the analysis of the nature of the order type of the natural numbers using the digit-pronunciation system in another base, such as base 16 (hexadecimal), or base 8 (octal), or base 2 (binary). Does using a different base affect the features of the overall order? Of course, in a different base, the numbers will be in a different order, but how much change occurs?

1.3 Explore the nature of the natural numbers in alphabetical order according to the digit pronunciation system using a language other than English. Is the order type the same? How much of the analysis given here carries over to your other chosen language? Is there a language where there is no final number?

1.4 Consider the natural numbers ordered alphabetically according to their usual representation written in base ten (or another base)—that is, using the numeral digits as the lexical sorting characters rather than writing the names of those digits as English words. So the number 123 would precede 123254 alphabetically, which would precede 13, and so on. Does every number have an immediate successor? Is there a least number? Can you say exactly what is the order type?

1.5 Write a short essay explaining the number nomenclature system of Conway and Guy and explaining how and why, in particular, it provides a definite number name for every natural number.

1.6 Write a short essay explaining the number-naming system proposed by John Gowers in the further reading section, and compare the advantages over the Conway-Guy system.

1.7 What can you say about the alphabetical order of the real numbers under the digit-pronunciation naming system, whereby π would be pronounced *three point one four one five nine* and so on? Is this a dense order? Is it complete? (Hint: Take care to consider how the word "point" figures in this order and your analysis.)

Further Reading

Conway and Guy (2012). A romp through fascinating number ideas, led by some leading mathematical minds. (Kindly note that despite the identical titles, this suggested reference is not the imaginary *Book of Numbers* that I consider in this chapter.)

Gowers (2020). A proposal for a more efficient number naming system.

Credits

This chapter is based on ideas that I had presented in both question and answer posts on Math.StackExchange at math.stackexchange.com/q/3047540/413, which grew out of a conversation that I had had over beers with Stefan Hoffelner and Stefan Mesken following a talk that I had given at the Logic Oberseminar at the University of Münster in 2019. The list of numbers up to one hundred, in alphabetical order, is from Hervé Graumann 1988 www.graumann.net/echo/graumann/aaa_pageshtml/dladnE.html, although I have used "one hundred" rather than the plain "hundred." The short-scale names were obtained from Wikipedia at en.wikipedia.org/wiki/Names_of_large_numbers. Question 1.4 was suggested by Nikita Danilov.

2 The Sand Reckoner

Abstract. How many grains of sand would you need to fill the universe? How much coffee is made each day in New York City? Would it fill the Statue of Liberty? If you stacked quarters to the Moon, how much would they be worth? Would they be more massive than the Earth? Would the coins fit in Central Park? If you gathered all the hot air that you have breathed in your life into a hot-air balloon, would it lift you and all your possessions?

2.1 A Universe of Sand

How many grains of sand would you need to fill the universe? This whimsical question is the principal concern of *The Sand Reckoner*, a remarkable treatise of Archimedes written in the third century BC. Ostensibly focused on the calculation of a definite numerical answer, the treatise also advances what is clearly a further underlying goal for Archimedes—namely, to showcase his number representation system and its capacity for handling extremely large numbers. This was powerful new number technology, transcending the comparatively primitive Greek number system of his time—it amounts presciently to a place-value system in base *myriad*, or 10,000. My reading of the text is that Archimedes was eager to show off his agility with large numbers, and the grains-of-sand problem was an entertaining example to exhibit his powers of calculation with numbers far larger than others had previously considered.

Archimedes sets out his project, dedicated to Gelon of the royal house of his native Syracuse, like this:

> There are some, king Gelon, who think that the number of the sand is infinite in multitude; and I mean by the sand not only that which exists about Syracuse and the rest of Sicily but also that which is found in every region whether inhabited or uninhabited. Again there are some who, without regarding it as infinite, yet think that no number has been named which is great enough to exceed its multitude. And it is clear that they who hold this view, if they imagined a mass made up of sand in other respects as large as the mass of the Earth, including in it all the seas and the hollows of the Earth filled up to a height equal to that of the highest of the mountains, would be many times further still from recognising that any number could be expressed which exceeded the multitude of the sand so taken. But I will

> try to show you by means of geometrical proofs, which you will be able to follow, that, of the numbers named by me and given in the work which I sent to Zeuxippus, some exceed not only the number of the mass of sand equal in magnitude to the Earth filled up in the way described, but also that of a mass equal in magnitude to the universe.

Thus Archimedes aims to provide a specific numerical upper bound for the number of grains of sand required to fill the universe. He jumps right into the discussion of the nature and size of the universe:

> Now you are aware that "universe" is the name given by most astronomers to the sphere whose centre is the centre of the Earth and whose radius is equal to the straight line between the centre of the Sun and the centre of the Earth.

Archimedes thereby takes the universe to be the sphere of the solar orbit.[1] Nevertheless, he immediately mentions a much larger universe concept, which he credits to Aristarchus of Samos, a heliocentric conception placing the Sun at the center with the Earth and the planets revolving around it, all inside the more distant sphere of fixed stars (the ancients did not realize that the stars were distant suns at vast but highly varying distances). Because Archimedes is interested in providing only an upper bound on the grains-of-sand question, he is perfectly willing to consider this more expansive Aristarchian conception, so after filling his own solar-orbit universe with sand, Archimedes also endeavors to fill the great Aristarchian universe, and his number system has names for all the requisite numbers:

> I say then that, even if a sphere were made up of the sand, as great as Aristarchus supposes the sphere of the fixed stars to be, I shall still prove that, of the numbers named in the *Principles* [a lost work of Archimedes], some exceed in multitude the number of the sand which is equal in magnitude to the sphere referred to, provided that the following assumptions be made.

He then proceeds to undertake the calculation in a careful sequence of steps. To begin, he claims that the circumference of the Earth is about 3 million stadia—1 stadion is a brisk run around the stadium track (about 185 meters by contemporary estimates):

> It is true that some have tried, as you are of course aware, to prove that the said perimeter is about 300,000 stadia. But I go further and, putting the magnitude of the Earth at ten times the size that my predecessors thought it, I suppose its perimeter to be about 3,000,000 stadia and not greater.

[1] According to Erhardt and Erhardt-Siebold (1942), the remark that "most astronomers" agree is not accurate for the astronomers of his day:

> Κόσμοζ, as a designation for the solar orbit or sphere is baffling; Archimedes' contention that "most" astronomers so understand the word contradicts the known facts. The astronomers of his time believed in a finite universe bounded by the sphere of the fixed stars. The word cosmos meant to them either this bounding sphere or the totality of things enclosed by it.

The von Erhardts go on to say, "The other alternative would be that Archimedes, who no doubt was well acquainted with the mathematical astronomy of his day, is not the true author of this unscientific statement." They emphasize that the *Sand-Reckoner* is "essentially an extract from an earlier mathematical work by Archimedes, which has not come down to us."

Next, he claims that the diameter of the Earth is greater than that of the Moon, and the diameter of the Sun is greater than that of the Earth. Furthermore, he claims that the diameter of the Sun is no more than 30 times the diameter of the Moon:

> It is true that, of the earlier astronomers, Eudoxus declared it to be about nine times as great, and Pheidias my father twelve times, while Aristarchus tried to prove that the diameter of the Sun is greater than 18 times but less than 20 times the diameter of the Moon. But I go even further than Aristarchus, in order that the truth of my proposition may be established beyond dispute, and I suppose the diameter of the Sun to be about 30 times that of the Moon and not greater.

Archimedes wants all his estimates, you see, to err on the side of a larger universe, so that his final number will be an upper bound on the number of grains of sand required to fill the universe.

By contemporary scientific accounts, Archimedes has overestimated the size of the Earth, which would be closer to 200,000 stadia, and underestimated the diameter of the Sun, which is slightly more than 400 times larger than the diameter of the Moon, although only 108 times larger than the Earth's diameter. These two errors tend fortuitously to cancel each other somewhat. But let us simply continue on Archimedes's own terms.

Next, Archimedes relates the size of the Sun to the size of the universe:

> The diameter of the Sun is greater than the side of the chiliagon [1000-sided polygon] inscribed in the greatest circle in the (sphere of the) universe. I make this assumption because Aristarchus discovered that the Sun appeared to be about 1/720th part of the circle of the zodiac, and I myself tried, by a method which I will now describe, to find experimentally the angle subtended by the Sun and having its vertex at the eye.

Archimedes explains how he used a long rod with a sliding disc attached, moving it out on the rod until it exactly occluded the Sun. By measuring the resulting angle on the device, he found it to be between 1/164th and 1/200th of a right angle, which would mean between 1/656 and 1/800 of a whole circle, and therefore in agreement with Aristarchus.

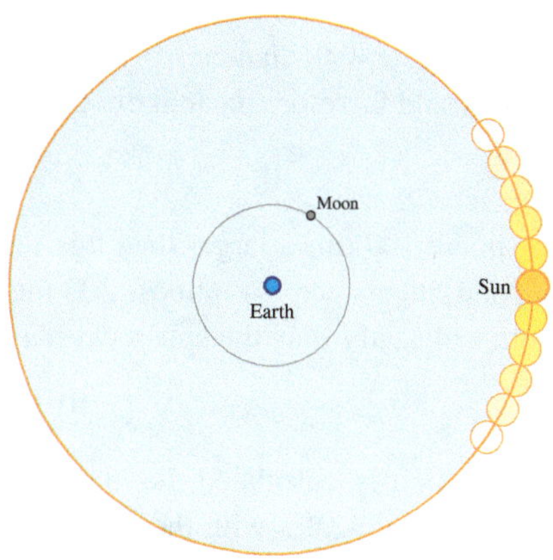

From these assumptions, it follows that the circumference of the universe is at most 1000 times the diameter of the Sun (the figure here is not to scale), which is at most 30 times the diameter of the Earth, so the circumference of the universe is at most 30,000 Earth diameters. Thus, since $\pi > 3$, the diameter of the universe is at most 10,000 Earth diameters—in short, the universe is a myriad Earths across. Since the Earth circumference is at most 3 million stadia, the Earth diameter is at most 1 million stadia (again since $\pi > 3$), so the diameter of the universe is at most 10^{10} stadia, to use modern exponential notation:

$$\begin{aligned} \text{universe diameter} &\leq 10{,}000 \text{ Earth dia } \times 1{,}000{,}000 \text{ stadia/Earth dia} \\ &= 10^{10} \text{ stadia} \end{aligned}$$

Turning now to the small scale, Archimedes considers how many grains of sand would fit inside the space of a poppy seed:

> Suppose a quantity of sand taken not greater than a poppy-seed, and suppose that it contains not more than 10,000 grains.

Now I admit that I find this statement rather odd, since in my experience with poppy seeds and grains of sand—I often enjoy a New York poppy-seed bagel with lox—I frankly find poppy seeds and grains of sand to be about the same size. To be sure, coarse sand can often have grains larger than a poppy seed, while fine sand grains would be smaller. If 10,000 grains of sand make the size of a poppy seed, that would be very fine sand indeed. But since Archimedes seeks only an upper bound, it is all right for him to allow such very fine sand since it can only result in a much larger number of grains required to fill the universe. So let us give Archimedes this estimate on the size of the very fine sand with which he aims to fill the universe.

Archimedes estimates that no more than 40 poppy seeds are required to make up the width of a human finger. And since volume scales with the cube of diameter, the small sphere with the diameter of a finger-width would contain no more than $40^3 = 64{,}000$ poppy seeds. Since each poppy seed has at most 10,000 grains of sand, this means that our little finger-width sphere of sand has at most 1 billion grains of sand:

$$64{,}000 \times 10{,}000 = 640{,}000{,}000 = 6.4 \cdot 10^8 < 10^9 = 1 \text{ billion.}$$

Archimedes now proceeds in a sequence of steps, increasing the size by a linear factor of 100 each time. So the number of grains of sand gets multiplied by the cube $100^3 = 10^6$, or 1 million, with each increase.

In a sphere with diameter 100 finger widths—almost 2 meters—the number of grains of sand would therefore be no more than

$$10^9 \times 10^6 = 10^{15}.$$

A sphere 100 times larger than this in diameter would would be 10,000 finger-widths, a myriad fingers across—almost 200 meters, or about 1 stadion. Therefore, the number of grains of sand filling the sphere with a diameter of 1 stadion would be at most

$$10^{15} \times 10^6 = 10^{21}.$$

We previously calculated the diameter of the Archimedean universe at 10^{10} stadia, and since volume scales with the cube, it follows that the number of grains of sand to fill the

Archimedean universe would be at most

$$10^{21} \times (10^{10})^3 \quad = \quad 10^{51}.$$

Thus, we have our final answer, the number of grains of sand sufficient to fill the universe, if we should use the original Archimedean conception of the universe, the sphere of the solar orbit: 10^{51} grains of sand will suffice.

But Archimedes also aims to fill the much larger Aristarchian universe, the sphere of fixed stars, with his very fine sand. He finds reason in the Aristarchus text to assume that the diameter of the sphere of fixed stars is to the solar orbit as the solar orbit is the diameter of the Earth:

$$\frac{\text{diameter of sphere of fixed stars}}{\text{diameter of solar orbit}} = \frac{\text{diameter of solar orbit}}{\text{diameter of Earth}}$$

Since we have already calculated the diameter of the solar-orbit universe as 10,000 Earth diameters, it follows that the sphere of fixed stars has a diameter that is 10,000 times larger than the solar-orbit universe, a myriad solar orbits across. The number of grains of sand required to fill the Aristarchian universe is therefore greater than that of the solar-orbit universe by the cube of this number. So the total number of grains of sand sufficient to fill the Aristarchian universe is

$$10^{51} \times (10{,}000)^3 \quad = \quad 10^{51} \times 10^{12} \quad = \quad 10^{63}.$$

Fantastically large. Shall we go to the beach?

2.2 Number Sense—Fermi Calculations

I place the Archimedean sand reckoner problem amongst the many other whimsical estimation problems considered in contemporary times. We use a process of estimation and approximation like Archimedes to calculate various unknown or even fantastical quantities: How much coffee is made per day in New York City? Would it fill the Statue of Liberty? If you made a stack of quarters to reach the distance to the Moon, what would the dollar-value be? If you piled those quarters in a heap, would they fit in Central Park? How much does the Empire State Building weigh?

At the heart of these questions is the idea that one can often provide a better estimate by breaking the problem into manageable pieces, undertaking what is known as a *Fermi* calculation, named for physicist Enrico Fermi (1901–1954) and his surprisingly accurate estimations made sometimes on the scantiest amount of data and observations. We do not just pull a final estimate out of the air; but rather like Archimedes, we make a series of simplifying assumptions and informed estimates for parts of the problem, thereby building in steps to a better estimate for the main quantity. Initially, we may have little idea what would be a reasonable answer, perhaps even within several orders of magnitude, but by the end, we gain some insight and find ourselves a little closer to the truth.

Math for Poets

Years ago, I had taught a Math for Liberal Arts class in New York—it was known informally as "Math for Poets"—a course for students in non-math majors to learn some math and thereby fulfill the college quantitative-reasoning requirement. I took the opportunity to have some math fun, covering elementary topics in game theory, graph theory, probability, and more. We spent several weeks on number sense and Fermi calculation—I tried to develop the student's ability to attack a technical problem by breaking it down into more manageable problems.

We began with some easy cases. How many seats are in the auditorium where the class was held? Without counting them explicitly, we might estimate approximately 12 seats per row and about 10 rows, so 120 seats in all. How old is one of the students, in seconds? Well, if he is 18 years old, we can simply multiply each year by 365 days, times 24 hours per day, times 60 minutes per hour, times 60 seconds per minute, to get

$$18 \times 365 \times 24 \times 60 \times 60 \text{ seconds.}$$

One student objected—what about leap days? The 365 should be 365.25 or so. But of course, we had already made far larger rounding assumptions. The student was not exactly 18 years old, for example, but 18 years old and some several additional months, whereas the leap days would amount to a difference of only 4 or 5 days in total. In these calculations, we should freely round the numbers to make the calculation easier, aiming always at a ballpark estimate rather than an exact answer. Indeed, in keeping with the number-sense goal, we didn't use a calculator for these problems, but rather simply estimated all the arithmetic, rounding some numbers up and others down to make it easier, like this:

$$\begin{aligned} 18 \times 365 \times 24 \times 60 \times 60 &\approx 20 \times 300 \times 25 \times 4000 \\ &= 6000 \times 25 \times 4 \times 1000 \\ &= 6000 \times 100{,}000 \\ &= 600{,}000{,}000. \end{aligned}$$

So the student was about 600 million seconds old.

Let me show you some of the other more whimsical absurd questions that we had considered, many of which had a New York theme.

Coffee in New York

How much coffee is made each day in New York City? Would it fill the Statue of Liberty?

First, let me say that I really do not have any definite information about how much coffee is made each day in New York, and I fear that my own coffee-obsessed perspective will lead me to overestimate the amount, but let's give it a try. New York City has a population of approximately 10 million people. Some of those people, like myself, drink a large amount of coffee each day, but many of the others do not drink coffee at all (like most children, for instance). I would think that a sizable percentage of the NYC population does drink coffee, and perhaps as much as a third or even half of them consume coffee daily. Many of those coffee-drinkers have more than one cup per day, and also surely more coffee is made than consumed. So it seems reasonable to me to estimate that we have approximately one medium cup of coffee per person on average per day in New York. Basically, we're saying that the heavy coffee drinkers and the made-but-not-sold coffee approximately makes up for those who abstain, making the average about one cup per person. So we are talking about 10 million cups of coffee per day. A medium cup of brewed coffee at Starbucks is about 12–16 ounces, I think, a little less than a pint, so let's say about 3 cups per liter. This amounts to roughly 3 million liters of coffee.

Would all that liquid fill the Statue of Liberty? The statue itself is, I estimate, about 20 stories tall, including the base, and if each story is 15 feet, or 5 meters, that would mean 100 meters tall, counting the base. But I think that the base is about half the height, so let's say 50 meters for the actual statue itself. Although I lived in New York for over 25 years, I have never actually been inside the statue, but my students say that it is about 10 meters across inside, a little more at the bottom than near the top. If we approximated it as a rectangular solid, that would give a volume of $10 \times 10 \times 50$ cubic meters, or 5000 cubic meters. But since the statue tapers as you go up, particularly in the arm holding the torch, it really is more like a cone than a rectangular solid, so we should divide by 3. But let's divide just by 2, because she isn't quite as tapered as a cone. So the Statue of Liberty has a volume of approximately 2500 cubic meters. One cubic meter can be thought of as a 10 by 10 by 10 array of little 10-centimeter cubes, and each of those is exactly 1 liter. So a cubic meter is 1000 liters, and therefore the Statue of Liberty has a volume of $2500 \times 1000 = 2.5$ million liters. But since we had 3 million liters of coffee, the answer that our calculation arrives at is:

> Yes, one day's worth of New York coffee would fill the Statue of Liberty!

Well, we do not have perfect confidence in our estimates and assumptions — for example, perhaps there are many fewer coffee drinkers in New York than we estimated, or perhaps we underestimated the volume of the Statue of Liberty. Since the estimated volumes were of basically similar magnitudes, we aren't really entitled to say that definitely the coffee would fill the Statue of Liberty. Rather, what we have come to know is that those two volumes are comparably similar in size; they are in the same ballpark.

Elevator Trips

While riding downtown one weekend with my son on the subway, a crowded 4 train, we overheard the group standing next to us talking about elevators.

One woman said,

> "My elevator company serves as many elevator trips in New York City in five days as the population of the entire world."

Her friends, impressed, nodded affirmatively in reply. But my own thoughts, upon hearing that, were to make a quick calculation. Suppose that all 10 million NYC residents rode an elevator 10 times every day, which is way too high (probably 1 trip per person per day is more reasonable, since many people live and work in buildings without elevators). Even in this extreme case of 10 trips per person per day, that would mean only 100 million trips total per day, or 500 million trips over 5 days. This is not even one-tenth of the world population, which is about 8 billion, so no way is that person's claim true, especially since there are also many elevator companies. I was tempted at the time to mention my calculation to those people on the subway, but decided against it. Walking out of the subway in the East Village, however, I asked my son (who was 14 years old at the time) whether he heard those people talking about elevators, and he replied, "Oh, yes, and when they said that, I calculated it in my head: No way is that true." He then proceeded to explain his calculation, which was similar to mine. Yay, Horatio!

The Chicago Marathon

In the run-up to the 2016 Chicago marathon, on a route that would wind through the windy city streets, *Newsweek* magazine reported, "Chicago Marathon organizers expect 1.7 million fans to line the route this year." The organizers had emphasized the economic impact of these spectators, many of whom would presumably be eating in Chicago restaurants and staying at Chicago hotels. But is this a reasonable number?

Let's calculate. A marathon is approximately 26 miles, and the route has two sides for spectators, so let's round it to 50 miles of spectator viewing spots. Each mile is about 1800 yards, so we have $50 \times 1800 = 90{,}000$ yards of viewing spots. Each spectator, standing shoulder-to-shoulder, with all their stuff, takes up about a yard of space. So if the marathon was lined on both sides for the entire route with spectators standing shoulder-to-shoulder, this would amount to about 90,000 spectators. To have 1.7 million spectators, therefore, they would have to be lined up behind each other. But even if the spectators were 10 people deep on each side for the entire route, which is a vast crowd, it would still amount only to 900,000 people. We would basically have to double this to get to 1.7 million. So, if the live event really had 1.7 million spectators lining the route, then it would mean that the race was lined 20 people deep on each side for the entire route.

No way is this number correct! I have never had the chance to attend the Chicago marathon, but at the New York marathon, which I assume is roughly comparable, I know that while there are thick crowds at the finish line in Central Park and at some of the other prominent or especially interesting race locations, most of the rest of the route is much thinner, and at the typical nothing-special location on the route, one could expect easily to have a front-row spot.

How Many Bricks on Campus?

The college campus at which I used to teach in New York is covered with some lovely woods and green meadows and has a number of brick Georgian buildings. Most of these are the same standard size as the mathematics department building, but there are also some larger buildings such as the library, the performing arts center, the campus center and the gymnasium. I posed the question for my students:

How many bricks are there on campus?

The students and I agreed that altogether it amounted to about 30 buildings of the size of the math building. This building is approximately 30 meters by 70 meters, which would make a perimeter of 200 meters, but because the building has wings and is not purely rectangular, let's add another 100 meters for the folds in the walls, so about 300 meters around the base of the building. The building is two stories tall, about 10 meters tall, making the area of the walls about 3000 square meters. Of course, there are windows and doors cut out of these walls, but let's say that they are approximately accounted for by the extra bricks used in the trimming flourishes at the corners and top of the building. So we have 3000 square meters of brick. Each brick face is approximately 10 cm by 30 cm, so 1 square meter of brick would have about 10 rows of a little more than 3 bricks across, or about 20 bricks per square meter. Now, there is the question whether the bricks are laid more than one layer deep. The newer larger buildings I expect have a steel frame, with only one layer of bricks, but perhaps in the older buildings, the bricks are used structurally, with multiple layers. So let's say on average that there are about 30 bricks per square meter. So each building-equivalent has on average about about $3000 \times 30 = 90{,}000$ bricks. And with 30 buildings, this amounts to $30 \times 90{,}000 = 2{,}700{,}000$ bricks in the buildings on campus. There are also some bricks in the central fountain, a few small brick walls and some bricks lining some of the walkways. So let's add another 10 percent for those extra bricks, arriving at about 3,000,000 bricks on campus in all.

Quarters to the Moon

How many quarters would you have to stack to reach the Moon? How much would it be worth? How much would it weigh? More than the Earth? More than the Empire State Building? If you put all those quarters into a pile, would it fit in Central Park?

The question is admittedly a little absurd, and there are all kinds of problematic issues: We couldn't ever actually make such a tall tower of quarters, as it would topple over; it doesn't make sense to build a tower to the Moon since the Moon is in orbit around the Earth and it is moving; the quarters would begin to orbit the Earth themselves, or fall back down to the Earth or to the Moon. But let's just try to ignore all those problematic issues and just try to answer how many quarters it would take to make a pile that high.

How far is the Moon from the Earth? I do not really know, and we could look it up to see what the astronomers say about it, but that would go against the back-of-the-envelope spirit of these problems. Let me simply use two facts that I happen to have picked up years ago: first, I know that the speed of light is about 300,000 km per second; and second, I happen to know that it takes radio signals traveling at the speed of light about 1 second to get to the Moon (8 minutes to the Sun), so the Moon is about 1 light-second away. I remember this fact from having learned it in my childhood because it was important in a science fiction story I read, in which radio communication from Earth to the Moon base had a 2-second delay: you send a message, it takes a second to get there, and then a second for the answer to come back. So the Moon is 1 light second away, and light travels 300,000 km in one second. So the Moon is about 300,000 km distant (OK, I looked up the actual value: 387,400 km).

Let's now stack up the quarters. By eyeballing a quarter that I have in my pocket, I'd say that each quarter is about 2 mm thick, or 10 quarters in a 2-centimeter stack, which seems about right. This makes 500 quarters per meter, or 500,000 quarters per kilometer. So altogether, we have

$$300{,}000 \times 500{,}000 = 150{,}000{,}000{,}000$$

many quarters in the stack to the Moon. That's 150 billion quarters to the Moon! It takes 4 quarters to make a dollar, so that is worth about $40 billion, which might be less than you would have expected initially. Elon Musk could have paid to stack quarters to the Moon instead of buying Twitter.

Let us now put those quarters in a big pile, packed as tightly as possible. Each quarter is a little over 2 cm in diameter, so the volume of the rectangular solid bounding the quarter is a little more than 2 cm by 2 cm by 2 mm, or about 800 cubic millimeters, which we can round up to 1 cubic centimeter. That makes sense because we can imagine folding a quarter into a little 1-centimeter cube. So 1000 quarters takes up about 1 liter. Thus, our quarters-to-the-Moon have a volume of about 150 million liters. There are 1000 liters in a cubic meter, so this is 150,000 cubic meters. Since a cube 10 meters on each side has a volume of 1000 cubic meters, we can fit all those quarters into 150 such cubes. I can easily imagine a giant art installation, with 150 such 10-meter cubes placed in Sheep Meadow in Central Park. They would definitely fit.

How much would the quarters weigh? Some of the students in my lecture had worked as cashiers, so they were familiar with quarter rolls from the bank, which fit in your palm and have $10 worth of quarters, or 40 quarters. I can imagine a quarter roll in one hand balancing the weight of a half-pound of Gouda cheese at the deli in my other hand. So 40 quarters weigh 0.5 pound, which makes 100 quarters weigh about 2.2 pounds, which is about 1 kg. We had 150 billion quarters, which is 1.5 billion times 100 quarters, so 1.5 billion kilograms.

Since we already imagined the quarters in Sheep Meadow in those 150 large boxes, clearly they weigh far less than the Earth, and they are perhaps comparable to the Empire State Building.

How Much Does the Empire State Building Weigh?

After posing this question in my lectures, I found out subsequently that it is also a famous Google interview question, and you can easily find other blog posts about it. But let me tell you how I thought about it. The Empire State Building is about 100 stories tall, and if we assume 12–15 feet per floor, or about 4 meters, this makes the total height (without the radio tower) about 400 meters. The CUNY Graduate Center, where I used to work, is across the corner on Fifth Avenue and 34th Street, and I have walked past the Empire State Building many times. I estimate that the base is about 75 meters on Fifth Avenue by 150 meters on 34th Street. But because of the step-backs, the middle part of the tower is considerably less than that—probably 30 by 80 near the top. Let's say that the average cross section of the building is 50 by 100 meters. Now, how much does that weigh? Of course, there is a lot of steel and masonry in the floors and walls, and concrete floors, and also all the contents of the building, with desks and paper files and books and interior walls and doors. Those things are heavy. I really have no idea how much those things weigh, but let me imagine that we build a virtual copy of a complete floor from the Empire State Building, without any of those floor beams or concrete or desks or walls or anything, and instead we flood that virtual room with water, until it has the same weight. Of course, the metal and concrete and masonary is much heavier than water, but the actual space is mostly empty air. How much water would it take to weigh the same? Well, let me just guess that we'd have to flood the virtual room about one-quarter deep with water to have the same weight as the actual materials. At 50 by 100 meters by 4 meters tall for each floor, this would mean a pool of water 50 by 100 meters, one meter deep, for a volume of 5000 cubic meters of water. We already said that 1 cubic meter of water is 1000 liters, each of which weighs 1 kg, so we are talking about 5000 × 1000 kg or 5 million kilograms per floor. With 100 floors, this is 500 million kilograms, or 500,000 metric tons. So according to this estimate, the Empire State Building with all its contents weighs about 500,000 metric tons.

So it weighs less than the quarters to the Moon (1.5 billion kilograms)—the Empire State Building weighs about one-third as much as the quarters to the Moon—but these two quantities are in the same ballpark.

How Many Babies are Born Each Day in New York City?

The population of New York is approximately 10 million people. And those people live about 70 years. Let's imagine spreading those births out uniformly over the 70-year period. That means 1 million people are born in 7 years. This is about 150,000 people born in one year, which is about 3000 births per week, or about 400 births per day. This estimate would be accurate if the city population were in a steady state, with births balancing deaths, but of course the population is increasing. On the other hand, most of that increase is from people moving here, not just being born here. But then again, the people moving to New York, I expect, are mainly young adults looking for a career, so they will contribute to a heightened birth rate as they have children. Shall we add 25 percent more to account for the fact that the city is growing and not in a steady state? In that case, our estimate is that there are about 500 births each day in New York City.

A Lifetime of Hot Air

If you collect all the hot air that you have breathed in your life, what would the volume be? If you made a hot-air balloon, would it be able to lift you and all your possessions?

To answer, let's start with the first part. How much do I breathe? If I imagine inhaling and then exhaling a deep, big breath, I figure that I could inflate a small paper bag, perhaps well over 1 liter, but probably not as much as 2 liters. But my passive resting breathing is probably much less than a deep, big breath. So let's figure a half-liter per ordinary, passive breath. How often do I breathe? Well, in the swimming pool, I can hold my breath under water for 1 minute or even 2 minutes (in my younger swim-team days); but if I hold my breath right now, I can say that it does start to feel a little unnatural, like I should take my next breath, even after just about 5 or 10 seconds, even though this impulse could be resisted longer. It seems to me that my body wants to take another breath in about that time. If we breathe every 5 seconds, that would mean 12 breaths per minute, so let's say 10 breaths per minute, which would mean a volume of 5 liters per minute. That makes $5 \times 60 = 300$ liters per hour, or $300 \times 24 = 7200$ liters per day. In a year, this would be 7200×365, which is less than 7000×400, which is 2,800,000 liters per year. Let's say 2.5 million liters per year of hot air. Multiplying that by 50 years old (my aspirational age) would make 125 million liters of hot air in all for me.

Now, each liter of air fills a cube 10 cm on a side, and 1000 of these fit into a cubic meter. So we have got 125,000 cubic meters of hot air. This is the same as a cube that is 50 meters by 50 meters by 50 meters. That is my hot-air balloon, filled with a lifetime of hot air.

This balloon is considerably larger, more than 100 times as large, as a typical recreational hot-air balloon, which I understand are usually under 1000 cubic meters. From this point of view, it would seem likely that it could lift me and all my possessions, although body temperature may be much less than is achieved in those balloons.

If the air was at my body temperature (98.6° F), then would it be able to lift me and all my possessions? Well, let's see how much it would lift. Hot air expands in proportion to temperature (from absolute zero). If it is a day that is like today, about 50° F, which is about a 50° F difference, and absolute zero is minus 460° F. So this is about a 10 percent increase in temperature. (Stated in metric terms, we have body temperature of about 37° C, and it is about 10° C today, so a difference of 27° C, and absolute zero is about minus 270, so about a 10 percent increase, as I said.)

So the heat of the hot air caused it to expand in volume by 10 percent. The buoyant force of the hot-air balloon is exactly the weight of this displaced air, by the Archimedean principle. Thus, the lifting force of my hot-air balloon will be equal to the mass of air filling 10 percent of the volume that we computed. How much does air weigh? I happen to remember from my high school science class that 1 mole of air at 1 atmosphere of pressure is about 22 liters (my teacher had a cube of exactly that size sitting on his desk to help us to visualize it). And I also know that air is mainly nitrogen, which forms the molecule N_2, and since nitrogen in the periodic table has an atomic number of 14, the molecule N_2 has a mass of 28 g per mole. So air weighs about 28 g per 22 liters, which is about 1.3 g per liter. Each cubic meter is 1000 liters, so 1.3 kg per cubic meter (this is much larger than I had expected—air weighs more than 1 kg per cubic meter! This will be useful to keep in mind for future such kinds of calculations). My hot air in total was 125,000 cubic meters, and we said that because of the temperature difference, the volume expanded by 10 percent, or 12,500 cubic meters. This expansion would displace an equal volume of air, which weighs 1.3 kg per cubic meter. Thus, the displaced air weighs $12{,}500 \times 1.3 \approx 16{,}000$ kg, or about 16 metric tons. So all my hot air, at body temperature in a giant hot air balloon on a chilly day, would have a lifting force that could lift 16 metric tons.

Would this lift me and all my possessions? Do I own 16 tons of stuff? Well, thankfully, I do not own a car, which would be a ton or more by itself. But I do own a lot of books, a piano, an oven, a dishwasher, some heavy furniture, paintings, and various other items, as well as a collection of large potted plants on my terrace. It seems likely to me that I could include most, if not all, my possessions within 16 tons. To gain a little confidence in this, let's estimate the mass of my books. My wife and I have about 12 large shelves filled with books, each about 2 meters, and then I have about 3 more such shelves filled with books in my offices at the university. If we count half of the home books as mine, plus my office books, that makes 9 shelves times 2 meters, for about 20 meters of books. If the books are about 25 cm tall on average, and 15 cm deep, that makes $20 \times 0.25 \times 0.15 = 0.75$ cubic meters of books. Let's round up to 1 cubic meter of books. How much does a cubic meter of paper weigh? Well, one ream of copy paper weighs about 2 kg, and that is a volume of 8.5 by 11 by 2 inches. And 1 meter is about 33 inches, so we could fill one cubic meter with a pile of reams of copy paper 3 by 3 by 17, which would be 153 reams, or about 300 kg. So not even a half-ton of books! So I can definitely include all my most important possessions within the 16 tons.

My final answer, therefore, is:

> Yes, if we filled a giant balloon with all the hot air I have breathed in my life, at body temperature, then it would lift me and all my possessions.

Questions for Further Thought

2.1 Essay suggestion: Learn the Archimedean terminology for large numbers and explain how it relates to the central calculation of *The Sand Reckoner*.

2.2 Suppose that someone left the water tap running in the kitchen. How long would it take to fill your home or apartment with water, assuming that no water leaked out?

2.3 How many doorknobs are there in your building?

2.4 How many subway tiles are there in the NYC subway system?

2.5 How many riders were on my crowded 4 train downtown this morning?

2.6 How many light switches are there on your university campus?

2.7 How much water does NYC use each day?

2.8 What are the odds that you drank a molecule of water that was once also drunk by Julius Caesar?

2.9 How many people have there been? What fraction of them are currently alive?

Further Reading

Archimedes (1897 [3rd century BC]). The classic text of Archimedes, translated by Thomas Heath (1897). Also available in Archimedes (2009).

Credits

Much of the material in this chapter was adapted from my blog posts Hamkins (2016b) and Hamkins (2016a), which were based on lectures that I had given on the topic as part of a Math for Liberal Arts course for students at the College of Staten Island of the City University of New York (CUNY). Thanks to the critical math commentary of Mark Iris (Chicago Tribune, Commentary, October 12, 2016) for bringing the Chicago Marathon example to light.

3 Zeno's Paradox

Abstract. An ancient puzzle leads ultimately to a remarkable observation about the malleable nature of infinite sums.

3.1 Zeno's Paradoxes of Motion

The Greek philosopher Zeno of Elea (c. 490–430 BC) argued in antiquity that all motion is impossible. It is impossible to walk through town or across the room, to go from here to there. What? We know, of course, that this *is* possible. We walk from here to there every day. And yet Zeno offers us proof that this is an illusion—we simply cannot do it.

Zeno argues like this. Suppose that it were possible for you to move from point A to another distinct point B:

Before you complete the move from A to B, however, you must have gotten halfway there:

But before you get to this halfway point, of course, you must get halfway to the halfway point! And before you get to *that* place, you must get halfway there:

And so on, ad infinitum. Thus, to move from A to B, or indeed to move anywhere at all, one must have already completed an infinite number of tasks—a *supertask*. It follows, according to Zeno, that you can never start moving—you cannot move any amount at all, since before doing that, you must already have moved half as much. And so, contrary to appearances, you are frozen, motionless, unable to begin. All motion is impossible.

Is the argument convincing? On what grounds would you object? Do you think, contrary to Zeno, that we can actually complete infinitely many tasks? How would that be possible?

It will be no good, of course, to criticize Zeno's argument on the grounds that we know that motion *is* possible, for we move from one point to another every day. That is, to argue merely that the conclusion is false does not actually tell you what is wrong with the argument—it does not identify any particular flaw in Zeno's reasoning. After all, if it were in fact an illusion that we experience motion, then your objection would be groundless.

Rather, to truly criticize Zeno's argument, one must engage more directly with it. What exactly might be the problem with his argument?

Achilles and the Tortoise

Consider the allegory of Achilles and the Tortoise. The Tortoise has challenged Achilles to a race around the stadium. How laughable! How could the slow Tortoise win against the swift warrior? Nevertheless, the confident Tortoise insists that he can win with only a modest head start, say, one-quarter of the way around. Achilles accepts, and the race begins. They line up at the start, and the Tortoise starts off running at a brisk pace (for a tortoise, that is), while Achilles waits patiently for him to reach the quarter-way-around mark at point *A*, at which time Achilles takes off at a fast clip from the start:

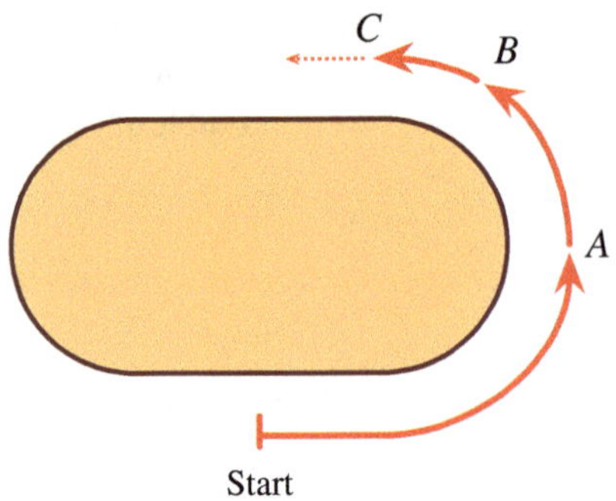

Why does the Tortoise think he will win? Between steps, panting heavily, the Tortoise explains his reasoning. He argues that Achilles will never overtake him. Why? Well, when Achilles had started running, the Tortoise had already had his head start, a quarter of the way around the stadium to point *A*. But by the time Achilles gets to that point *A*, the Tortoise will naturally have moved on to some further point *B*. So, to catch him, Achilles also will need to move to this further point *B*. But again by the time Achilles gets to point *B*, the Tortoise will have moved on to some further point *C*. And so on, ad infinitum. Every time Achilles catches up to where the Tortoise had been, the Tortoise has moved on to a new further location, still beyond Achilles. Thus, the Tortoise reasoned, Achilles shall never catch up!

3.2 An Infinite Sum with a Finite Value

Does it make sense ever to add up infinitely many numbers? Consider the following sum:

$$\frac{1}{2}+\frac{1}{4}+\frac{1}{8}+\frac{1}{16}+\cdots$$

Since we are adding up infinitely many positive numbers, should one expect the sum to be infinite?

No—in fact, this infinite sum adds up altogether to a finite value. What I claim is that this summation sums to 1:

$$\frac{1}{2}+\frac{1}{4}+\frac{1}{8}+\frac{1}{16}+\cdots \quad = \quad 1.$$

To see this, consider a line segment of unit length 1, as follows:

Let us take the first half, like so:

And next we take also half of what remains, which will add a quarter more:

And then an eighth and a sixteenth more:

At each stage, we take half of what remains, which will be half as much as we had just added at the previous step, and in this way, we exhaust more and more of the original unit interval:

At each stage, the current sum remains less than 1, but the difference between it and 1 decays to zero. The total sum, which means the limit value of the finite partial sums, will therefore be exactly 1. The length of the original interval is exhausted by the infinite sum, which therefore sums to 1:

$$\frac{1}{2}+\frac{1}{4}+\frac{1}{8}+\frac{1}{16}+\cdots \quad = \quad 1.$$

So we have an infinite sum with a finite value.

A Different Perspective on Zeno

Perhaps this analysis provides an alternative perspective on Zeno's paradox. Let us consider a forward-oriented version of his argument. That is, let us argue, like Zeno, that to get from one point to another, we must first traverse half the distance, and then *after* that, traverse half the remaining distance, and then half the still-remaining distance, and so on, ad infinitum. This is exactly the process that we had used in these diagrams to analyze the $\frac{1}{2} + \frac{1}{4} + \frac{1}{8} + \cdots$ and so on. We can imagine an alternative Zeno arguing that it is impossible to arrive at the destination, since one must traverse separately all these infinitely many disjoint intervals. Such a forward-oriented version of his Zeno's argument seems to have much in common with his original ideas. Yet, meanwhile, the paradox seems to be at least partly addressed by our analysis that although there are infinitely many distinct segments, nevertheless the total length is finite, so one can indeed sometimes traverse infinitely many segments in a finite time. This observation seems to pour some cold water on the idea that it is inherently impossible to traverse infinitely many distinct intervals. Does this lead the way to a resolution of Zeno's paradox?

An Alternative Illustration

Here is an alternative way to analyze the same sum. Consider the unit square shown here, with area 1. We have successively divided it into smaller pieces: a rectangle of area $\frac{1}{2}$, a square of area $\frac{1}{4}$, a rectangle of area $\frac{1}{8}$, and so on. Each rectangle is followed by a square of half the size, and each square is followed by a rectangle of half the size. So the sum of all the pieces is $\frac{1}{2} + \frac{1}{4} + \frac{1}{8} + \frac{1}{16} + \cdots$, and since they exhaust the unit square in the limit, the sum is 1, as claimed.

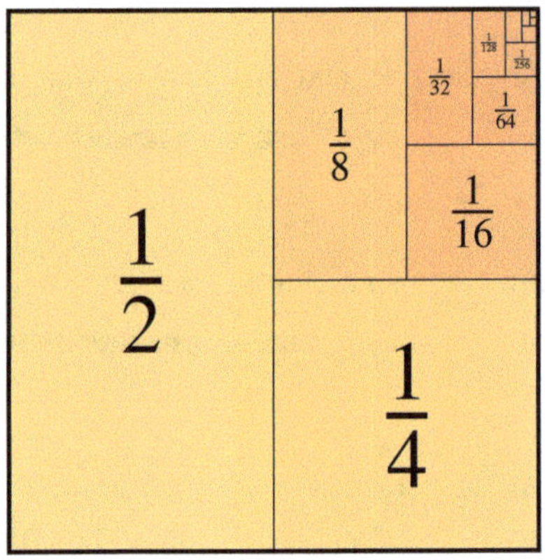

Such a sum is known as a *geometric* series—each successive term exhibits the same ratio with the previous term, getting half as large at each step.

3.3 The Most Contested Equation in Middle School

Consider the equation that has spurred countless hours of contested argumentation and debate in our middle schools and far beyond. I refer, of course, to the following identity:

$$0.999999999\ldots \quad = \quad 1.$$

It is true? Is 0.99999... fully *equal* to 1, or is there some minute or infinitesimal difference between them?

Yes, indeed, it is true. Let us prove it. Here is one common argument. Let $x = 0.999999\ldots$, whatever value that may be. It follows that $10x = 9.999999\ldots$, and we may compute the difference $10x - x$ simply by lining up the decimal and performing the subtraction:

$$\begin{array}{rcl} 10x & = & 9.999999999999999999\ldots \\ -\quad x & = & 0.999999999999999999\ldots \\ \hline 9x & = & 9.000000000000000000\ldots \end{array}$$

On the left we have $9x$, and on the right, all the 9s in the decimal parts cancel each other out, leaving just 9. So $9x = 9$, which implies that $x = 1$. So we seem to have proved that indeed $0.9999999999\ldots$ is equal to 1.

Here is another proof. Many of us know the decimal expansion of one-third to be point three repeating, like this:

$$\frac{1}{3} = 0.33333\ldots$$

This can be easily seen if one simply performs long division. But in this case,

$$0.9999999\ldots = 3 \times 0.3333333\ldots = 3 \times \frac{1}{3} = 1,$$

which shows the desired identity.

Perhaps both of these proofs can be criticized, however, on the grounds that they assume that $0.99999\ldots$ and $0.3333\ldots$ are meaningful expressions in our real number system. The first argument was really of the following form: If $0.9999\ldots$ is a meaningful expression, then its value is 1. But is it a meaningful mathematical expression? Yes, it turns out that these expressions are both instances of the geometric series. So let us discuss that before returning to this topic.

3.4 What our Decimal Notation Really Means

Our familiar place-based decimal notation for the natural numbers provides a means of describing every number in terms of the number of ones, tens, hundreds, thousands, and so on. Thus, the number 7547 is written that way precisely because it is the value of:

$$7547 = 7 \text{ thousands} + 5 \text{ hundreds} + 4 \text{ tens} + 7 \text{ ones.}$$

In other words,

$$7547 = 7 \cdot 10^3 + 5 \cdot 10^2 + 4 \cdot 10 + 7.$$

The decimal notation for the real numbers works similarly, except that fractions of a unit are considered. So the number 5.385 means:

$$5.385 = 5 + \frac{3}{10} + \frac{8}{100} + \frac{5}{1000}.$$

But now the point is that this same idea applies to infinitary decimal representations, whether they are repeating or not. Every infinite decimal expression represents an infinite sum. The decimal number 5.171717... is simply a way of writing the infinite sum:

$$5.17171\ldots \quad = \quad 5 + \frac{1}{10} + \frac{7}{100} + \frac{1}{1000} + \frac{7}{10000} + \frac{1}{100000} + \cdots$$

The number $\pi = 3.14159265358979323846\ldots$ is a notation for a certain sum:

$$3.14159265358979323846\ldots = 3 + \frac{1}{10} + \frac{4}{100} + \frac{1}{1000} + \frac{5}{10000} + \cdots$$

To have digit d in the kth place after the decimal point means that we are adding $d/10^k$ to the sum.

From this point of view, the expression 0.99999... refers to the following infinite sum:

$$0.99999\ldots = \frac{9}{10} + \frac{9}{100} + \frac{9}{1000} + \cdots + \frac{9}{10^k} + \cdots$$

This is an instance of what is called a *geometric* series.

Geometric Series

A *geometric series* is an infinite summation whose successive terms all stand in the same ratio to the previous term. That is, you get the next term always by multiplying by the same factor. For example, the series considered earlier is a geometric series:

$$\frac{1}{2} + \frac{1}{4} + \frac{1}{8} + \frac{1}{16} + \cdots$$

because each term is half as large as the previous term. In general, a geometric series has the form

$$a + ar + ar^2 + ar^3 + \cdots$$

where a is the starting value and r is the constant ratio of successive terms. Each term is obtained by multiplying the previous term by r.

Let us consider the simple case $a = 1$, which is nevertheless general enough since every geometric series is obtained from this case by multiplying through every term by factor a. So we consider

$$1 + r + r^2 + r^3 + r^4 + r^5 + r^6 + \cdots$$

Such an infinite series is said to *converge* to a limit value L, if the sequence of finite sums

$$1 + r + r^2 + \cdots + r^n$$

gets as close as desired to L as n increases. That is, for any positive degree of accuracy $\varepsilon > 0$, there is a number N of terms in the series, such that for all larger $n \geq N$, the finite partial sum is within ε of L. Such a kind of limit-value analysis is a core conception of the calculus, and some of my readers may be familiar with that.

It is clear that if $r \geq 1$, then the series cannot converge since the individual terms are at least 1 and the finite sums will grow without bound. A similar analysis works if r is negative, with $|r| \geq 1$.

So let us assume that r is smaller than 1 in magnitude, meaning that $|r| < 1$. We shall calculate the exact value of the finite partial sums. Let x be the finite partial sum

$$x = 1 + r + r^2 + r^3 + \cdots + r^n,$$

taking the terms up to r^n for some fixed n. Observe that

$$\begin{aligned} x + r^{n+1} &= (1 + r + r^2 + r^3 + \cdots + r^n) + r^{n+1} \\ &= 1 + (r + r^2 + r^3 + \quad \cdots \quad + r^{n+1}) \\ &= 1 + r(1 + r + r^2 + \quad \cdots \quad + r^n) \\ &= 1 + rx. \end{aligned}$$

So $x + r^{n+1} = 1 + rx$, which can be easily solved for x to obtain

$$x = \frac{1 - r^{n+1}}{1 - r}.$$

So this is the value of the terms in the geometric series up to r^n. The thing to notice about this expression is that because $|r| < 1$, it follows that the r^{n+1} term becomes increasingly small, toward zero. So the limit of the expression is exactly $1/(1 - r)$. Thus, we have found the value of the geometric series!

$$1 + r + r^2 + r^3 + \cdots = \frac{1}{1 - r}, \qquad \text{provided that } |r| < 1.$$

In the general case, we have

$$a + ar + ar^2 + ar^3 + \cdots = \frac{a}{1 - r}, \qquad \text{provided that } |r| < 1.$$

Application to our Initial Examples

In the special case of our initial geometric series, we have $a = \frac{1}{2}$ and $r = \frac{1}{2}$, leading to

$$\frac{1}{2} + \frac{1}{4} + \frac{1}{8} + \frac{1}{16} + \cdots = \frac{\frac{1}{2}}{1 - \frac{1}{2}} = 1,$$

which agrees with our earlier analysis of the value of that geometric series.

The case of 0.9999... is the geometric series with $a = \frac{9}{10}$ and $r = \frac{1}{10}$, so

$$0.999\ldots = \frac{9}{10} + \frac{9}{100} + \frac{9}{1000} + \cdots = \frac{a}{1 - r} = \frac{\frac{9}{10}}{1 - \frac{1}{10}} = 1,$$

which again agrees with our earlier analysis.

And similarly, 0.3333... is the geometric series with $a = \frac{3}{10}$ and $r = \frac{1}{10}$, so

$$0.333\ldots \quad = \quad \frac{3}{10} + \frac{3}{100} + \frac{3}{1000} + \cdots \quad = \quad \frac{a}{1-r} = \frac{\frac{3}{10}}{1 - \frac{1}{10}} = \frac{3}{9} = \frac{1}{3},$$

as expected.

Harmonic Series Diverges

And what about the following series, known as the *harmonic* series:

$$1 + \frac{1}{2} + \frac{1}{3} + \frac{1}{4} + \frac{1}{5} + \frac{1}{6} + \cdots$$

Does it converge? The individual terms $\frac{1}{n}$ are vanishing to zero, which is a necessary requirement for convergence of the series. But is there a finite limit value for this series?

No, in fact there is not. The harmonic series diverges to infinity. To see this, suppose that we have considered the terms of the series up to $\frac{1}{n}$, like this:

$$1 + \frac{1}{2} + \frac{1}{3} + \cdots + \frac{1}{n}.$$

Let us consider doubling the number of terms, like this:

$$\left(1 + \frac{1}{2} + \frac{1}{3} + \cdots + \frac{1}{n}\right) + \left(\frac{1}{n+1} + \frac{1}{n+2} + \cdots + \frac{1}{2n}\right).$$

Notice that we have added n extra terms in that second grouped expression, and each of them is at least as large as the final term $\frac{1}{2n}$. So the extra amount added by the second group is at least $n \cdot \frac{1}{2n} = \frac{1}{2}$.

In other words, we can always add $\frac{1}{2}$ to a finite partial sum simply by doubling the number of terms appearing in it. By doing this twice (taking four times as many terms), we can add 1 more. By doing that twice, we can add 2 more to the value. And so on. We can make the value as big as desired, simply by doubling the number of terms that we take a sufficient number of times. This series therefore diverges to infinity.

Alternating Harmonic Series

A curious variation on the divergent harmonic series is the *alternating harmonic* series, whose terms are just like in the harmonic series, except that they alternate in sign from positive to negative to positive again:

$$1 - \frac{1}{2} + \frac{1}{3} - \frac{1}{4} + \frac{1}{5} - \frac{1}{6} + \cdots$$

Does it converge? Yes, indeed it does. This can be seen by considering the nature of the graph of the finite partial sums as the series progresses:

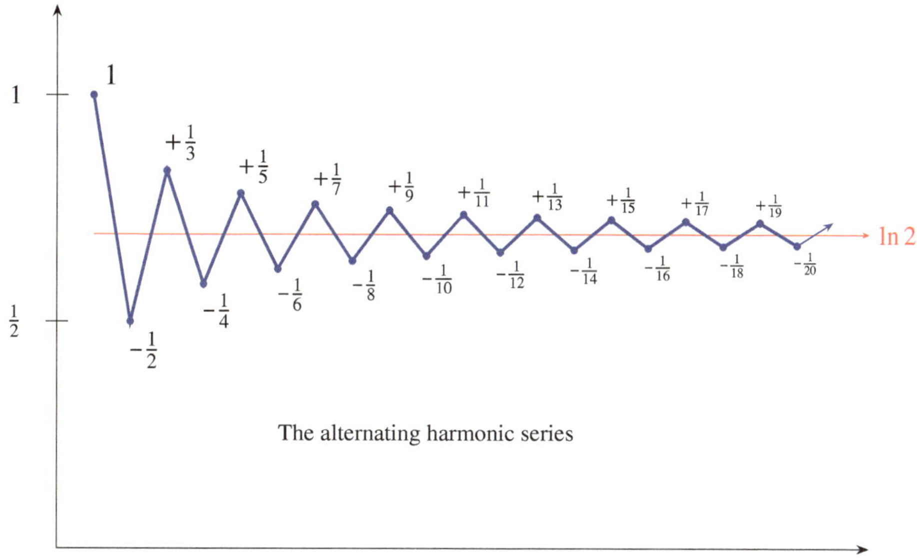

The points on this graph represent the value of the running tally as we add successive terms to the sum as indicated. Notice how the sum jumps up when we have just added a positive term and down when we have just added a negative term. And because these jumps also become smaller as the series progresses, the upper sums are progressively descending and the lower sums are progressively ascending. Furthermore, those approximations are getting closer to each other—as close as desired—because the magnitude of the jumps is $\frac{1}{n}$, which tends to zero. And so the series will converge to some limit value, which is simply the infimum of the upper approximations, which is the same as the supremum of the lower approximations. In fact, using methods from calculus, one can show that the limit value is precisely $\ln 2$, the natural logarithm of 2, which is about 0.693. So the alternating harmonic series converges to $\ln 2$.

3.5 Riemann Rearrangements

The alternating harmonic series is an example of a *conditionally convergent* series. Namely, it is convergent, with limit value $\ln 2$ as we have mentioned, but if one makes all the terms positive, then it becomes the harmonic series, which diverges to infinity.

Bernhard Riemann (1826–1866) proved a remarkable theorem about such conditionally convergent series. Namely, for any conditionally convergent series, we can rearrange the terms of the series to give the limit sum whatever value we desire. Yes, you read that correctly—rearranging the terms of a conditionally convergent infinite series can cause the final sum to have a different value, and indeed, suitable rearrangements can realize any desired sum. In this respect, the basic arithmetic of infinite sums can be quite different from what we used to with finite sums, which, of course, are invariant under rearrangement.

As an instance of the rearrangement theorem, we can rearrange the terms of the alternating harmonic series to make the limit value whatever we want. In the usual order, we have mentioned that the sum is ln 2, which is about 0.693. Suppose that we take a new target value of 0.9 for a rearranged version. Can we realize it?

Yes, indeed we can. The idea of the proof is that we build the rearrangement by taking positive terms only until they meet or exceed the target, and then negative terms until it drops below the target, after which we go back to positive terms until it reaches or exceeds the target again:

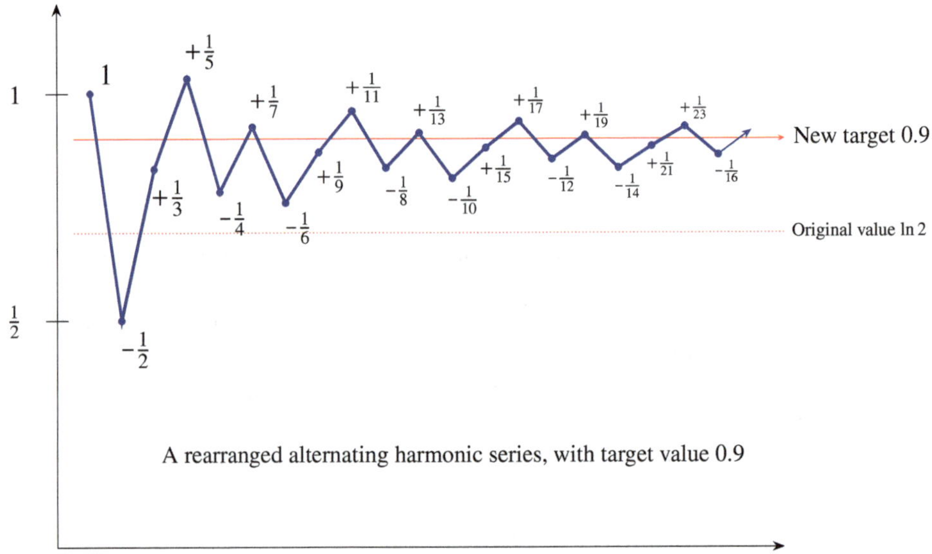

A rearranged alternating harmonic series, with target value 0.9

Notice how a few times, we took positive values twice in succession since the sum hadn't yet exceeded the new target value. The graph shows that for this target value, we should use the rearranged alternating harmonic series beginning with

$$1 - \frac{1}{2} + \frac{1}{3} + \frac{1}{5} - \frac{1}{4} + \frac{1}{7} - \frac{1}{6} + \frac{1}{9} + \frac{1}{11} - \frac{1}{8} + \cdots$$

This is no longer strictly an alternating series—it doesn't simply alternate from positive terms to negative to positive again. Rather, it uses positive terms at a little greater frequency than the negative terms. Sometimes there will also be negative values twice or more in succession.

The main point is that this procedure will produce a rearrangement realizing exactly the target value as the desired sum. Since the positive parts sum to infinity and the negative parts to minus infinity, we will always be able to zigzag back and forth across the target as the construction proceeds. And since the individual terms in the series are becoming very small, they provide a bound for how far the finite partial sums will be from the target. So the rearranged sum converges exactly to the target, as desired.

The argument is fully general and works with any conditionally convergent series and any desired target value, whether positive or negative. Simply by taking additional positive values when the current partial sum is below the target and negative values when the current sum is above the target, one can in this way attain any desired target value as the limit of the rearranged series. Question 3.7 invites readers to realize an infinite sum, or negative infinity, by a similar method, as well as a series whose partial sums oscillate wildly. It is truly remarkable.

Questions for Further Thought

3.1 Provide an analysis of Zeno's various paradoxes of motion. What is the flaw, if any, in Zeno's argument that all motion is impossible? Will Achilles overtake the Tortoise? If so, what is the flaw in the Tortoise's argument?

3.2 Is it possible for us actually to complete a supertask?

3.3 Do the limit conceptions of calculus (in particular, the ideas surrounding the convergence and divergence of series), provide an adequate resolution of Zeno's paradox? Why or why not?

3.4 Does Zeno's paradox apply just as much to time as to space? That is, mightn't Zeno argue that time passing is an illusion because to get from time A to time B, we must first get halfway there, and before that we must get halfway to the halfway point, and so on, ad infinitum.

3.5 Can you prove that every positive real number r is the value of a suitable subseries of the harmonic series? That is, for every $r > 0$, there is a set $A \subseteq \mathbb{N}$ for which $r = \sum_{n \in A} \frac{1}{n}$. Is A unique for a given r? If not, how many sets A are there?

3.6 Similarly, can you prove a generalization of Riemann's rearrangement theorem showing that if a series is conditionally convergent, then for any target value r, there are many different rearrangements of the series having sum r?

3.7 Prove that every conditionally convergent series admits a rearrangement that diverges to infinity, another that diverges to $-\infty$, and a third that oscillates wildly.

Further Reading

Huggett (2019). A summary philosophical account of Zeno's paradoxes of motion.

Credits

The red-orange figure on page 28 showing the sum of the geometric series $\frac{1}{2} + \frac{1}{4} + \cdots$ was adapted from Hamkins (2020).

4 The Method of Exhaustion

Abstract. Archimedes calculates the area of a parabolic segment by systematically filling it with triangles—exhausting the area—thereby anticipating several core ideas of the integral calculus by nearly two millennia. Here, we shall exhaustively explore the method of exhaustion, including several paradoxical failures.

4.1 Archimedes and *Quadrature of the Parabola*

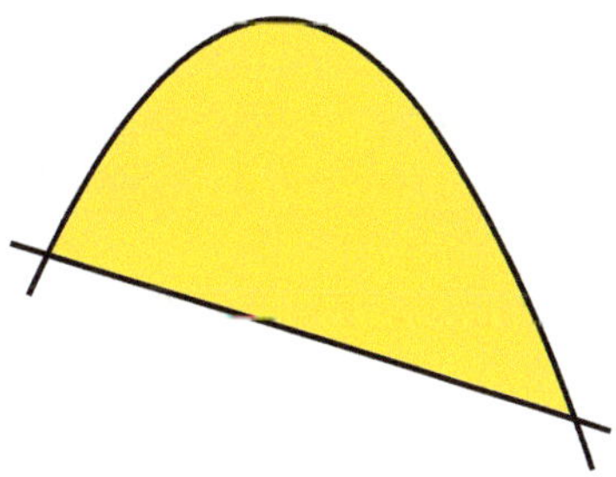

Archimedes (third century BC) set himself the task of calculating the area of a parabolic segment, the region between a parabola and a linear chord crossing it, as shown in the figure here. His treatise, *Quadrature of the Parabola*, develops the geometric theory of parabolas and contains two proofs of his solution.

What would it mean in ancient times to calculate this area? In contemporary mathematics, we might express the area of a parabolic segment with a formula involving a definite integral such as

$$\int_a^b p(x) - \ell(x)\,dx,$$

expressing the area between parabola $p(x)$ and line $\ell(x)$ as the difference of two curves.

In ancient times, however, there were no integrals, no calculus, no algebraic formulas, no analytic geometry—not even an equality sign. Archimedes's achievement was a prescient anticipation of some of those later developments, particularly his use of the method of *exhaustion*, which foreshadowed several central ideas underlying the integral calculus. The methods in his treatise reveal how very close he was to developing calculus nearly two millennia before Isaac Newton and Gottried Wilhem Leibniz.

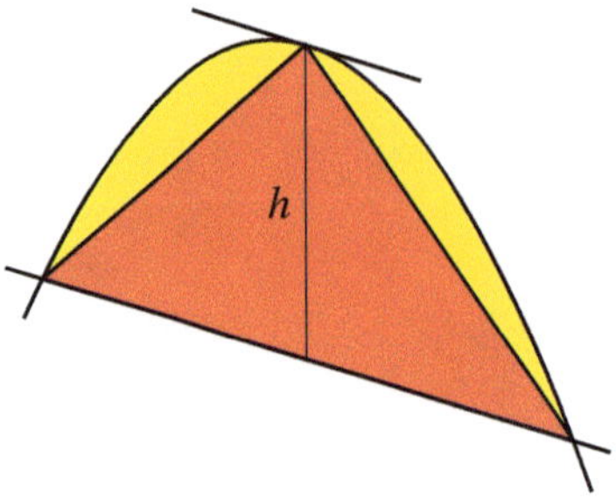

How does Archimedes express the result of his calculation? He simply states that the area of a parabolic segment is 4/3 the area of the inscribed triangle with the same base and height, as shown here. The red triangle is inscribed from the chord to the point where the tangent to the parabola is parallel to it. That point also maximizes the vertical height of the triangle, which is the vertical distance between the parallels. Furthermore, as Archimedes proves, this maximal height occurs exactly halfway along the chord. This is clear from a contemporary perspective since the difference function between the parabola and the line is a quadratic that vanishes on each end of the segment and is consequently maximized exactly in the middle.

Exhaustion

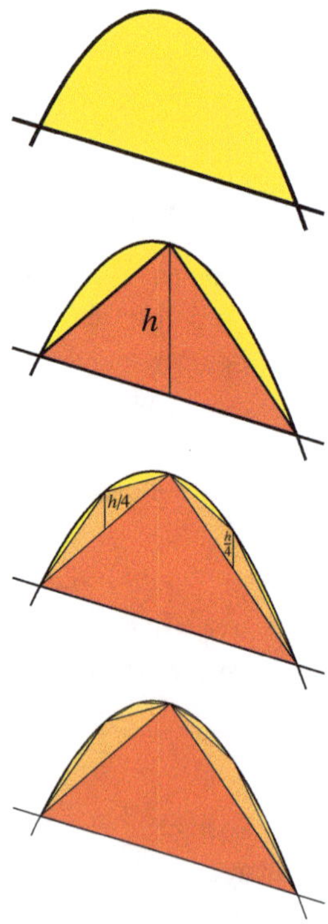

The key idea of Archimedes's geometric proof is the method of *exhaustion*, by which he inscribes further triangles to carve out progressively more of the area. Namely, after placing the main red triangle, we have in effect created two additional chords with the upper sides of the triangle. We may place two more triangles on them, shown in orange, in just exactly the same manner as the red triangle. Each of these orange triangles in turn creates two new chords, upon which we may place further triangles, and so on ad infinitum.

At each step of the construction, the heights of the triangles occur above the midpoints of their bases. Therefore, the purely horizontal widths of the triangles are individually half as big at each step. It follows that their heights (the vertical height from the midpoint of their chord base to the parabola) are quartered at each step. This is clear with contemporary analytic methods by arguing that the difference function between the parabola and the chord is a parabola with the same coefficient on x^2 as the original. And then, since doubling the horizontal distance from the center peak will quadruple the height, it follows that halving the width will divide the height by 4, as we claimed. Since the combined total width of the next-level triangles remains the same, but the height is $1/4$ as much, it follows that at each stage of iteration, the combined area of the subsequent triangles is 1/4 as much as the previous stage.

If the original inscribed red triangle has area T, therefore, then the two orange triangles have combined area $T/4$, and the four triangles at the next stage have combined area $T/16$, and so on:

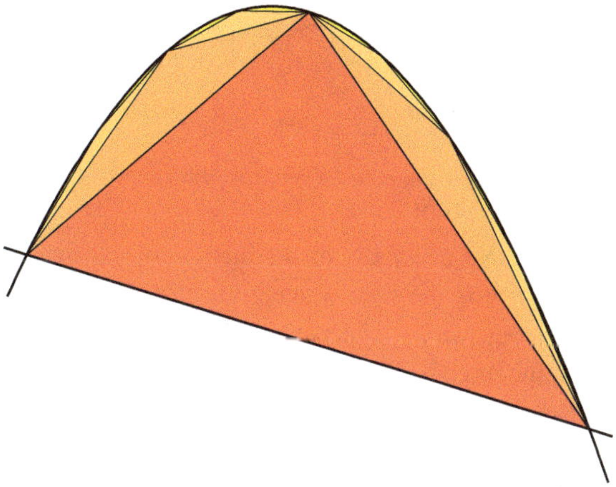

Therefore, the total area of all the inscribed triangles is the infinite series:

$$\text{Total area} = T + \frac{T}{4} + \frac{T}{16} + \frac{T}{64} + \cdots = T\left(1 + \frac{1}{4} + \frac{1}{16} + \frac{1}{64} + \cdots\right).$$

We can calculate this sum just as Archimedes did by using the figure shown here, in which the unit square is successively partitioned into shaded subsquares. The first orange subsquare at the lower left has area $\frac{1}{2} \times \frac{1}{2} = \frac{1}{4}$, the next has $\frac{1}{4} \times \frac{1}{4} = \frac{1}{16}$, and so on, making the total shaded area here equal to the infinite sum

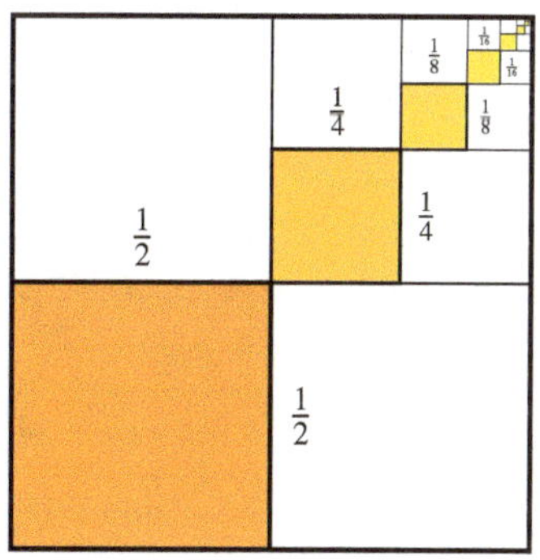

$$\frac{1}{4} + \frac{1}{16} + \frac{1}{64} + \cdots$$

The truly insightful trick of this ancient figure, however, is the observation that comes next. Namely, we may observe that exactly one-third of the whole square is shaded, precisely because each shaded square is one of exactly three squares of that same size, which altogether exhaust the unit square.

Thus, the total shaded area is 1/3, and we therefore conclude that

$$\frac{1}{4} + \frac{1}{16} + \frac{1}{64} + \cdots = \frac{1}{3}.$$

By simply adding 1 to both sides, we realize the Archimedean conclusion

$$1+\frac{1}{4}+\frac{1}{16}+\frac{1}{64}+\cdots \quad = \quad \frac{4}{3}.$$

So the total area of all the inscribed triangles in the parabolic segment is 4/3 the area of the initial inscribed triangle. One can also make this deduction by using our analysis of geometric series in chapter 3, since this series has geometric ratio $r = 1/4$, making the sum

$$\frac{1}{1-r} \quad = \quad \frac{1}{1-\frac{1}{4}} \quad = \quad \frac{1}{\frac{3}{4}} \quad = \quad \frac{4}{3}.$$

Two Parts to Proof by Exhaustion

I should like to emphasize that every use of the method of exhaustion has two essential parts. First, of course, there is the core observation that the exhaustion itself provides a lower bound for the target quantity. In our case, the area of the parabolic segment is at least the total area of the inscribed triangles, simply because those triangles fit entirely inside that region without any overlap. Since we calculated the total area of all the inscribed triangles as 4/3 the area of the initial triangle, it follows that the parabolic segment has at least 4/3 the area of that triangle.

But for an equality claim to be made, there is a critical second step to the argument. Namely, it is necessary to provide the converse relation showing that the given region is not strictly larger than that of the exhausting elements. We must establish that the inscribed triangles truly exhaust the parabolic segment. For this, we may observe that any point in the interior of the parabolic segment is eventually encompassed within one of the inscribed triangles. This is because any point in the interior of the parabolic segment has some nonzero distance to the parabola above it. But since the heights of the inscribed triangles will eventually descend below that amount—they descend as close to zero as desired—it follows that the triangle whose chord it lies above will encompass it by that stage if it has not already been captured, since the heights of the triangles are the maximal vertical distances from their chords to the parabola. Thus, the triangles do truly exhaust the entire interior region of the parabolic segment, so we have established Archimedes's quadrature of the parabola.

Anticipating the Integral Calculus

I should also like to emphasize something that I find truly remarkable about Archimedes's proof—namely, the sense in which Archimedes had anticipated one of the central ideas of the integral calculus, which nevertheless took another 2000 years to come fully to fruition. In calculus, we calculate the area of an arbitrary region between two functions using the definite integral, defined as the limit of the associated Riemann sums, finite sums of increasingly thin rectangles fit into the region as in the illustration that follows.

In calculus we usually have an analytic representation of the function, and we can thus find an explicit account of the desired area as a limit of finite sums. In this way, the Riemann sums of calculus implement a version of the ancient method of exhaustion. The trapezoidal method of Riemann sums, which have tall, thin trapezoids rather than rectangles, are an even tighter fit with Archimedes—if we halve the widths of the trapezoids at each stage, then the cumulative region after $n + 1$ steps of the trapezoid construction is literally identical to performing the Archimedes construction for n steps, although the manner of calculating this area differs because Archimedes has triangles rather than tall, thin trapezoids.

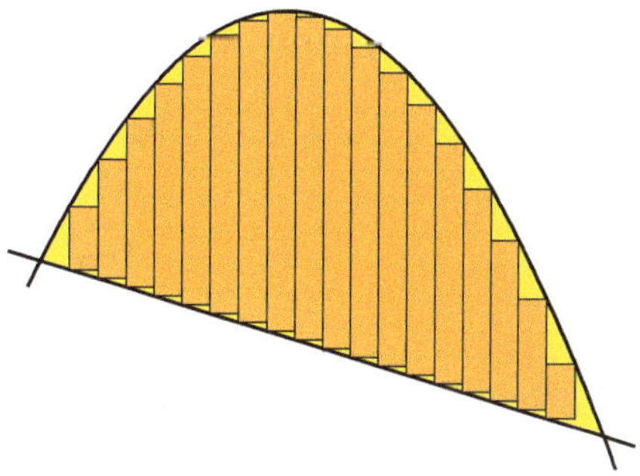

Did Archimedes thus invent calculus 2000 years before Newton and Leibniz? Give your answer in the questions at the end of this chapter.

4.2 The Cantor Set

Let us form the *Cantor set*, introduced by Georg Cantor (1845–1918), by aggressively cutting out the open middle thirds of all the line segments in our possession. We begin with the unit segment from 0 to 1, as shown on the orange top line of the figure here—what a beautiful, elegant segment it is, stretching across the page:

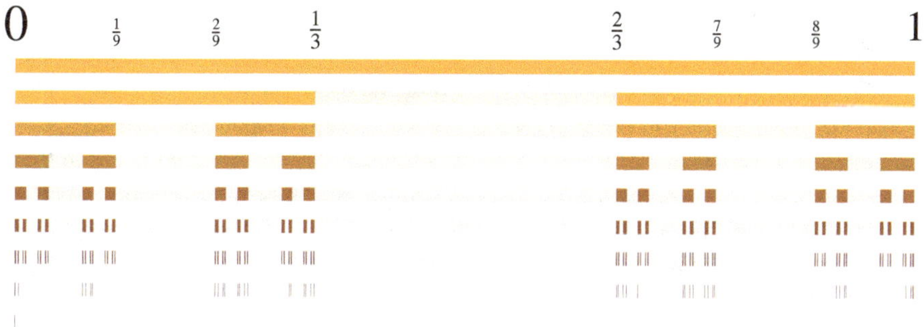

But we shall regrettably attack with knives at the center, cutting out the middle-third segment and removing the open segment between $\frac{1}{3}$ and $\frac{2}{3}$, leaving as disfigured remains the two segments, from 0 to $\frac{1}{3}$ and from $\frac{2}{3}$ to 1, respectively, shown on the next line. Continuing our attack, we cut again, removing the middle thirds of both of those segments, leaving behind four short segments on the next line. Let us continue like this forever, systematically excluding the middle third of every segment that remains at a previous stage, iterating infinitely.

The *Cantor set* is the scattered, disconnected, and discouraged set of points that remain, the points that have survived every omission—these are the numbers that do not find themselves in the interior of any of the omitted middle-third segments.

The Cantor Set Has Many Points

Are there any points remaining? Well, sure: At the very least, the endpoints of the middle-third intervals themselves survive, since we excluded always only the open middle-third intervals, keeping the endpoints. Those endpoints will never again find themselves in the interior of a subsequent middle third, so they will survive all the subsequent carnage.

But there are many more points in the Cantor set besides just the endpoints of the omitted segments. One can see this by observing first that the Cantor set is a closed set, since every time that we omitted a point from it, we also omitted an entire open neighborhood of that point. Because of this, we may consider any path proceeding downward in the diagram, choose first the left interval at a level and then the right interval, staying within the surviving points at each stage. The limit of this path will be a point in the Cantor set since it is a limit of surviving points, but it moves systematically away from every interval endpoint that it encounters. Indeed there are many paths like that do not converge to any of the endpoints of the omitted intervals.

Another way to see that there are many points in the Cantor set is to think in base 3, where we use the digits 0, 1, and 2 only. The number 0.1, for example, means 1/3 in base 3, and 0.2 means 2/3. The key observation is that the numbers in the first middle-third segment are exactly those numbers that can be represented with digit 1 in the first place after the point in base 3. And the numbers in the second two middle-thirds segment are the numbers with digit 1 in the second place. Thus, the first digit in base 3 tells you exactly which of the three one-third segments the number is in. And so on with every place—the numbers omitted at stage n of the construction are exactly those not yet excluded that have digit 1 in the nth place of their base 3 representation.

The conclusion is that the Cantor set consists precisely of those real numbers in the unit interval that have a representation in base 3 using only digits 0 and 2 and never using digit 1. For example, the number $0.220002202022020\ldots$ in base 3 is in the Cantor set. (A technical point: Although 1/3 has a representation as 0.1 in base 3, which uses the forbidden digit 1, we can also represent it as $0.022222\ldots$, which would be a base 3 analogue of the most-contested equation in middle school, the fact that $0.9999\ldots = 1$ in decimal notation; in this way, the endpoints of the intervals all have a representation without using the forbidden digit 1. For example, 7/9 is 0.21 in base 3, being $\frac{2}{3} + \frac{1}{9}$, but this can also be written as $0.202222\ldots$) So there are many points in the Cantor set. In this sense, the Cantor set is large.

The Cantor Set Has Few Points

Nevertheless, there is another sense in which the Cantor set contains almost no points. Let us simply calculate the total length of the omitted segments. At the first stage, we omit the middle-third segment, which has length 1/3. At the next stage, we omit two segments, each of length 1/9, or 2/9 together. After that, we omit four segments each of length 1/27, making 4/27 altogether. The total sum of the lengths of the omitted segments is an infinite series:

$$\text{Omitted length} \quad = \quad \frac{1}{3}+\frac{2}{9}+\frac{4}{27}+\cdots+\frac{2^n}{3^{n+1}}+\cdots$$

This is a geometric series with initial value $a = 1/3$ and ratio $r = 2/3$, so by our analysis in chapter 3, the sum is consequently

$$\frac{a}{1-r} \quad = \quad \frac{\frac{1}{3}}{1-\frac{2}{3}} \quad = \quad \frac{\frac{1}{3}}{\frac{1}{3}} \quad = \quad 1.$$

In other words, the sum of the lengths of the omitted intervals is 1, the same as the length of the whole unit interval. We have omitted an entire unit interval worth of length!

Wait, what? How can that be, if we still have points remaining? Indeed, this is precisely the sense in which the Cantor set is also small. It has measure zero, length zero as a set of numbers, since the complement of the Cantor set, the points that have been cast out of it, have the full measure of the unit interval.

Another way to describe the situation is that those open middle-third segments exhaust the unit interval in terms of length—they add up to the same total length as the whole interval—and the scattered, disconnected remains that constitute the Cantor set should therefore be seen as very small, with measure zero, despite it having so many points.

The Cantor Set as a Probability Zero Event

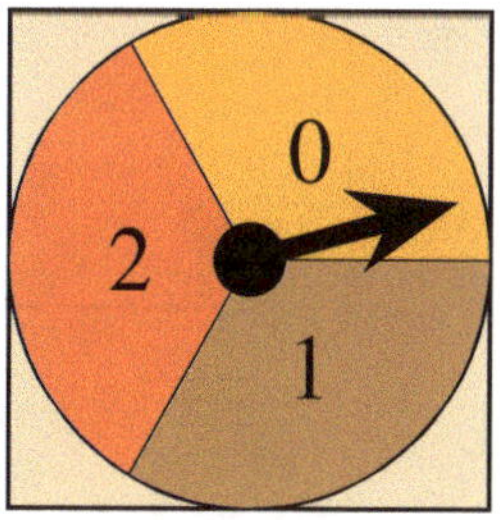

Let me next describe a probabilistic characterization of the Cantor set, realizing it as a probability zero event in a certain random process. Suppose that we have a three-way spinner with three equal segments. We shall use the spinner to determine the successive digits of a real number in base 3, which is the same as determining successively during the Cantor-set process which of the three intervals the point is in. So the successive digits will be 0, 1, or 2, each with equal one-third chance for each spin, corresponding to the first third, the middle third, or the last third. In the end, our random process produces a real number x, represented in base 3 as

$$x \quad = \quad 0.d_1d_2d_3d_4d_5\ldots,$$

where each trinary digit d_i is determined by the spinner at stage i.

How likely is it that you will never spin 1 as a digit? That is, how likely is it that you never choose the middle third? Although it is logically possible for you never to have spun 1 as a digit, this is very unlikely—a probability zero event. The probability of not getting 1 in the first 10 digits, for example, is $(2/3)^{10}$, which is about 0.017 (less than 2 percent), and the probability of not getting 1 on the first N spins is $(2/3)^N$, which goes to zero as N becomes large. So the probability of never getting 1 as a digit is zero—almost surely you will get a 1 at some stage.

The numbers in the Cantor set are exactly the lucky survivors in this process, the possible ways to never spin 1. As we said, this is a measure zero set—you are almost surely not going to land in it—but nevertheless there are indeed a lot of logically possible ways to never spin 1. You could spin 0.2200222022202... or 0.0020002002220... and so on. So although there are lots of numbers in the Cantor set, you are very unlikely to be in it by choosing randomly.

Devil's Staircase

Cantor uses his set to construct the *Devil's staircase*, a certain function exhibiting some surprising paradoxical properties:

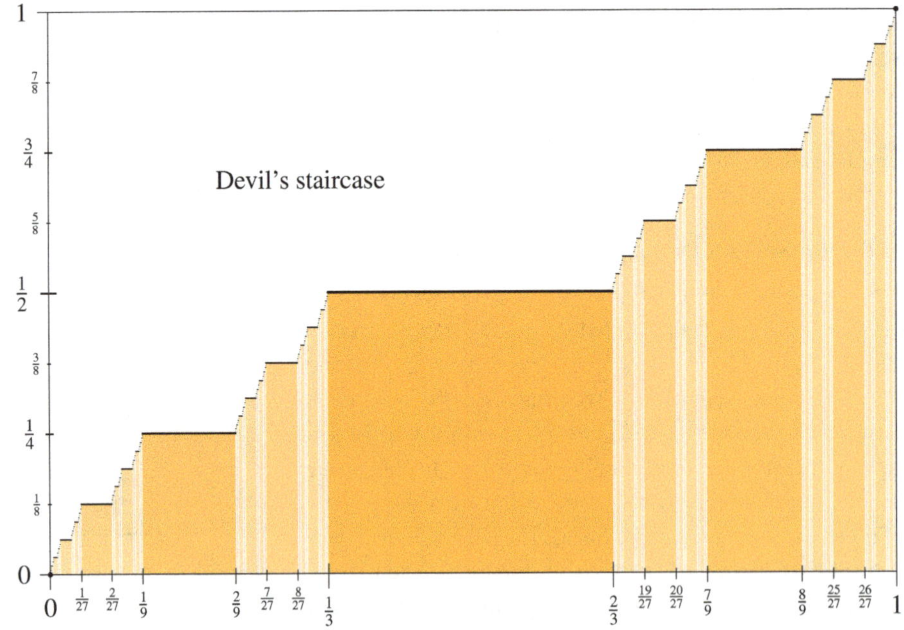

The Devil's staircase serves as a counterexample to what might have been a certain natural extension of the fundamental theorem of calculus, since it is a continuous function from the unit interval to itself, which has a zero derivative at almost every point with respect to the Lebesgue measure—it is almost everywhere perfectly flat—and yet it rises from 0 to 1 during the interval. This is how the Devil ascends from 0 to 1 while remaining almost always motionless.

To construct the function, one starts with value 0 at the left and 1 at the right of the unit interval. Interpolating between these, one places the constant value $\frac{1}{2}$ on the entire middle-third interval; this leaves two intervals remaining, one on each side. Next, one places the interpolating values $\frac{1}{4}$ and $\frac{3}{4}$ on the middle-thirds of those intervals, and so on, continuing successively to subdivide. This defines the function on the union of all the resulting middle-thirds sections, indicated in orange in the previous figure, and because of the interpolation, the function continuously extends to the entire interval. Since the function is locally constant on each of the middle-thirds intervals, the derivative is zero there, and those intervals add up to a full measure of 1, as we calculated earlier. The function is locally constant on points outside the Cantor set.

4.3 Filling a Square with Balls

What does it mean exactly to *exhaust* a region? To get a better grip on this question, let us imagine filling the unit square with circles, iteratively choosing always the largest possible circle that fits into the remaining space. Such a process is called a *greedy* algorithm, for it maximizes the area that is covered by the circles at each step, regardless of whether a more reserved strategy might lead to a more favorable overall covering proportion.

So we begin with the cental inscribed circle, shown in gold here, pressing against all four sides. Next, we place four orange circles neatly into the corners, tangent to the central circle and braced by the sides. On either side of those corner circles, we can fit more circles, and then still more in the gaps between those that we have already placed. We might enjoyably continue in this way forever, continually placing at each stage the largest possible circles that fit into the remaining gaps. I can recommend as a relaxing art project or doodle to create such a largest-circle decomposition of a given shape—why not make one the next time that you are at a meeting that has gone on too long? The resulting figures are known as *Apollonian gaskets*, when one has iteratively placed the largest possible disc into what remains of a given region, with the later circles generally placed tangent to the circles placed earlier.

In our case here, do the circles collectively exhaust the square? What does that mean exactly? Let me give two different answers. One way to answer is yes, these circles do exhaust the square, in the sense that the total area of all the circles is equal to the area of the original square. This is not obvious, and indeed it is a challenging little mathematical problem in geometry and analysis, but it is true.

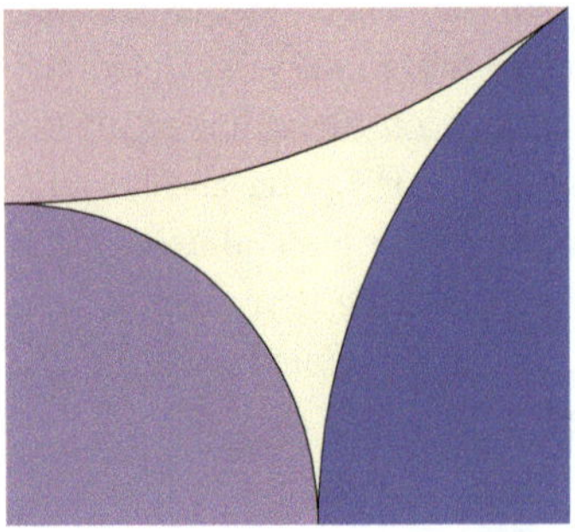

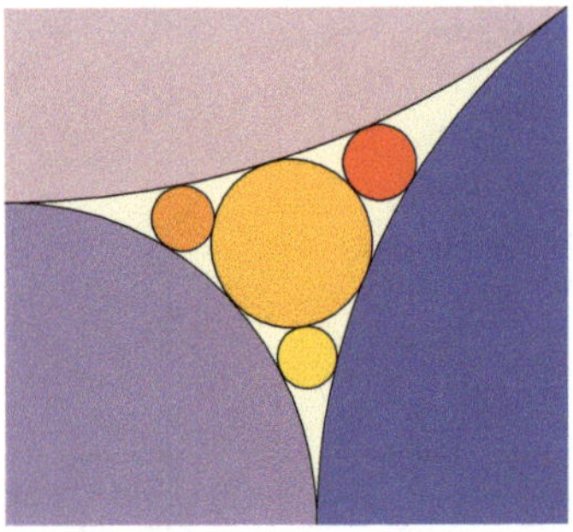

Let me describe how one can prove it. Consider an open space arising between the circles, as shown here with the yellow region between the purple circles. We will eventually place circles in this gap, of course, as shown with the orange circles placed into the yellow region. The main observation that I should like to make about this is that these circles will collectively cover a certain proportion $r > 0$ of the empty space that had been left. That is, after placing any finite number of circles, then by placing more circles, we shall always be able to cover proportion r of whatever area remains.

The long thin sections between two large circles, for example, can be eventually covered to a fixed proportion r, as illustrated here, if we simply use enough circles. Even covering merely $r = 10$ percent will be good enough for this argument, provided that we can always cover at least this proportion, and we have certainly covered far more than 10 percent in the figures shown here.

To see that covering any fixed proportion $r > 0$ is sufficient to know the circles are exhaustive, observe that if the total area of the circles were less than the area of the whole square, then at some finite stage of the covering process, we would be close enough to that limit so that covering proportion r of what remained would bring us over the limit value, which would be impossible. So if we can always cover at least a fixed positive proportion of the remaining space, then the total area of the circles will exhaust the area of the square, as claimed.

But let me now give a different answer to the question of whether the circles exhaust the square. In fact, they do not fully exhaust the square in the sense that every point in the interior of the square is part of some circle. The fact of the matter, on the contrary, is that some points in the interior of the square are not encompassed as a part of any circle, neither in the interior nor on the boundary of any circle that is placed. This is quite different than what occurred with the parabolic segment of Archimedes, since in his construction, every point in the interior of the parabolic segment is eventually encompassed within one of his triangles.

Let me briefly sketch how to show that indeed some points must remain uncovered by any circle. Consider the pseudo-triangular regions between three circles, such as the yellow region between the three purple circles considered previously. By placing the orange circle in the midst of that region, we in effect divided it into three smaller pseudotriangular regions. Let us now make an infinite sequence of hops from a psuedotriangular region to one of the three subregions, and then to a further subsubregion, and so on, like falling down a rabbit hole into these ever-smaller regions. At each stage, let us choose specifically to hop into the subregion that is furthest from the currently largest circle abutting the previous region. By continuing our hops in this manner, we shall stay ultimately within each of the successively chosen tinier regions, so our sequence will converge to a limit point. This point, however, cannot lie on any particular circle since that circle would have some finite size, but when the inscribed circles get small enough, we would specifically have hopped away from that circle and remained bounded away from it by that amount. So the limit point cannot lie on any particular circle. Thus, there are points in the square that are not a part of any of the circles. Because this hopping limit construction is extremely flexible, it follows in fact that there are many points like this, just as with the Cantor set, although the collective area or measure of these extra points will be zero, precisely because the area of the circles adds up to the total area of the square.

4.4 Paradoxical Failures of Exhaustion

Let me conclude this chapter by mentioning a few somewhat paradoxical failing instances of exhaustion.

Fat Cantor Set

Let us form a version of the Cantor set by cutting out somewhat less than a middle third at each stage. Suppose that we begin by cutting out the middle fifth, leaving two larger segments remaining. And then we cut out a segment of length 1/25 from each of them, and then cut a $1/125$ segment from the remaining segments. At stage n, we shall have 2^n segments remaining, and we cut out a segment of size $1/5^{n+1}$ from each of them. The points that survive constitute what is known as a *fat* Cantor set:

The set indeed seems to be fatter than the Cantor set, if one judges by the amount of color remaining in the later stages of the construction in comparison with the earlier figure for the classical Cantor set (page 41). To see that this set is indeed fat, let us calculate the total length of the omitted intervals, which form the infinite sum

$$\frac{1}{5} + \frac{2}{25} + \frac{4}{125} + \cdots + \frac{2^n}{5^{n+1}} + \cdots$$

This is a geometric series with $a = \frac{1}{5}$ and $r = \frac{2}{5}$, so the total value is

$$\frac{a}{1-r} = \frac{\frac{1}{5}}{1-\frac{2}{5}} = \frac{\frac{1}{5}}{\frac{3}{5}} = \frac{1}{3}.$$

Therefore, the fat Cantor set is missing only one-third of the interval! Somehow the fat Cantor set retains two-thirds of the unit interval content, even though it is totally disconnected and has no remaining intervals to omit, even very tiny ones. Much fatter Cantor sets arise by omitting a smaller amount than 1/5 and its powers at each stage. By omitting less, we can make fat Cantor sets with measures as close to 1 as desired.

Notice that we are not omitting middle fifths at each stage—we do not omit a constant proportion of what remains. Earlier, in our discussion of circles exhausting the square, we had observed that when one omits a constant positive proportion of what remains, then the omitted portion inevitably has full measure. In this fat Cantor set construction, rather, we omit 1/5 on the first step, and then two 1/25 sections in the next stage. This is strictly less than a fifth of what remains, which would have been 1/5 of 4/5, or 4/25. At the next step, we omit four 1/125 sections, which is an even smaller proportion of what remains.

The fat Cantor set does not contain any interval, and it is completely disconnected since any two points are eventually separated by an omitted middle segment. Nevertheless, one can view this situation as a failure of exhaustion, because the omitted sections add up collectively to less than the measure of the whole unit interval. The paradoxical situation is that we have cut out only 1/3 of the interval, and yet there is nothing more to cut.

Maximal but Thin Exhaustions of the Square

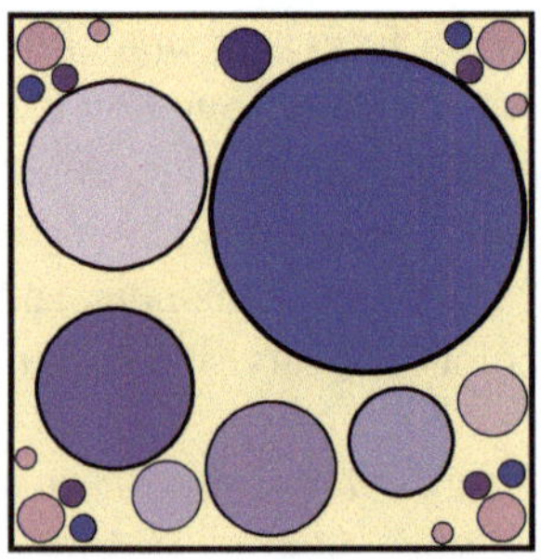

Let us similarly consider the exhaustion of the unit square by circles, but drop the greedy algorithm requirement that we use the largest possible circles at each stage. That is, let us simply place circles of any desired size and location (without overlap) until no space remains in which to place another. Perhaps we have started to place the circles as illustrated here—there is plenty of room for more! We should continue placing circles infinitely until we have a maximal collection of circles, with no possibility of adding another. The partition of the square into circles is maximal overall, in the sense that after adding all infinitely many circles, no further circle of any size will fit into the remaining space; but it is not necessarily maximal at each stage since we didn't necessarily place the largest circle that fits at a given stage. In particular, perhaps none of the circles are tangent to each other or to the square.

Will our circles constitute an exhaustion of the square? We had observed earlier that when we systematically place the largest possible circles at each stage, then the total area of the circles adds up to the total area of the square. In that sense, such a decomposition of the square by circles would be an exhaustion, although we had also observed that not every point in the interior of the square will occur on one of the circles.

When we drop the requirement to place the largest possible circles, however, then the areas of the circles may in fact add up to considerably less than the area of the square, even when we have achieved a maximal configuration. To see this, imagine that we initially place a circle of area 1/8, then a circle of area 1/16, then a circle of area 1/32, and so on. By placing these circles at each step as far from the others and the edges of the square, however, we can ensure that they are distributed ever more finely over the square, and thereby leave no room for further circles to be placed. Meanwhile, the total area of these circles will be merely

$$\frac{1}{8} + \frac{1}{16} + \frac{1}{32} + \cdots \quad = \quad \frac{1}{4}.$$

We will have achieved a maximal configuration of circles while using only one-quarter of the area of the square. The points that remain and are not on any of the circles will constitute three-quarters of the area content of the square, even though we cannot place any additional circles without overlap.

A similar idea will allow us to find maximal circle configurations with as small a total area as desired. Such configurations amount to a paradoxical failure of exhaustion akin to the fat Cantor sets, since we will have placed so many circles that no more circles fit, and yet the total area of the circles that have been placed might be very tiny.

Questions for Further Thought

4.1 Did Archimedes invent calculus? How close was his method to what we have today with the integral calculus? Which key calculus ideas are missing from the ancient method?

4.2 Use Archimedes's square diagram to show that

$$\frac{3}{4} + \frac{3}{16} + \frac{3}{64} + \cdots + \frac{3}{4^n} + \cdots = 1.$$

Explain how this can be interpreted as the assertion that 0.3333... in base 4 is equal to 1. What do you think about 0.7777... in base 8? Or 0.1111... in binary? Can you find the general phenomenon?

4.3 Discuss the senses in which the unit square is exhausted by the maximal inscribed circles and the senses in which it is not. Do you think Archimedes would accept this as an exhaustion? How does your analysis apply to the Cantor set and its complement?

4.4 Mount an analysis of the Cantor set arising by omitting middle fourths at each stage—that is, removing from each segment remaining at each stage the middle one-fourth of it. Is this an exhaustive process? What is the total length of the omitted segments? (Hint: Use the constant-proportion argument that I describe in this chapter.) Are there points that survive all these cuts? What about omitting middle fifths, or middle sixths in this way?

4.5 Develop analogues of the Cantor set in two and three dimensions. Describe your constructions, with pictures, explaining how they proceed. What can you say about them?

4.6 What are the higher-dimensional analogues of the circle decomposition of the square? In what senses should we consider them as exhaustions or not?

Further Reading

Swain and Dence (1998). A nice account of the Archimedean quadrature of the parabola.

Credits

The material on the Devil's staircase was adapted from Hamkins (2021b, p. 69). I learned the argument about the inscribed circles exhausting the area of the square, when choosing always of the largest size, from my CUNY colleague Joseph Maher.

5 Supertasks

Abstract. Is it perfectly sensible or fundamentally incoherent to speak of completing a task with infinitely many steps?

5.1 Thomson's Lamp

You are sitting comfortably in your study reading in the twilight. It is becoming a little dark, so you turn on the light. But after a little while, you realize that the brightness of the lamp has ruined the twilight mood, so you switch it off. But now it is becoming truly too dark to read, so you turn it back on again. But of course, it is still too bright, so you quickly turn it off, and then on again, and off, over and over, ever more rapidly in the vacillating panic of your indecision:

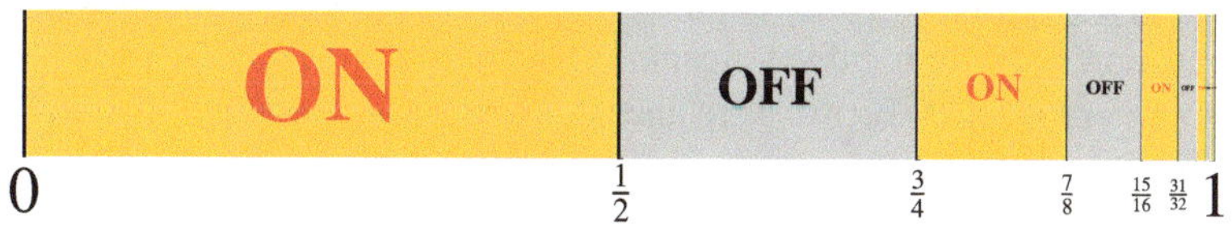

Thus you have found yourself undertaking the supertask of *Thomson's lamp*, introduced by James F. Thomson (1954) for the purpose of considering the nature and intelligibility of supertasks—tasks, that is, with infinitely many steps. The lamp is turned on at time $t = 0$, off at $t = 1/2$, on again at $t = 3/4$, and so on.

Question. What is the ultimate state of the lamp when the supertask is completed? Specifically, at time $t = 1$, is the lamp on or off?

One can make various physics-based objections to the realizability of the puzzle. The fuse could blow; it is not possible to flick the switch so rapidly toward the end since each instance would take some minimum length of time; the electric current flows at some finite speed, and so it would take some minimum time for the filament to come to light;

photons are discrete, and there aren't infinitely many available for the separate intervals of brightness; and so on. Obviously, we cannot undertake this supertask as an actual physical process, and from this point of view, the thought experiment of the puzzle may be a failure.

Nevertheless, to my way of thinking, all those objections are beside the point. The implicit question of interest here is not whether the task can be physically undertaken in a universe obeying our physical laws, but rather whether the very concept of this supertask is fundamentally intelligible or unintelligible. Does it make sense to undertake a supertask of this form even as a logical possibility? And furthermore, if it is intelligible, does the supertask procedure determine the ultimate state of the lamp, at time $t = 1$?

On the one hand, one might try to argue that the lamp should be off at $t = 1$, since every time it had been turned on before that moment, it was subsequently turned off. So there is no reason for it to be on at $t = 1$. But of course, one can argue symmetrically that at every moment before $t = 1$ when it had been turned off, it was subsequently turned on, so there is no reason for it not to be on at $t = 1$.

Perhaps it is clear that either outcome, with the lamp being on at $t = 1$ or off at $t = 1$, is logically compatible with the description of the state of the lamp before $t = 1$. The prior lamp activity simply does not determine the state of the lamp at $t = 1$. The on/off state would be a discontinuous function of time. From this point of view, the puzzle of Thomson's lamp is not so very puzzling after all.

5.2 Ordinary Tasks Seen as Supertasks

In the discussion of Zeno's paradox in chapter 3, we had considered the forward-oriented version of his argument, where to walk from here to there, from point A to another point B, you must first get halfway there, but after that, you must get halfway again from there to the destination, and then halfway again, and so on. To complete the simple task of walking from here to there seems to require that you complete each of these infinitely many subtasks:

If it is sensible to complete the task of walking from here to there, and if completing this task involves completing these infinitely smaller tasks, then it would seem to be sensible to speak of completing a supertask.

The Always-On and Always-Off Thomson Lamps

We can seem to divide almost every ordinary task into infinitely many subtasks in this same manner. Is nearly every ordinary task actually a supertask? Consider, for example, the always-on Thomson's lamp—we turn a lamp on for one full minute. But this task is the same as the supertask of leaving it on for the first half-minute, and also on for the next quarter-minute, and on for the next eighth of a minute, and so on. It is on the whole time:

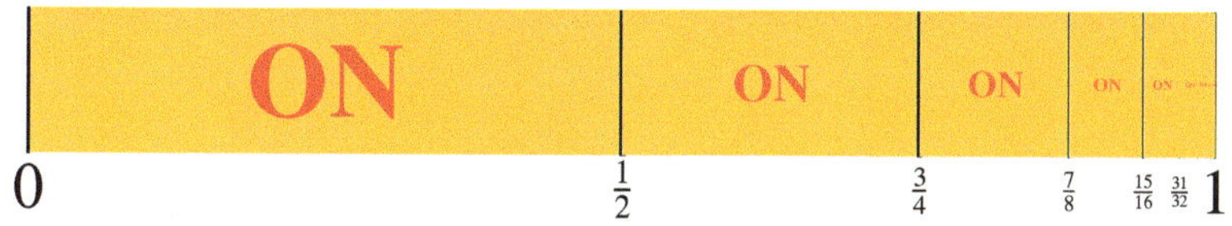

And it is similar for the always-off Thomson lamp:

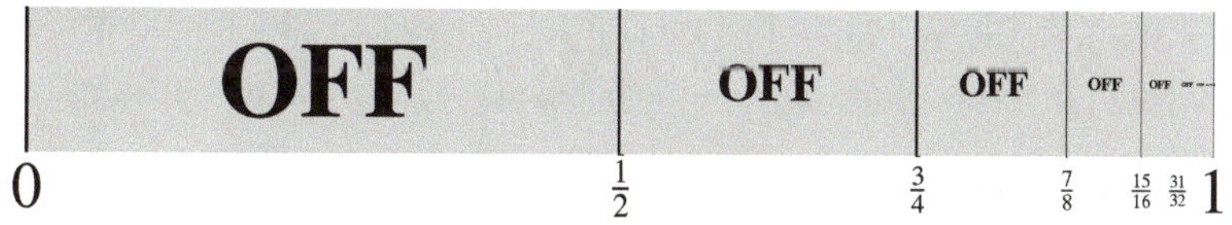

Do these tasks count as supertasks? In each case, an ordinary task is divided into progressively tinier subtasks in such a way that completing the whole task amounts exactly to completing all the constituent subtasks.

5.3 Infinite Sums as a Supertask

Consider the summation of an infinite series as a supertask:

$$\frac{1}{2}+\frac{1}{4}+\frac{1}{8}+\frac{1}{16}+\frac{1}{32}+\cdots \quad = \quad 1.$$

Is it intelligible to think of an infinite sum as being actually completed? Can we complete the task of adding up infinitely many numbers?

A Reduction of Supertask to Task via Potentialism

The ordinary account of the meaning of an infinite sum, after all, can be taken as *potentialist* in nature, as opposed to *actualist*. That is, by definition, an infinite series sums to a particular value if no matter how close to that value you want to get, for any $\epsilon > 0$, there is a stage N in the series such that all the finite partial sums beyond that are indeed within ϵ of that target value.

In other words, the standard account of infinite sums never refers directly to the process of adding up infinitely many numbers; but rather, it defines the infinite sum entirely in terms of the asymptotic limiting behavior of the finite partial sums. This definition in effect reduces the meaning of the supertask of adding up all the numbers to the nature and asymptotic behavior of the finite tasks of adding up only finitely many of the terms.

To my way of thinking, there are two directions in which one can take this observation to a philosophical conclusion about the corresponding infinite-sum supertask. On the one hand, the reductionist observation can be seen as proof that supertasks are perfectly intelligible—indeed, because the very definition of the infinite sum is reductive, our understanding of the infinite-sum supertask is no more problematic than considering the asymptotic behavior of the ordinary finite-sum tasks. So there is nothing particularly mysterious about the infinite-sum supertask because the limit definition tells us exactly what it means.

On the other hand, it seems, one could also reasonably take the reductionist observation as illustrating the basic incoherence of the actual supertask, since one cannot seem to define the sum of an infinite series except by speaking of the finite partial sums leading to it. If it were intelligible to speak of completing the process of an infinite sum, then the definition of the value of a series would refer directly to that conception and not have to resort to the reductive account in terms of the asymptotic behavior of the finite partial sums.

5.4 Deal with the Devil

Because of a shrewd investment, you carry on your person infinitely many $1 bills, numbered 1, 3, 5, and so on. You enter a dimly lit underground bar and encounter the Devil, sitting at a table piled high with money. He notices you immediately and beckons you to join him. He explains to you his admiration for your particular bills and is willing to pay you a premium to buy them from you. Specifically, he is willing to pay $2 for each of your $1 bills. To carry out the exchange, he proposes an infinite series of transactions, in each of which he will hand over to you $2 and take from you $1. The first transaction will take a leisurely half-hour, with drinks, the second a quarter-hour, the third an eighth of an hour, and so on, so that after an hour, the entire exchange will be complete. The Devil takes a sip of whiskey while you mull it over. Should you accept his proposal? Perhaps you think you will become richer, or perhaps you think with infinitely many bills, it will make no difference. At the very least, you think, it will do no harm, so you sign the contract and the procedure begins. How could the deal harm you?

It appears initially that you have made a good bargain, because at every step of the transaction, he pays you $2 and you give up only $1. The Devil is particular, however, about the order in which the bills are exchanged. The contract stipulates that in each sub-transaction, he buys from you your lowest-numbered bill and pays you with higher-numbered bills. Thus, on the first transaction he accepts from you bill number 1 and pays you with bills numbered 2 and 4. On the next transaction he buys from you bill number 2 (which he had just paid you) and gives you bills numbered 6 and 8. Next, he buys bill number 3 from you with bills 10 and 12, and so on. When all the exchanges are completed, what do you discover?

Think about it...

You have no money left at all! The reason is that at the nth exchange, the Devil took from you bill number n, and never subsequently returned it to you. So while it seemed as though you were becoming at least no poorer with each exchange, in fact the final destination of every dollar bill in the transaction is under the Devil's control. The Devil is a shrewd banker, and you should have paid more attention to the details of the supertask transaction to which you agreed. Similarly, when we analyze other supertask algorithms, we should take care not to make inadvertent assumptions regarding them, which may be true only for finite processes.

5.5 Balls in a Sack

Having lost a bet with your friends, you stand with humiliation holding a large, empty wool sack near an endless supply of billiard balls, anxious with the intent of undertaking for their amusement a certain absurd and hopeless but laborious supertask procedure:

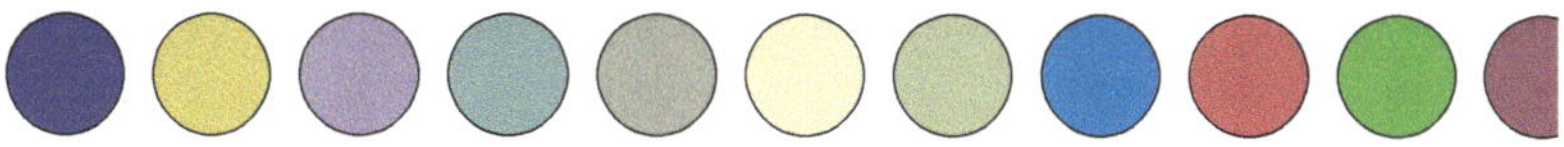

You begin by placing two billiard balls in the sack, but then immediately remove one of them and discard it. Next, you place two additional balls into the sack—so it now has three balls—and then remove one of them, tossing it aside. And so on, with ever more urgency as your friends look on. At each stage, you must add two fresh balls to the sack but may then remove one. You must undertake the steps faster and faster in such a way that you are able to complete all the steps in finite total time—the first step takes a half-minute, and the next a quarter-minute, and so on, so that after 1 minute, the entire supertask is completed, as your friends erupt with laughter.

The hopeless, impossible challenge they have set for you is: To achieve an empty sack at the end. Can you do it?

Think about it…

The sack is clearly getting heavier at each step with one additional ball, so after n stages, it holds n balls. Perhaps this makes it obvious that the sack will have infinitely many balls in it at the end—they had set you an impossible task, just to watch you fail. Indeed, if at each stage, the ball that you remove is one of the two balls just added, then the prior balls in the sack will never be removed, so you will be steadily accumulating balls in the sack that are never removed. With this procedure, therefore, the sack will have infinitely many balls at the end.

Nevertheless, I claim that you can turn the tables on your friends and achieve the impossible goal. You can undertake the process in such a way that the sack will be empty at the end. Yes, empty!

What? How can you make it empty? Let me explain. I am going to be particular about which balls you are to remove at each stage. Let us imagine that the balls are numbered like the positive integers:

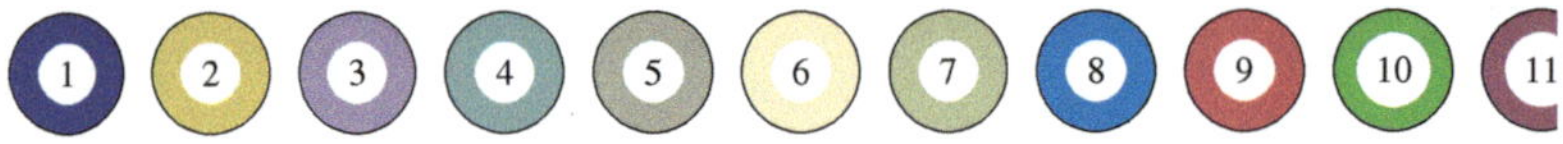

At each stage, you add two new balls to the sack, but I insist that the ball to be removed should always be the *lowest-numbered* ball currently in the sack. You should always choose and discard the lowest-numbered ball available. For example, if you add the balls in order, then you begin by adding balls 1 and 2, and removing ball 1; next add balls 3 and 4 and remove ball 2; add balls 5 and 6, remove 3; and so on.

With this procedure, I claim, there can be no ball in the sack at the end. Every ball has some finite number, but ball n is removed at stage n and discarded. So none of the balls can be in the sack at the end—every conceivable ball that could be there was already dealt with and ultimately discarded along the way. Your friends stand stupified at your success with the task that they had thought impossible.

The procedure works just as well no matter how you add balls to the sack, so long as you always remove and discard the lowest-numbered ball available in the sack. If there were any balls left at the end, there would have to be a lowest-numbered such ball, and that ball would have been the lowest-numbered ball available at some finite stage, since there are only finitely many balls with lower numbers, so it would have been already removed—a contradiction. So there can't be any balls left at the end.

The balls in a sack puzzle is fundamentally similar, if you think about it, to the Deal with the Devil puzzle considered earlier, since the surprising outcome with an empty sack was essentially the same idea that the Devil had employed to get your money. Both puzzles show that one must sometimes take care with the details of an infinitary process to ensure that it is not underspecified in a way that allows for the possibility of a surprising or unanticipated outcome.

A Stochastic Supertask

Let us consider a stochastic variation of the balls-in-a-sack supertask, one for which the ball chosen for discarding at each stage is chosen at random uniformly amongst the balls currently in the sack. What outcome should we expect when this supertask is completed?

We began by placing two balls in the sack. For any particular one of them, the probability that it survives the first discarding is 1/2. At the next stage, it will find itself amongst three balls, and the probability that it survives again (given that it has survived the first stage) is 2/3. And at the next stage, given that it has not yet been discarded, the probability that it survives again is 3/4. And so on. The chance that it survives n rounds is the chance that all those situations come to pass, which therefore has probability

$$\frac{1}{2} \cdot \frac{2}{3} \cdot \frac{3}{4} \cdots \frac{n}{n+1}.$$

Notice how almost everything here cancels from one term to the next, like so:

$$\frac{1}{\not{2}} \cdot \frac{\not{2}}{\not{3}} \cdot \frac{\not{3}}{\not{4}} \cdots \frac{\not{n}}{n+1} \quad = \quad \frac{1}{n+1} \to 0,$$

so this converges to 0 as n becomes large. In other words, the probability that this ball is never picked at any stage is 0. Thus, almost surely—this is simply a phrase that means *with probability 1*—it will eventually get discarded. We should expect that that first ball eventually will get discarded.

A similar calculation works with any other particular ball. Consider any fixed ball in the sack when there are $k+1$ balls in the sack. The probability that that ball is not chosen right then is $k/(k+1)$. And given that it survived that selection, the probability that it survives again at the next stage is $(k+1)/(k+2)$, and so on. The probability that it survives for n steps in this way is

$$\frac{k}{k+1} \cdot \frac{k+1}{k+2} \cdots \frac{k+n-1}{k+n} = \frac{k}{k+n} \to 0,$$

which again (because of the cancellation) converges to 0 as n becomes large. So any particular ball added at any stage has a probability 1 chance of being discarded.

It follows that for any particular stage, all the balls in the sack at that stage will almost surely get discarded eventually. Since this is true for every stage, it follows that the expected outcome of the whole process is that the sack is empty at the conclusion—almost surely every ball will be discarded. So this would have been another way for you to succeed against your friends.

Two Issues in the Philosophy of Probability

To my way of thinking, the probability calculation that we just undertook leads immediately to two interesting philosophical issues in the philosophy of probability that I should like to discuss briefly.

First is the fundamental distinction between a probability zero event and a logically impossible event. Based on our calculation, after all, we had said that the probability that a given ball survives undiscarded is zero, and yet, it seems logically possible that a given ball is simply never chosen for discarding. But shouldn't probability zero mean that the event is logically impossible?

The distinction is a fundamental feature of probability theory, particularly in infinite probability spaces. For an event to have probability zero means that the likelihood of that event is smaller than any positive number. But this doesn't mean it is logically impossible. Consider the problem of selecting a point uniformly at random from a disk—we throw a dart at the target to select a point. The probability of hitting a certain region of the target is proportional to the area of that region, and since points have zero area, the probability of hitting any given point is zero. For any given point, you are very unlikely—probability zero—to hit *exactly* that point with the center of the dart tip. But these are all logically possible events, and indeed, you must hit *some* point.

A dual version of this same issue concerns probability 1 events. A property or event holding *almost surely* means just that it holds with probability 1, and this is not the same as being logically necessary. Almost surely, no ball is left at the end in the sack, but it is logically possible for a ball to be left at the end. With the dartboard, almost surely you will miss a particular fixed point on the target—probability 1—but it is logically possible that you do *not* miss that point.

A second philosophical issue that I should like to mention is the move we had made after saying that every particular ball will almost surely be discarded to the conclusion that almost surely every ball will be discarded. We had in effect mounted the argument:

1. Every ball will almost surely be discarded.
2. Therefore, almost surely every ball will be discarded.

Is this legitimate? What is the justification? This conclusion is indeed a standard part of the standard theory of probability—but why?

The situation may be truly puzzling, in light of the following similar deduction in the dartboard case, which is definitely not part of the standard theory of probability:

1. Every point on the dartboard is almost surely missed by the dart.
2. Therefore, almost surely every point on the dartboard is missed by the dart.

What explains the difference? Why does the standard theory of probability accept the first deduction and not the second? The reader is invited to discuss the matter in the questions at the end of this chapter. For now, I aim merely to open the question on the matter.

5.6 The Chocolatier's Game

Consider the *Chocolatier's Game*, an infinite game played between two players, the Chocolatier and the Glutton. On each move, the Chocolatier serves up finitely many unique, exquisite chocolate creations on a platter, and the Glutton selects one to eat. Play continues in just that way—at each stage, the Chocolatier adds finitely many chocolates to the platter, and the Glutton consumes one of those available. Uneaten chocolates, if any, simply accumulate, and the Glutton may select them on a later round.

After infinitely many rounds, the Glutton wins if he has eaten all the chocolate—that is, if every chocolate that was ever served was also eventually consumed. Otherwise, if there is any chocolate left uneaten, the Chocolatier wins.

What? How could the Glutton possibly win? After all, the Chocolatier can be placing huge numbers of chocolates on the platter each round, but the Glutton can eat only one. Won't he simply get further and further behind?

And wait a minute—what do I mean anyway by saying that the winner is determined "after infinitely many rounds"? How can there be anything after infinitely many rounds? Well, this game is a supertask, a game with infinitely many turns, and just as with Zeno and the billiard-ball supertasks we have considered thus far, we may imagine completing all infinitely many steps of the game in a finite amount of time.

Question. Who wins? Does either player have a winning strategy?

Think about it...

The Glutton has a simple winning strategy. Namely, the Glutton can simply imagine the chocolates on the serving platter as forming a queue, with the new chocolates added always at the back of the queue but the chocolate to be eaten is chosen always from the front. This is the same as the first-in first-out perishable goods management strategy commonly used in restaurants—to minimize spoilage, they always use the oldest ingredients first and place new deliveries at the back of the cupboards:

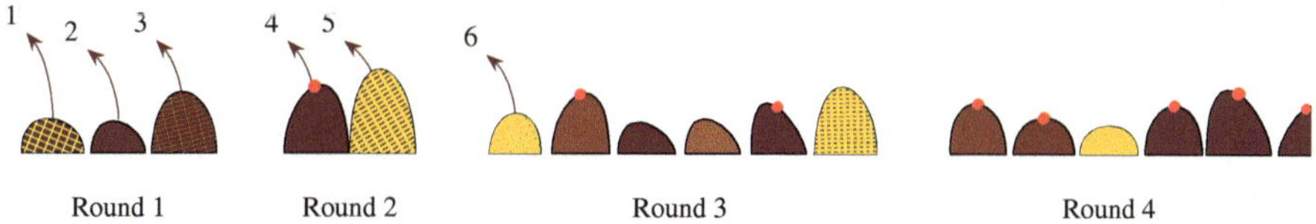

To see that this will be a winning strategy for the Glutton, consider any particular chocolate that is served. At the moment that it is served, the Glutton places it in the queue in a certain position, and there are only finitely many chocolates ahead of it in the line. Since no new chocolates will ever be placed in front of it in the queue (but only far behind, at the back), we already know exactly which round the given chocolate will be eaten during. So every chocolate that is served will be eaten at some stage, and the Glutton will win.

Infinite Chocolates Each Round

Consider a variation of the game in which the Chocolatier is allowed to serve an *infinite* row of chocolates each turn. We can imagine the chocolates on each turn numbered with positive integers, 1, 2, 3, and so on.

This variation of the game immediately ruins the chocolate-queue strategy described for the original case, since if the second-round and later chocolates are to be added to the end of the queue, the Glutton will never get to that part since he will spend all his turns eating the first-round chocolates only. The later chocolates will remain uneaten.

Can the Glutton win the variation allowing infinitely many chocolates each round? Readers of a certain age may be tempted to imagine an infinitary version of Lucy and Ethel in the chocolate factory sketch on *I love Lucy*, with chocolates emerging ever more rapidly on the conveyor belt.

Think about it...

Yes, indeed he can. Let the Glutton arrange the chocolates not in a linear queue as before, but in an infinite array, placing the first-round chocolates on the top row, the second-round chocolates on the next row, and so on. The Chocolatier is not obliged to serve infinitely many chocolates each round, and if only finitely many are served, then some rows in the array may remain only partially filled, which will be fine. The key idea of the strategy is that the Glutton should follow a winding eating path through the array, starting with the chocolate at the upper left, and then eating the chocolates in the order that they appear on the winding path, back and forth, on the finite diagonals as shown in the figure:

If a location happens to be empty when the path comes to that place, the Glutton can simply choose another chocolate for eating. This is a winning strategy for the Glutton in the infinite-chocolates variation, I claim, since every chocolate is placed in the array, and the winding path specifies a turn on which it will be eaten (if not earlier). Notice the critical detail that the winding path never visits a row before that row has been played—it takes at least n steps to get to any chocolate on row n.

Memory-Free Strategy

The strategies that we have described thus far for the Glutton depend not only on the set of chocolates being offered but also on the history of how those chocolates were served up. It is interesting to inquire whether there could be a *memory-free* winning strategy for the Glutton—one that depends at each stage only on the assortment of chocolates on the serving platter, and not on the history of how chocolates were added to the platter.

One consideration is that if the Chocolatier is allowed to repeat chocolates (that is, if he is allowed to place identical copies of earlier chocolates again at a later stage), then it is easy to see that there can be no such memory-free winning strategy for the Glutton. The reason is that the Chocolatier can simply start with any assortment of at least two chocolates and let the Glutton make his choice—perhaps the most delicious-looking one—but then the Chocolatier simply replaces that chocolate with an identical chocolate on the next move and all subsequent moves. If the Glutton is following a memory-free strategy, then the menu will always be the same, and the Glutton will always make the same most-delicious choice. The slightly less enticing chocolates will never get eaten, so the Glutton will lose.

It will therefore make for a more interesting game if we require the Chocolatier always to play new distinct chocolate types, without repetition. The Chocolatier must be increasingly creative to offer always only new chocolate types. For this version of the game, there are some interesting things to say.

How creative can the Chocolatier be? Let us suppose that the set of all possible chocolates that the Chocolatier can make—his complete menu of possible chocolates—can be formed on a list ordered like the positive integers. So there is chocolate type 1, type 2, type 3, and so on. At each stage, he is choosing to make some of these possible chocolates from his repertoire and then serving them up. The menu order may be totally unrelated to the order of preparation and serving of the chocolates.

In this case, I claim, the Glutton has a memory-free winning strategy, which works even if the Chocolatier serves up infinitely many chocolates each round. Namely, the Glutton should at each stage simply select the chocolate that appears earliest in the menu order. If chocolate 1 is available, then he should select that with the highest priority, but if not, look for chocolate 2, and so on. If he always eats the lowest-numbered chocolate offered, then there can't be any chocolate left uneaten at the conclusion because if there were, there would have to be a lowest-numbered chocolate at that time. But since there are only finitely many chocolates ahead of this chocolate in priority, it would have been the lowest-numbered chocolate at some stage, when it would have been eaten. Thus there can be no chocolates left at the end and the Glutton has won.

There is actually much more that one can say about the Chocolatier's Game. For example, what if the menu space of possible chocolate types has the character of the real numbers, rather than the positive integers—can the Glutton still win? When the menu of possible chocolates is uncountable, things get mathematically interesting.

Questions for Further Thought

5.1 The standard definition of the value of an infinite sum $\frac{1}{2} + \frac{1}{4} + \frac{1}{8} + \cdots$ does not refer to the process of adding up infinitely many numbers; rather, it is reductive to finite sums, defining the final value of the infinite sum in terms of the asymptotic behavior of the finite partial sums. What evidence, if any, does this observation provide as to the intelligibility or unintelligibility of the supertask of an infinite summation?

5.2 Is there a manner of undertaking the balls-in-a-sack supertask so as to leave exactly one ball at the end? Exactly two balls? Which sets of balls can be realized as the limit set by a suitable undertaking of the process?

5.3 Consider a variation of the random-selection balls-in-a-sack supertask, where we add considerably more balls at each stage. Suppose that we double the number of balls at each stage, in such a manner that we are selecting one ball from 2^n at stage n. Should we still expect the sack to be almost surely empty?

5.4 Is the deduction from *Every ball is almost surely discarded* to the conclusion *Almost surely every ball is discarded* legitimate? And what of the similar deduction in the dartboard example, from *Every point is almost surely missed by the dart* to the conclusion *Almost surely every point is missed by the dart.* What, if anything, distinguishes these cases?

Further Reading

Laraudogoitia (1996). A beautiful account of the beautiful supertask.

Hamkins (2002). An introduction to supertask computability.

Credits

Thomson's lamp appears in Thomson (1954). My account of the deal with the Devil was adapted from Hamkins (2002); I first heard of the allegory from Peter Vallentyne in 2001. The balls-in-a-sack paradox is also known as the *Ross-Littlewood* paradox, due to John Edensor Littlewood, appearing in his 1953 article, and later developed further by Sheldon M. Ross. I introduced the Chocolatier's Game on MathOverflow in my answer to a question of Richard Stanley's at Hamkins (2021a), with a follow-up question of my own Hamkins (2021d), and the account here was adapted from my book-in-progress on infinite games, *Infinite Games: Frivolities of the Gods*, Hamkins (2025a). The *I Love Lucy* connection was suggested by my copyeditor Susan.

6 The Infinite Coastline Paradox

Abstract. How unexpected difficulties in a practical measurement problem challenged fundamental conceptions of length and led to a new theory of dimension, while revealing mathematical monsters lurking at small scales.

A century ago, Lewis Fry Richardson (1881–1953), a noted polymath with prizes in metereology and scientific peace research now named in his honor, had a mathematical theory for predicting the likelihood of war between two countries based in part on the length and nature of their common border. Aiming to test his theory, he looked into the details of various borders in Europe but found some strange disagreements—the Portuguese reported their border with Spain to be 1214 km, while the Spaniards thought it was 987 km long; the border between the Netherlands and Belgium was reported as 380 km or 449 km; and similar disagreements held for many other borders. Curiously, when two countries shared a common border, it was typically the smaller country reporting a larger value. Why should there be such a level of disagreement about such basic measurements, especially one of vital importance for issues of sovereignty?

Richardson pondered the manner in which one calculates the length of a border or coastline. Naturally, one lays off unit segments on a large, detailed map—perhaps the map is 50 km to the centimeter and one lays these off as unit segments, counting their number to find our estimate. Using smaller unit segments would presumably give a more accurate result, right?

What Richardson found, however, was that smaller units often led to a huge increase in the estimated length. And using still smaller unit segments would make another unexpected increase. The total estimated length of the border did not always stabilize; rather, in practice it was often sensitive to the scale of measurement. Richardson began to realize a basic incoherence of the concept of a determinate length for a given border, and he introduced an invariant for calculating how the estimated length would scale larger as the unit of measurement became smaller—a deeply insightful contribution anticipating later developments of Benoit Mandelbrot. Infinite border lengths remained a theoretical possibility.

6.1 Measuring the Length of a Coastline

How long is the coastline of Great Britain? Consider the measurements illustrated in figure 6.1. At the coarsest scale, in the top map, we use a comparatively crude unit scale of 150 km, finding the estimated coastline length to be 2610 km. In the middle, with a considerably finer measurement unit of 4.4 km, we resolve many more of the bays and river inlets, more than doubling the measured coastline length to 6220 km. At the finest end of the scale, with a measurement unit of only 50 m, we find far more detail, nearly doubling the estimate again to 11500 km.

The accompanying graphs show how the length estimate varies with the measurement unit, generally producing longer estimates with smaller scales. Since this is a logarithmic scale, the basically linear appearance of the graph indicates an exponential growth relationship between the length estimates and the measurement scale. The slope of that line is essentially the invariant that Richardson found—different coasts and borders will exhibit their own characteristic slopes, depending on the topographical and geographic nature of each border.

Notice particularly the zoomed-in region of the coast in the Scottish Highlands—it is the coastal region from Loch Alsh to Loch Hourn—which clearly shows the phenomenon in play. The crude measurement unit of necessity passed completely over the details of this region, while the medium-fine scale was able to follow some of the ins and outs of the coastline. The finest measure resolved a considerably longer coast, precisely because it was able to follow the river inlets and bays much further. This is precisely the process by which a finer measure leads inevitably to a longer coastal estimate.

Benoit Mandelbrot (1967), building on Richardson's ideas, mounted the theory of *fractals* to explore the strange new theory of this entirely new class of curves and geometrical objects, introducing the *fractal dimension*, building on the Richardson invariant anticipated earlier.

The subject of fractals thus finds its origin in a practical problem—measuring the length of a border or coastline—but gives rise to a new mathematical theory, one concerned with the purely mathematical features of the abstract mathematical objects of our imagination. The puzzling theoretical examples shed light on our core conceptions of the nature of the length of a mathematical curve, and even our conceptions of dimensionality.

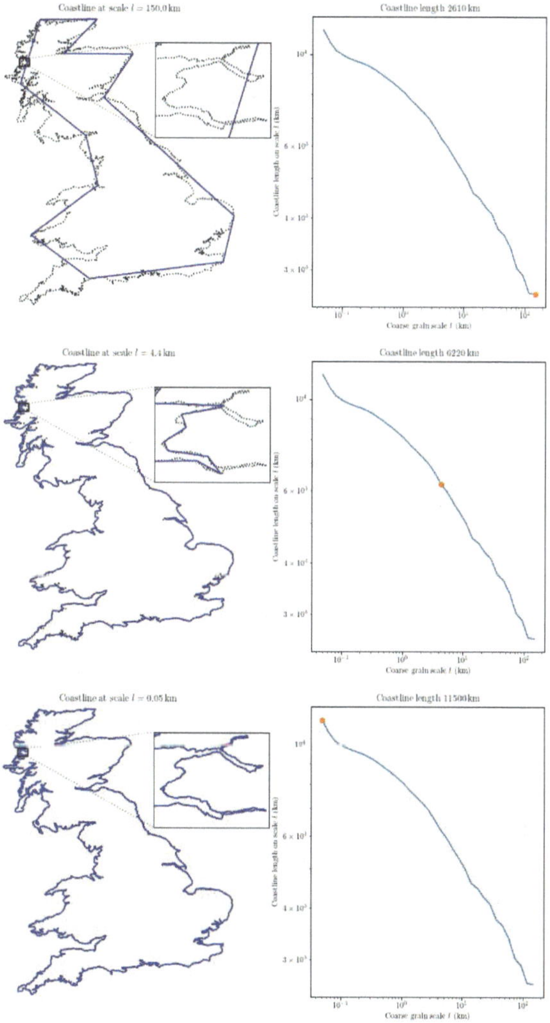

Figure 6.1: Great Britain coastline, at various scales of measurement.

6.2 Paradoxes of Perimeter

Let us adopt this coastline measurement way of thinking to consider a few natural attempts to approximate the values of $\sqrt{2}$ and π.

Length of a Diagonal

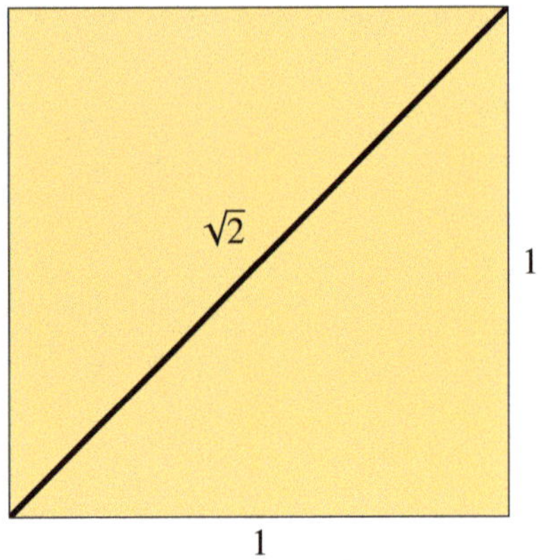

Consider the diagonal of a unit square, as shown in the figure here. The sides have unit length 1, so the diagonal has length $\sqrt{2}$ according to the Pythagorean theorem. Let us try to calculate approximations to $\sqrt{2}$ by constructing increasingly fine approximations to the diagonal using simple finite combinations of linear segments in such a way that we can easily calculate their total length. Consider, for example, the series of increasingly tight staircases illustrated in the figure that follows. As the number of steps becomes very large, it will become difficult to distinguish the staircase approximation from the diagonal itself:

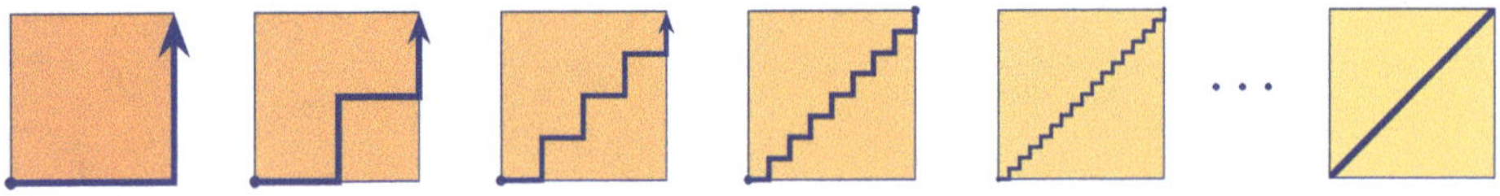

How well did we do? Let's calculate the lengths of the approximations and observe how well they approximate $\sqrt{2}$ as the resolution becomes very fine. The very first staircase at left, following the bottom edge and then the right-side edge of the square, has total length $1 + 1 = 2$. Crude, but perhaps an acceptable start. The second staircase has 4 segments, each of length 1/2, which makes again a total length of 2. Hmmnn. . . it didn't change. The third staircase approximation has 8 segments of length 1/4, again making a total length of 2. Wait, what? Why isn't it converging to $\sqrt{2}$? The next approximation has 16 segments of length 1/8, again with total length 2. Indeed, in fact all the staircase approximations, no matter how fine the resolution, will have a total length of 2 because for each staircase, the sum of all the horizontal segments will be 1 and the sum of the vertical segments will also be 1, making the total length 2. The strategy of approximating $\sqrt{2}$ in this way is a total failure. What is going on? Although the staircases seemed to approximate the diagonal to an increasingly fine degree, they all have length 2, whereas the diagonal has length $\sqrt{2}$, and we know that these are not the same. How can one explain this paradox?

Circumference of a Circle

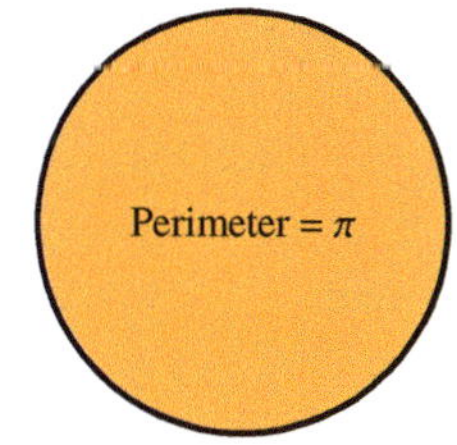

Here is another similar instance—we shall try to calculate an approximation to π. Start with a circle of diameter 1, which consequently has a circumference of exactly π. We place the circle in a square box, as shown next on the left, and proceed successively to remove corners from the box to make the boundary of the enclosing box more and more closely approximate the circle. Keep trimming the corners to make the bounding box as close to the circle as you can. Eventually, with a very fine resolution, the bounding box will become indistinguishable from the circle:

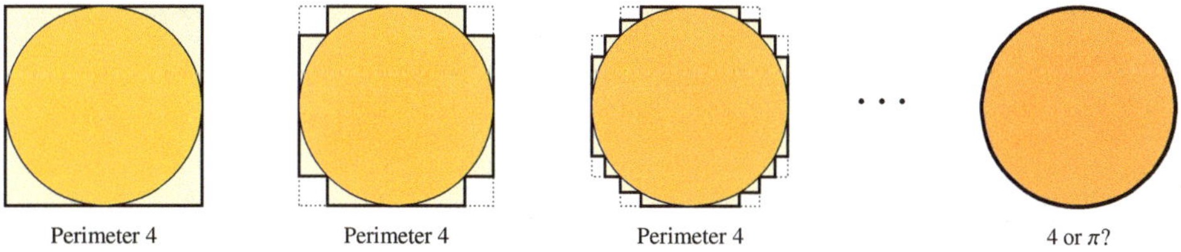

How well do the approximations approximate π? The initial square has four sides of unit length, which makes a perimeter of 4. A crude estimate, to be sure, but perhaps acceptable as a first approximation. For the next box, let's consider how long the perimeter is. How shall we calculate it? Do we need to know further details about exactly which points were used when pushing in the corners? No, we do not, because the process of removing a corner from a box actually preserves the perimeter, since it simply replaces a vertical/horizontal arrangement with a horizontal/vertical arrangement of exactly the same length. So the second box actually also has perimeter 4. But wait, hmmmm. The very same argument shows that the third box also will have perimeter 4, and indeed all the bounding boxes will have the same perimeter of 4, since the process of pushing in a corner does not change the perimeter. Unfortunately, this method of approximating π has also been a total failure. We know that π is definitely less than 4, simply by considering the very first box. What is going on?

6.3 The Length of a Curve

The problem with the previous paradoxical examples is that they rely on a subtly faulty notion of what it means to have an approximation to a curve. With the right concept, we shall realize that we were mistaken about the staircases and bounding boxes—we should not regard them as approximating the curves in a way that is useful for estimating the length.

So what does it mean to have an approximation to a curve or to speak of the *length* of a curve? To be sure, we know how to conceptualize and measure the length of a straight line segment. We can use the distance formula, for example, which is really just another way of thinking about the Pythagorean theorem. But when a curve is *really* curvy, what does it actually mean to refer to the length? Consider the elegant curve here, climbing gracefully from point A to point B. What does it mean to speak of the length of this curve?

We can provide a robust conception of the length of a curve by considering its piecewise linear approximations, joining finitely many points on the curve with linear segments:

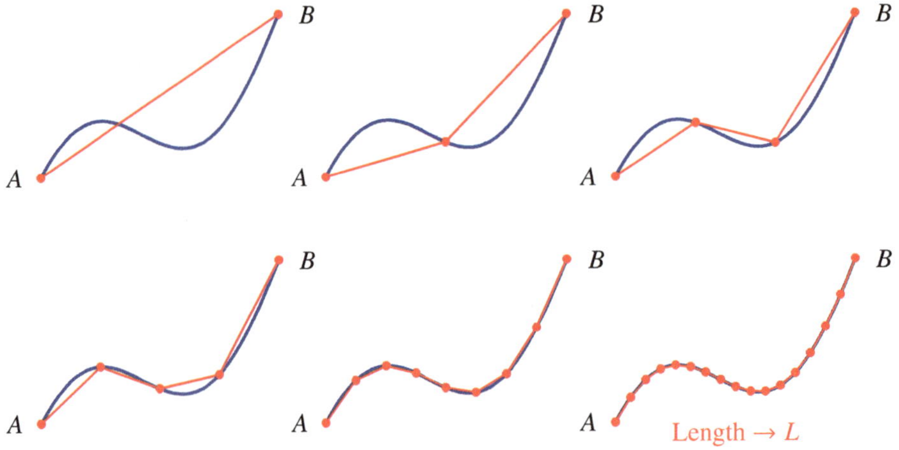

As we take more and more points, these linear approximations may converge to a limit value L, and in this case, we say that the length of the curve is L. Since the curve is at least as long as any piecewise linear approximation, we can simply define the length of the curve to be the supremum of the lengths of its piecewise linear approximations. A curve is *rectifiable* when this limit exists—in this case, the curve is well approximated by linear segments and it has a well-defined length.

OK, so what is the difference between this rectifiability idea and the approximations that we had considered earlier? The difference is that to fulfill rectifiability, we consider linear approximations consisting of segments joining points *on the curve*. But the staircase and bounding box approximations, in contrast, were criss-crossing the curve, often making turns at points not on the curve. These inefficient excursions away from the curve caused wasteful extra length in the approximations, making them systematically too long, and this is why we do not want to count them as piecewise linear approximations. Indeed, those deviations from piecewise linearity explain why those approximations each had a longer length than expected, thereby resolving the paradoxes.

Archimedes Approximates π

Mathematicians often regard rectifiability and the limit process described previously as amongst the core ideas underlying the differential calculus. And yet, nearly two millenia before Newton and Leibniz and the calculus, Archimedes famously used the method of piecewise linear approximation to calculate some prescient early approximations to π. Perhaps this shows again how close Archimedes was to discovering the calculus.

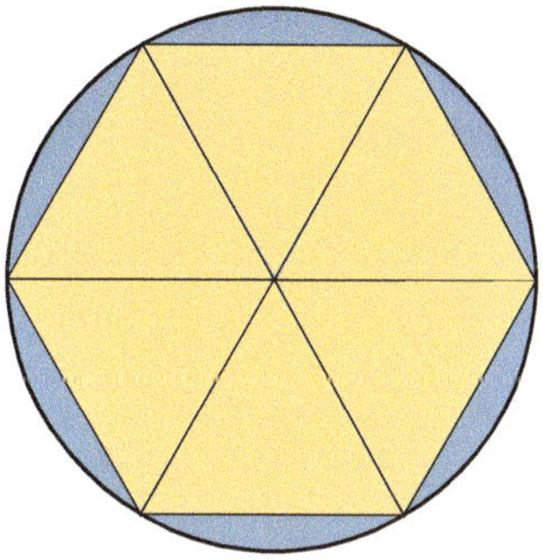

Consider the unit circle shown here, with radius 1. We inscribe a regular hexagon within it, consisting of six equilateral triangles, so each side of the hexagon is the same as the radius of the circle. Thus, the perimeter of the inscribed hexagon is exactly 6, while the perimeter of the circle is 2π. Since the straight sides of the hexagon are shorter than the curved segments of the circle—they form a piecewise linear approximation of the circle in the sense of rectifiability—it follows that $6 < 2\pi$, and therefore, for our first approximation, $3 < \pi$.

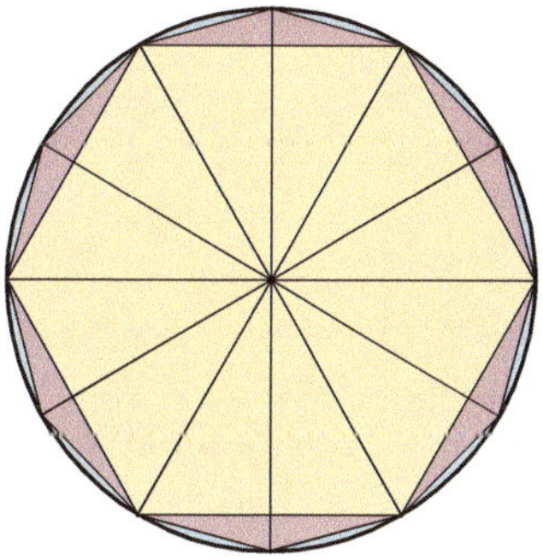

Archimedes established better bounds on π by extending the inscribed hexagon to an inscribed 12-gon, and then subsequently an inscribed 24-gon, and so on. Let us consider how one might argue with just the regular 12-gon, shown here, formed by bisecting the equilateral triangles of the inscribed hexagon each with a radius of the circle, making 12 angles of 30° at the center. If you consider the 12 small purple triangles that are added around the circle as a result, the leg bordering on the hexagon has length $1/2$ since it is half of the original equilateral-triangle side; the other very small leg on the radius has length $1 - \sqrt{3}/2$, since it is what remains of radius 1 after the equilateral-triangle altitude of $\sqrt{3}/2$. So according to the Pythagorean theorem, the perimeter segment of the 12-gon, which is the hypotenuse of the small extra dark triangle, has length

$$\sqrt{\left(\frac{1}{2}\right)^2 + \left(1 - \frac{\sqrt{3}}{2}\right)^2},$$

which some simple algebraic manipulation shows is equal to $\sqrt{2 - \sqrt{3}}$.

Since there are 12 sides and the perimeter of the circle is 2π, we may conclude that

$$6\sqrt{2-\sqrt{3}} \quad < \quad \pi.$$

We may calculate the square-root expression to have value 3.1058 . . . , so in particular, this shows $3.1 < \pi$.

The inscribed 24-gon has a perimeter of approximately 3.132, it turns out, and the inscribed 48-gon has a perimeter of approximately 3.139, and one thereby begins to see how to obtain increasingly better approximations to π this way. Using regular 96-gons, Archimedes ultimately established that $223/71 < \pi < 22/7$, using the inscribed polygon for the lower bound as we have done here, but using a circumscribed outer polygon for the upper bound. In decimal notation, the bound establishes

$$3.1408 < \pi < 3.14286.$$

6.4 What Can Go Wrong—Nonrectifiable Curves

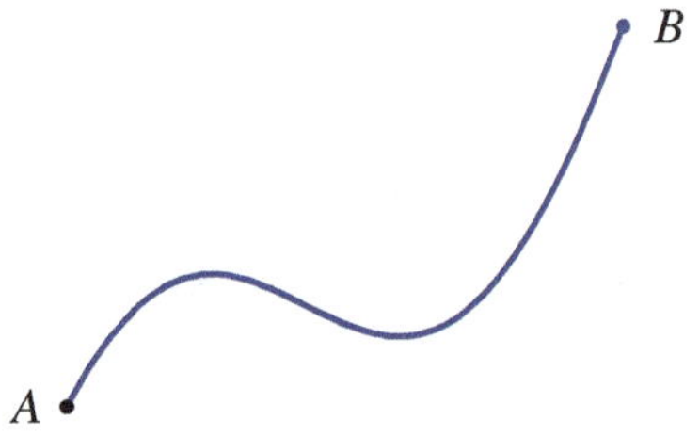

Let us consider again our elegant curve, climbing gracefully from point A to point B. Is it rectifiable? Does this curve have a well-defined length? It certainly seems to have a definite length, and earlier we had drawn a series of piecewise linear approximations, which seemed to be converging nicely. If indeed those approximations do converge, then this would be a rectifiable curve with a well-defined length. We seem to have a gentle, innocent, well-behaved curve.

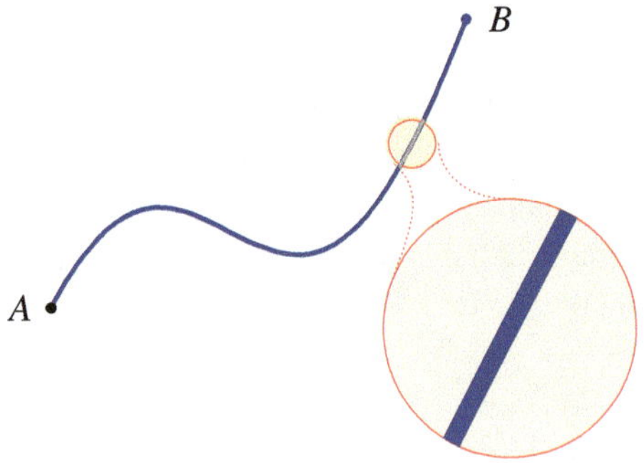

To be more certain, however, perhaps we should inspect the curve a little more closely. Shall we look at it under a microscope to make sure that there is nothing untoward? Here, I have zoomed in on a particular part of the curve. Does it look acceptable? Does it still seem rectifiable? Yes, indeed, everything seems to be in order—the curve seems completely fine and well behaved. What could go wrong?

Our friend, who as a child was anxious about monsters hiding under the bed, insists that we look closer. After all, he says, didn't you notice above that under the microscope, the curve seemed weirdly thicker than we might have expected? What could be causing that? Our friend has a more powerful microscope (he carries it everywhere), and with it, we can zoom in a bit closer on that part of the curve. What does it reveal?

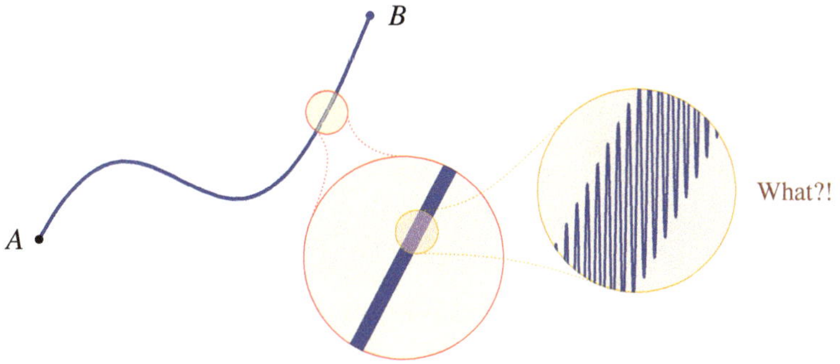

OMG! What is this?! We were completely mistaken in our judgement about the nature of the curve. What we had thought was a gentle, upward-sloping rise in the curve was actually a chaotic scramble of violent upward and downward jerks. This explains why the curve seemed a little thicker than expected—there was indeed a monster in the small scale.

Notice the drastic effects of the violent up-and-down motion on the length of the curve. What we had thought to be an extremely short segment of curve, if we were measuring with a piecewise linear approximation at the previous scale, is actually considerably longer by a large multiple. Each short segment of the original curve with this behavior is at least dozens of times longer than expected because of all the up-and-down motion. Although the absolute magnitude of those up-down jerks are minuscule, they cause a multiplicative factor in the length because of their number and density. Indeed, if the up-down zigzags were more tightly packed in together, the actual length of the curve could have been much longer than expected by an enormous factor, perhaps millions of times as long as expected.

Suppose that we zoom in even closer? We might look closely at one of those violent zigzags. Perhaps at a tighter scale, it will look fine, like the narrow parabola shown on the green background here:

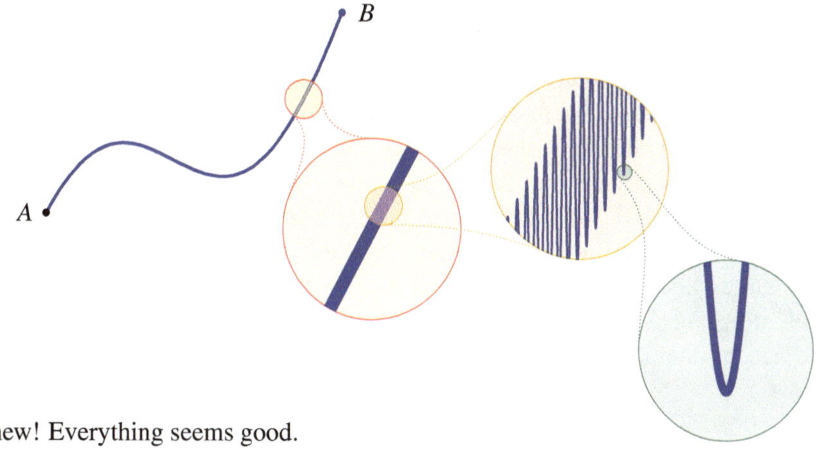

Whew! Everything seems good.

But wait—our anxious friend insists that we look still closer again:

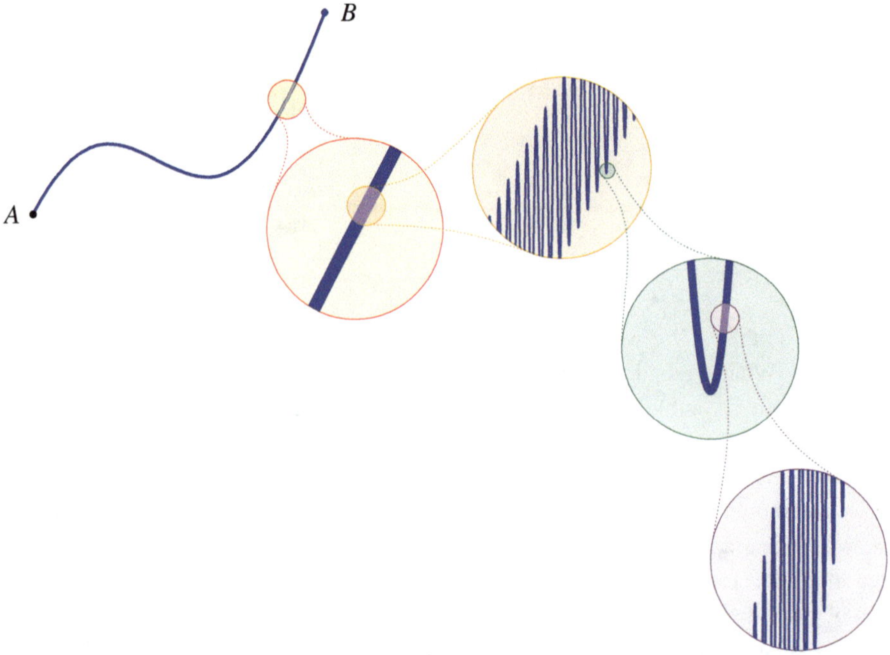

Again, monsters in the small scale!

Oh, no! We have found a deeper, tinier realm of violent action, occurring at a much smaller scale of resolution. The curve overall appears nice and smooth at some scales, and yet it is chaotic and scrappy at other scales.

The main point that I want to make is that essentially any curve that we observe might exhibit this kind of monstrous behavior at small scale. Are we surrounded by mathematical monsters? Perhaps all the functions we have been plotting and graphing and analyzing are like this when we zoom in close enough. And if it occurs often enough or badly enough, then the curve will not have a rectifiable length. The increasing multiplicative factors could cause the total length of the curve from A to B to become infinite.

Let me point out briefly, for those readers who are familiar with the calculus concept of the derivative, that because of the monstrous behavior at very small scales, the elegant function that we considered from A to B has a truly monstrous derivative, and not just at small scales. Although initially the slope of the curve looked quite tame, perhaps with a slope everywhere bounded by about 2, nevertheless the small-scale violent behavior that we observed causes the actual slopes to be much higher—the chaotic up-and-down motion is extremely steep where it occurs, making the values of the derivative swing from enormous positive values to enormous negative values and back again. The derivative of this function is not nice.

6.5 Fractals

The curves known as *fractals* often illustrate this kind of behavior in a systematic, self-reflective manner. Let us explore a few examples next.

Koch Snowflake

Consider the Koch snowflake curve, which was introduced by Niels Fabian Helge von Koch (1870–1924). Despite its complexity, it is generated as the limit of a sequence of approximations according to a surprisingly simple underlying rule:

Namely, for the Koch snowflake curve, we begin with an equilateral triangle, and then at each stage, we generate the next approximation by replacing the straight line segments, at whatever scale they appear in the current approximation, with a corresponding four-segment notched version as shown here. Thus, each segment gets a little notch in it, and then the notches get tinier notches, and so on ad infinitum:

Can you see how this rule transforms each snowflake approximation into the next? At each stage of generation, these approximations will be piecewise linear approximations to the final snowflake curve, which is the limit of these approximations.

Notice that the notched replacement configuration consists of four segments, each one-third of the length of the original segment. The length of the replacement is therefore 4/3 times as long as the original segment—the total length of the approximation grows by a factor of 4/3 at every stage of approximation.

This is a disaster for rectifiability. To see why, consider the original triangle, with length 3, the three unit sides. The next iteration has total length $3 \cdot \frac{4}{3} = 4$, and then $3 \cdot (\frac{4}{3})^2 = 16/3$, and $3 \cdot (\frac{4}{3})^3 = 64/9$, and so on. After n iterations, the total length is $3 \cdot (\frac{4}{3})^n$. Since this exponential grows without bound—it can be made as large as desired—the conclusion is that the Koch snowflake has an infinite perimeter! In other words, this is a nonrectifiable curve. An island in the shape of the Koch snowflake would have an infinite coastline.

Sierpiński Gasket

The Sierpiński gasket, introduced by Polish mathematician Wacław Franciszek Sierpiński (1882–1969), is another fractal generated by a simple replacement rule.

The unit operation replaces each segment with a three-segment arrangement, alternating the orientation. These images were generated by an algorithmic process that literally performed that process iteratively.

The resulting limit curve has infinite length.

6.6 Fractal Dimension

The fractal curves just considered open the door to a fascinating observation about fractional dimensions. Some curves are best described as having a particular numerical dimension, strictly between 1 and 2. What does that mean?

Consider how scaling works in familiar dimensions. If we scale a linear segment by a linear factor of 2, it becomes twice as long—we get two copies of the segment. If we scale a square up by a linear factor of 2, however, meaning that each side becomes twice as long, then we get a square with four times the area (four copies of the original). Scaling a cube by a linear factor of 2 gives us eight copies of the original. In general, in dimension d, scaling by a linear factor of s results in an object that is s^d times as large in the d-dimensional sense.

This formula also works when scaling downward—that is, when $s < 1$. Scaling a square down by a factor of $s = 1/3$ (making the side lengths one-third the size), makes the area $s^2 = 1/9$ as large—we have divided the square into a 3×3 array of nine smaller squares, each one-third the linear size.

How does this dimensional analysis apply to fractals such as the Koch snowflake and the Sierpiński gasket? If you inspect the final Sierpiński gasket shape, you will see that it exhibits a self-similarity by which the entirety of the figure consists of three copies of itself, each scaled down by a linear factor of 2. In other words, doubling the linear scale caused an increase in size of 3. But what dimension is like that? It must be somewhere between 1 and 2 since for those dimensions, we would get an increase in size of 2 or 4, respectively. We can find out the exact dimension of the Sierpiński gasket by solving the equation $3 = 2^d$. Taking logarithms of both sides, we see that $\log 3 = \log 2^d = d \log 2$, so $d = \log 3/\log 2$, which is about 1.5849. Thus, the dimension of the Sierpiński gasket is between 1 and 2, slightly closer to 2. Perhaps this fits with appearances, since in a sense the gasket begins almost to fill out a two-dimensional area in the plane, but not quite fully.

For the Koch snowflake curve, notice how each component exhibits self-similarity by which it consists of four copies of itself, each scaled down by a linear factor of 1/3:

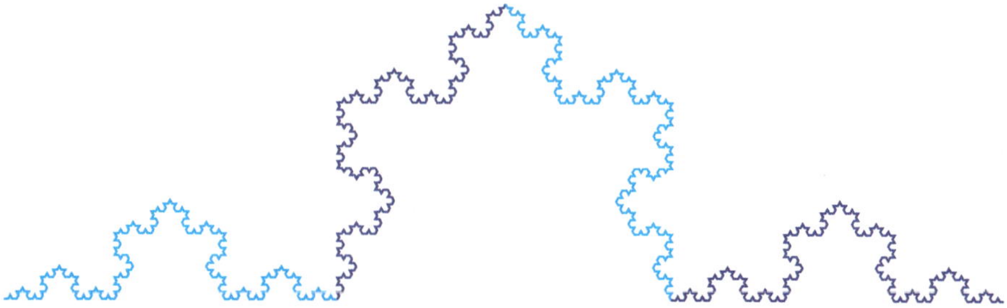

Thus, tripling the linear scale caused an increase overall of four. This corresponds to a fractal dimension of d solving the equation $4 = 3^d$, which again, by taking logarithms, has the solution $d = \log 4/\log 3$, which is about 1.262. So the snowflake curve also has a dimension strictly between 1 and 2, closer to 1 than 2 and less than the Sierpiński gasket, which perhaps fits with appearances since the snowflake curve seems much less two-dimensional than the Seirpinski gasket. And yet it is more than merely 1 dimensional.

6.7 Space-Filling Curves

What other monsters might there be in the small scale? How strange and unwieldy can a curve be? Very strange, indeed. Consider the phenomenon of *space-filling curves*, curves that entirely fill a solid region of the plane or space in which they reside. Giuseppe Peano was the first to describe such a curve, but let us explore David Hilbert's simplified variation.

Like the fractals we saw earlier, the Hilbert curve is defined by a limit process. One starts with the crude approximation next at left, consisting of three line segments, and then each approximation is successively replaced by one with finer detail, and the final Hilbert curve itself is the limit of these approximations as they get finer and finer.

The first six iterations of the process are shown here:

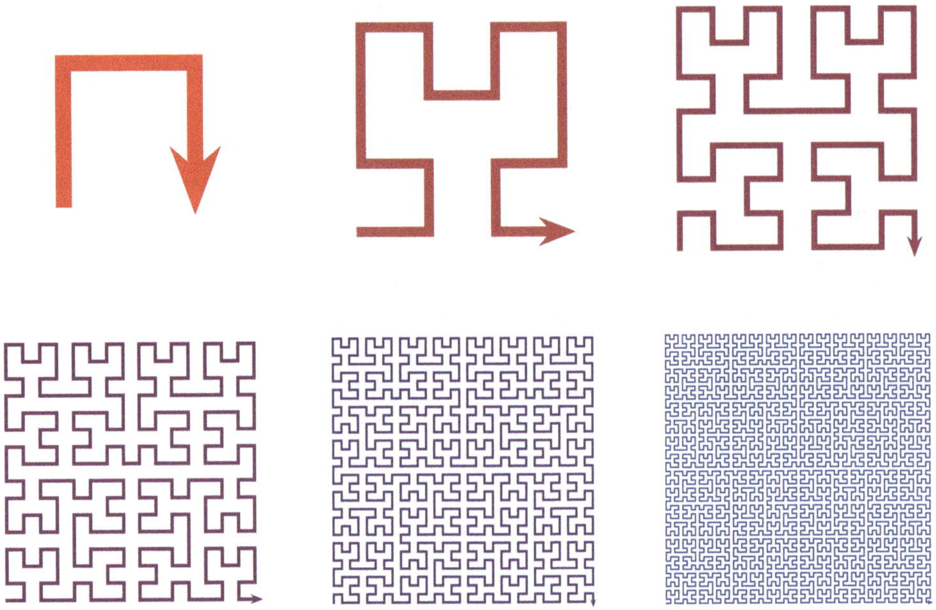

Each approximation curve traces a path in the plane—let us imagine they describe the position of a figure skater performing an exhibition on ice in the town square. In each case, she enters the stage at the lower left, gives an exhilarating performance with many twists and turns, and then exits at the lower right. With each successive performance, her skill increases, and she undertakes ever more intricate twists and turns. In the limit performance, her routine follows the fully space-filling figure-skating curve, visiting every point on the ice at some point during her performance. The limit path $h(t)$ is a point describing her position at time t, which is the limit of the time-t positions as the approximations proceed.

One can begin to see this property already in the approximations to the Hilbert curve here. At the halfway point of each performance, the skater's path crosses from the left side of the ice to the right side—there is only one such crossing with each approximation; can you find it? These crossing points are converging to the very center of the square, and the halfway value $h(\frac{1}{2})$ of the limit path h will therefore be the point $(\frac{1}{2}, \frac{1}{2})$. Because the wriggling twists and turns are increasingly local, the limit curve will be continuous; and because the wriggling gets finer and finer and ultimately enters every region of the square, it follows that every point in the square will occur on the limit path. So the Hilbert curve fills space. To my way of thinking, this example begins to stretch, or even break, our naive intuitions about what a curve is or what a function (or even a continuous function) can be.

Questions for Further Thought

6.1 What is your explanation for why it was generally the smaller country with the larger estimates for the length of a common border with another country?

6.2 Is the infinite coastline paradox about actual coastlines? Or is it instead only about the purely mathematical existence of curves and functions that behave differently from those in our ordinary experience?

6.3 Can you make a graph estimating the derivative of the elegant function from A to B considered in this chapter? What does it look like if the function is as well behaved as one might hope for? What does the derivative look like if the function has the small-scale monstrous nature discussed in the chapter?

6.4 Does the existence of nonrectifiable curves, including space-filling curves, challenge the coherence of the intuitive mathematical ideas we might have about the concept of the length of a curve?

6.5 Explain the meaning of Richardson's quotation, "Big whirls have little whirls that feed on their velocity; little whirls have lesser whirls and so on to viscosity." (1922, p. 66)

Further Reading

Mandelbrot (1982). Mandelbrot's classic treatise on fractals in nature.

Credits

Information about Lewis Fry Richardson was found in Stoa (2019). The map of Europe at the opening of the chapter was cropped from a map of user Alexrk2, CC BY-SA 3.0, via Wikimedia Commons. The maps of the Great Britain coastline in figure 6.1 were extracted stills from a GIF video of user Tveness, CC0 via Wikimedia Commons. The figure on approximations of π is adapted from Hamkins (2020, exercise 6.7) and the related discussion in Hamkins (2021c). The material on space-filling curves was adapted from Hamkins (2021b, section 2.6).

7 The Paradox of Giants

Abstract. Galileo argued that the giants of folklore, taking human form but at much larger scale, are physically impossible. Similar ideas will lead us to many other paradoxes of dimension.

According to legend, giants once roamed the Earth. Odysseus encounters the Cyclops, the one-eyed son of Poseidon, living in a huge cavern, hungry and angry—he grasps men and sheep in one hand, devouring them whole. In the days of King Arthur, a clever young boy earns the title of "Jack the Giant Killer," using his sharp wit to outsmart and slay the various giants plaguing the land. Another Jack—or perhaps it is the same Jack—famously plants some magic beans and climbs the resulting beanstalk to a castle in the clouds, cleverly outwitting the giant residing there. Jonathan Swift's character Gulliver travels to lands afar, encountering the Lilliputians, a society of tiny people to whom he appears as a giant, as well as the Brobdingnagians, a society of giants to whom he appears Lilliputian, thereby experiencing giantism from both perspectives, without himself ever changing size.

All these giants of legend generally have human form, but at larger scale, and they undertake generally human activities—the giants walk about, stomp, dance, carry heavy loads, run, climb ladders, throw stones, and so forth. They act and move about in a human manner, but at scale.

7.1 Galileo's Paradox of the Giant

Galileo Galilei (1564–1642) argues in *Dialogues Concerning Two New Sciences* (1638) that this folklore understanding of the nature of giants is fundamentally flawed—the very idea of a humanlike creature, acting and moving in a humanlike manner, but at much larger scale, is simply impossible physically. In short, giants are impossible creatures.

To begin his argument, Galileo calls for us to imagine a structural beam of sturdy oak. The beam might be used to support a heavy load—perhaps a load of bricks or a great stone:

Let us now imagine a much larger oak beam, scaled up in size but with the same proportions and material. A thicker, solid oak beam, of course, will naturally support more than a slender beam of the same wood. But how much more can it support? Will the larger beam be able to support the same load, but also scaled up in size? Should we expect the larger beam to support a scaled-up load of bricks or a great scaled-up stone?

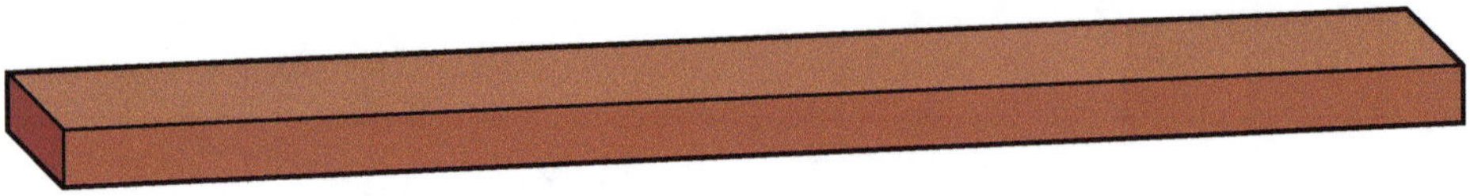

Galileo ingeniously argues no—the larger beam will not support a similarly scaled-up load. Indeed, he argues that at a certain sufficient scale, the beam will no longer support even its own weight! His argument relies on a certain subtle observation concerning how scaling works in different dimensions.

Galileo observes first that the load-bearing strength of a beam is connected with its cross-sectional area since a failure of the beam involves it breaking across a cross section. Since area scales as the square of the linear factor, a 10-times larger beam, with length, width, and depth each scaled by a factor of 10, will have a cross-sectional area 100 times larger than before. In other words, a 10-times larger beam is 100 times stronger! It will be able to support 100 times the load as the smaller beam.

That may seem initially very good. But the problem is that the weight of the load, for a given density of material, is determined by its volume, and volume scales with the *cube* of the linear factor. Scaling up a load of bricks or a great stone by a linear factor of 10, therefore, will cause a 1000-fold increase in the volume—a 10-times larger stone weighs 1000 times more.

Galileo brings these observations to their natural collision. A 10-times-larger beam is 100 times stronger, yes, but the similarly scaled-up load became 1000 times heavier. If the smaller beam had been carrying the optimal load, therefore, then the scaled-up beam will not be able to support the scaled-up load—not even close! It would support only one-tenth of it.

Indeed, Galileo argues that any given beam will have a certain sufficiently scaled-up size at which it will no longer support even its own weight. The strength of the beam scales with the square, but the mass of the beam itself scales with the cube, so at a sufficient scale, the beam will simply be too heavy for its own strength.

Galileo then draws the connection to giants by considering that the bones of the giant serve in effect as structural beams supporting the giant's body mass, his flesh and muscles. But if the giant's bones are made of the same stuff as ordinary men, his strength has not scaled the same as his mass.

The conclusion is catastrophic for the giant:

> Clearly then if one wishes to maintain in a great giant the same proportion of limb as that found in an ordinary man he must either find a harder and stronger material for making the bones, or he must admit a diminution of strength in comparison with men of medium stature; for if his height be increased inordinately he will fall and be crushed under his own weight. Galileo Galilei (1914 [1638], p. 131)

Similarly, scaled-up versions of wooden ladders would not support a scaled-up giant human climbing them; scaled-up swords would be too heavy for such giants to lift; scaled-up armor would be too heavy to wear; scaled-up wooden houses would not support their own roofs; a scaled-up wooden chair would not support even itself, let alone a giant sitting in it; and scaled-up dining glasses would be too weak to hold the volume of wine or water within them. In short, the entire world of giants as pictured in folklore does not make physical sense.

Animal Architecture

Galileo's argument explains many of the principal differences in body plan and architecture that we can easily observe in large as opposed to small animals. Namely, very large animals such as hippopotamuses, rhinoceroses, elephants, and dinosaurs tend to be stocky, with proportionally thicker, sturdier bones than smaller animals. Elephants, for example, have much thicker legs relative to their size than do dogs and cats. This is explained by Galileo's observation that the strength of bones scales with the square of the linear scale, since it depends on the cross-sectional area of the bones (and perhaps the pulling strength of muscles also depends on cross-sectional area), but the weight of the animal scales with the cube. So the bones themselves need to become proportionally thicker to support the increased weight.

A dual effect occurs when scaling down in size. Very small animals, for example, are typically slender and lithe. Insects typically have thin limbs, which would be entirely inadequate to support the animal if scaled up proportionally to much larger size. Small insects can often jump many times their own height. Insects can walk on walls and ceiling—they weigh so little that electrostatic forces are strong enough to hold them fast to the surface. If one were to encounter a fly that was 10-times normal size, but with the same proportions and body design, it would weigh 1000 times as much, and the tiny electrostatic forces would be insufficient to allow it to stick to walls. It would certainly not be able to fly, and probably it could not even support itself standing on those toothpick legs. Water bugs can walk on the surface of water, because at that scale, the surface tension of the water is proportionally strong enough to support the meager mass, but larger bugs would be too heavy. The laws of physics and the basic nature of chemical and physical reality simply do not scale uniformly with size.

The Paradox of the Miniature Human

Galileo's paradox applies just as much to the folklore conception of the miniature human, such as with the Lilliputians or the Hollywood films *Downsizing*, *Ant-Man*, or *Honey, I Shrunk the Kids!* The Lilliputians of *Gulliver's Travels* are reportedly about six inches tall, and the bones of such a 10-times scaled-down human would be 100 times weaker, while weighing 1000 times less. Such a person would therefore not walk about as normal but could probably jump very high (relative to his or her height) and carry correspondingly huge loads. For example, at normal size, a human can often carry another human on his or her back, but not much more. But the one-tenth-size humans would be able to carry much more:

> Whereas, if the size of a body be diminished, the strength of that body is not diminished in the same proportion; indeed the smaller the body the greater its relative strength. Thus a small dog could probably carry on his back two or three dogs of his own size; but I believe that a horse could not carry even one of his own size. Galileo Galilei (1914 [1638], p. 131)

Meanwhile, however, the tiny humans would find themselves subject to all manner of forces that are more prominent at small scale—they would stick to walls and other things more easily because of the electrostatic forces, and water would be a big problem, seeming much stickier at small scale. In short, the folklore vision of Lilliputians or other tiny humans walking about as humans ordinarily do is fundamentally flawed and ultimately just as physically impossible as giants are.

Evolution and Body Size

Galileo's observations enable us to see easily how evolutionary forces might work on body size. Perhaps for some organisms, there are comparatively few genes that control the overall body size in such a way that the typical size of an animal will be subject to evolutionary forces of natural selection.

We have clear evidence of species of animals greatly changing in size over evolutionary time scales. Horses used to be much smaller—the size of small dogs—and indeed a residue of this can be seen today in the minature horse breeds. Before dinosaurs and birds, dragonflies were kings, with wingspans of several feet, and we have fossils of huge cockroaches. Early mammals were very small.

The main point is that a given body architecture may have different levels of fitness with respect to a given environment depending on overall size and the varying effects of physical phenomenon that change with scale, including the ratio of strength to mass. Thus, evolutionary forces can be expected to act on those plans, perhaps changing size over time to improve survival. The evolutionary forces in play would also include, of course, the fact that small animals may be less able to defend themselves from predation. Becoming larger helps with that, of course, but imposes its own costs from increased mass, increased energy requirements, and lower relative strength, even when there is greater absolute strength.

7.2 The Paradox of Gabriel's Horn

Consider next the sonorous conundrum of Gabriel's horn, sounding out its sublime deep multitones from heaven, the secret chord that pleased the Lord:

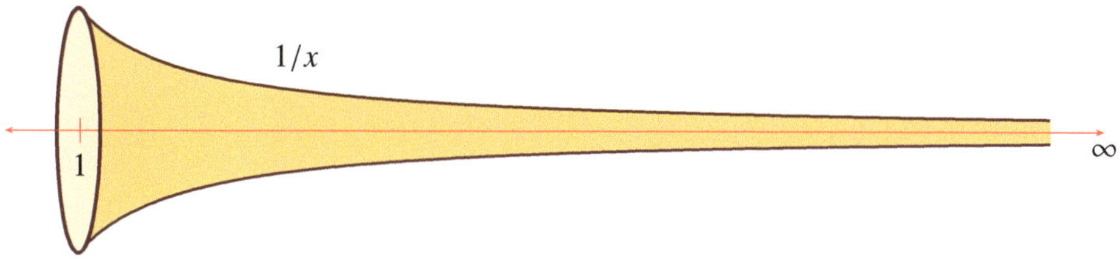

The horn is formed by revolving the function $f(x) = 1/x$ on the infinite interval $1 \le x < \infty$ about the x-axis. The result is a hollow, circularly symmetrical horn shape, flared open at left and stretching out along the x axis ever more thinly to infinity.

The horn has an infinite extent. But what is the volume? Let us calculate it. The cross sections are circles with radius $1/x$, making an infinitesimal volume element $dV = \pi(\frac{1}{x})^2\,dx$, which leads therefore to a total volume of

$$V = \int_1^\infty dV = \int_1^\infty \frac{\pi}{x^2}\,dx = \frac{-\pi}{x}\Big|_1^\infty = 0 - (-\pi) = \pi.$$

The total volume of the horn is finite! An infinite physical object with finite volume.

A further paradoxical feature of Gabriel's horn concerns the total surface area. The area calculation is a little more complex than the volume integral, but we can compute it like this: Slice the horn into bands, each forming the frustum of a slightly sloped cone. The band at x has average circumference $2\pi\frac{1}{x}$ and edge length ds, where

$$ds = \sqrt{dx^2 + dy^2} = \sqrt{1 + \left(\frac{dy}{dx}\right)^2}\,dx = \sqrt{1 + \left(\frac{-1}{x^2}\right)^2}\,dx = \sqrt{1 + \frac{1}{x^4}}\,dx,$$

which makes the infinitesimal area element $dA = \frac{2\pi}{x}\,ds$. Thus, the total surface area is

$$A = \int_1^\infty \frac{2\pi}{x}\sqrt{1 + \frac{1}{x^4}}\,dx \ge 2\pi \int_1^\infty \frac{dx}{x} = 2\pi \ln x \Big|_1^\infty = \infty.$$

Thus, the surface area is infinite. The paradoxical situation here with Gabriel's horn is that it has finite volume but infinite surface area.

Wait, what? How can that be?

Extended Real Numbers

Before considering that paradox further, let us make a brief diversion to discuss the use of ∞ in this calculation. This ∞ is the infinity of the *extended real numbers*, the number system $\bar{\mathbb{R}} = \mathbb{R} \cup \{\infty, -\infty\}$ formed by simply adding new, ideal objects $\pm\infty$ to represent positive and negative infinity, forming a new extended number system in a way that enables practical calculation with infinity for the sort of calculations that often arise in calculus, such as we had here. The underlying idea is simply that we introduce ∞ as a new special symbol or object and then simply specify as a matter of definition how it behaves in calculations with real numbers.

In the extended real numbers, we can add any finite quantity to infinity and remain infinite, such as $\infty + 5 = \infty$ and also $\infty + \infty = \infty$ and $-\infty - \infty = -\infty$, but $\infty - \infty$ is not defined. We can similarly multiply infinity by any positive quantity, as with $\infty \cdot 3 = \infty$, and also with negatives $\infty \cdot (-\pi) = -\infty$ in the expected manner. We also have $1/\pm\infty = 0$, but $0 \cdot \infty$ and ∞/∞ are not defined. The extended real numbers are thus a way of treating infinity ∞ as a kind of number, subject to natural rules that respect all the usual identities of the real numbers, such as commutativity, associativity, and distributivity, which continue to hold for extended real numbers in any instance where all the expressions are defined.

Meanwhile, division by zero $(1/0)$ is not defined in the extended real numbers, in part because the function $1/x$ approaches both ∞ and $-\infty$ as $x \to 0$, depending on which side of 0 one approaches. In contexts where one is considering only positive values, however, then it is reasonable and convenient to define $1/0 = \infty$ for that context.

Ultimately, the infinity ∞ of the extended real numbers $\bar{\mathbb{R}}$ need not be seen as especially heavy philosophically—one can take the situation simply that we have introduced a new symbol and specified some new arithmetic operations with it. The situation is fundamentally similar to how one may view complex numbers $a + bi$ simply as ordered pairs of real numbers (a, b), with addition and multiplication operations $(a, b) + (c, d) = (a + c, b + d)$ and $(a, b) \cdot (c, d) = (ac - bd, ad + bc)$, defined exactly to implement the effects of $i^2 = -1$. By doing so, one interprets complex numbers as pairs of real numbers, defined with a curious multiplication operation, so there need be no terrifying mystery of the square root of -1. Similarly, in the extended real numbers, we have simply introduced a new symbol and defined how to calculate with it, and there need be no terrifying mystery to this formulation of ∞.

The Painter's Paradox

Returning to the paradox of Gabriel's horn, let us imagine that we should simply fill it with paint. Because it has finite volume, we can completely fill it to the brim with a finite amount of paint altogether. Having done so, paint would thus be touching every point of the surface on the inside of the horn, so we will have seemed to have painted the infinite surface area with a finite volume of paint. How is that possible? How can we paint an infinite surface area with a finite amount of paint?

Well, this is cheating, of course, because the paint is spread ever more thinly inside the horn as one moves out on the x-axis, as the diameter of the horn becomes thinner and thinner. We painted an infinite surface with a finite amount of paint simply by spreading it ever more thinly. An honest approach to painting a surface, in contrast, would insist that every part of the surface was covered with a certain minimal uniform thickness of paint.

Criticizing the Paint-Based Account of Surface Area

That much is how one often hears the painter's paradox described in connection with Gabriel's horn. But I should like to consider the matter a little more carefully, providing a fuller analysis of the paint-based account of surface area. Indeed, I shall criticize the paint-based account, which is more subtle and problematic than one might have expected, even when there is no issue of cheating as before. The central idea of the paint-based approach to surface area, of course, is the following principle:

Paint-based area criterion. *A surface has finite area if and only if we can coat it to a minimal uniform thickness with a finite total volume of paint.*

Is this principle correct? No—this criterion turns out to be subtly wrong, and indeed, I shall give counterexamples on both sides of it—the criterion is neither necessary nor sufficient. Consider first the case of a single infinite line, which of course has infinite length but zero area. And yet we cannot coat it to uniform thickness with a finite amount of paint, since for any thickness $\varepsilon > 0$, the volume surrounding the line will be infinite.

Well, perhaps the line doesn't count as a surface at all. A more satisfactory but ultimately similar counterexample is obtained simply by modifying Gabriel's horn to use a more sharply tapered function, one converging to zero more rapidly to zero than $1/x$. For example, using $1/x^2$ makes a more sharply tapered horn with both finite volume and finite surface area:

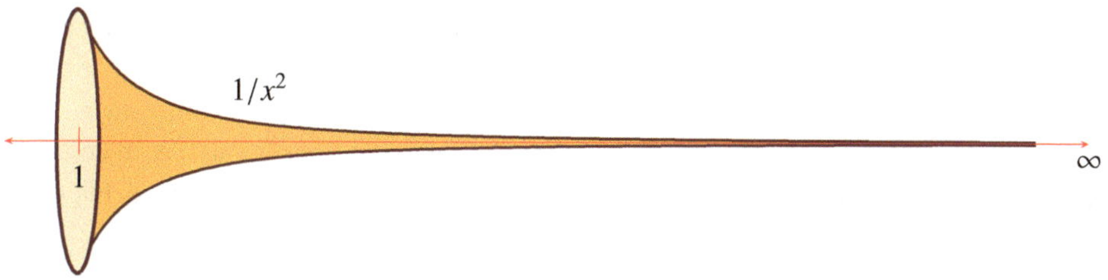

Do you see how this horn is more sharply tapered than Gabriel's? Unlike Gabriel's horn, however, this horn has finite surface area, and yet it cannot be covered to uniform thickness with a finite amount of paint, since we would need at least as much as used previously to paint a line to uniform thickness, and this is already infinite. So this is a case of a finite surface area that nevertheless cannot be coated to uniform thickness, no matter how thin, with a finite amount of paint, contrary to the paint-based area criterion.

But let me also provide another kind of counterexample, one of the converse type, a case of an infinite surface area that can nevertheless be painted with a finite amount of paint. Consider the surface formed by extending a section of the Koch snowflake curve, stretching it orthogonally to form a surface, like a corrugated section of paper, as depicted here:

All the cross sections of this corrugated surface are the same as the original Koch snowflake curve segment.

Because the Koch snowflake has infinite length, this corrugated surface area is infinite. But the whole surface is contained within a bounded region, so we might dunk the whole thing deep into a vat of paint. This would be a finite total amount of paint, and in this way we could ensure that every point on the surface was covered in paint to a desired thickness ε. Thus, the surface area is infinite, and yet we can paint it to uniform thickness with a finite amount of paint.

Perhaps one way to resolve the failure of these two kinds of counterexamples is to take account issues of curvature. In the case of sharply tapered horn, which has a finite area and yet can't be painted, the surface area has a progressively tighter curvature, which in order to be covered with paint means that a lot of extra paint is being used merely to make the turn around the axis. In particular, even though the original surface area was finite, the surface area of the outer surface of paint would be infinite. In the Koch curve surface, in contrast, the surface area is infinite because of various contours and wiggles of the surface, but these various tight crevasses in effect allow the same bit of paint to be part of the painting cover of multiple parts of the surface. The outer surface of the paint will have finite surface area, even though the underlying painted surface itself is infinite.

Finite and Infinite Spirals

Let me quickly mention a similar issue arising at a lower dimension—namely, the question of the nature of various kinds of spirals. Some turn infinitely around the origin and yet have finite length, while others have infinite length:

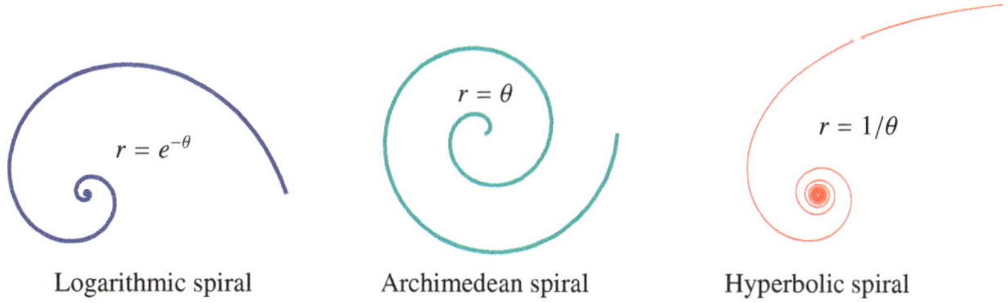

The logarithmic spiral turns around the origin infinitely many times, but it converges rapidly enough that the total length is finite. The Archimedean spiral, in contrast, has only finite turning and finite length from any point to the origin. The hyperbolic spiral has infinite turning and infinite length as it falls into the origin.

7.3 Paradox of High Dimension

Let us explore a few more paradoxes in high dimension.

Volume of the Unit Sphere in Dimension n

It may sound strange at first, but in which dimension does the unit sphere have the greatest volume? Of course, I mean to use the dimension-appropriate conception of volume, using length in dimension one, area in dimension two, volume in dimension three, and in general, n-dimensional volume in dimension n. This is sometimes referred to as *hypervolume* in dimension four and higher, but let us just call it n-dimensional volume, with this generalized understanding.

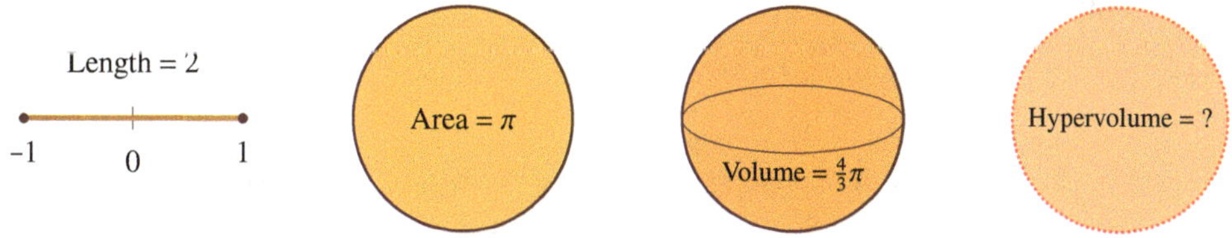

The one-dimensional unit "sphere" is simply a line segment sticking out in each direction by 1 unit, for a total length of 2. The two-dimensional unit sphere is the unit circle, with area π. The ordinary three-dimensional sphere has a volume of $\frac{4}{3}\pi$.

Note that so far the volumes are increasing, since

$$2 \quad < \quad \pi \quad < \quad \frac{4}{3}\pi.$$

Does this increasing pattern continue into higher dimensions? If not, in which dimension does the unit sphere have the greatest volume?

It turns out that one can calculate the volume of the n-dimensional unit sphere exactly. If v_n is the n-dimensional volume of the unit sphere in dimension n, one can use integration in spherical coordinates to establish the following recursion relation:

$$v_n = \frac{2\pi}{n} v_{n-2}.$$

Since $v_1 = 2$ and $v_2 = \pi$, as we observed previously, this leads directly to the following table of values:

v_1	$=$	2			
v_2	$=$	π	$\approx$	3.14159	
v_3	$=$	$4\pi/3$	$\approx$	4.189	
v_4	$=$	$\pi^2/2$	$\approx$	4.935	
v_5	$=$	$8\pi^2/15$	$\approx$	5.264	← Maximal hypervolume
v_6	$=$	$\pi^3/6$	$\approx$	5.168	
v_7	$=$	$16\pi^3/105$	$\approx$	4.725	
v_8	$=$	$\pi^4/24$	$\approx$	4.059	
$\vdots$					

After the first two values, one simply applies the recursion formula.

Since the recursion formula multiplies v_{n-2} by the factor $\frac{2\pi}{n}$, this will be increasing provided that $\frac{2\pi}{n} > 1$, which occurs exactly when $n < 2\pi$. Starting in dimension $n = 7$, therefore, this factor is less than 1, so $v_7 < v_5$ and similarly, $v_n < v_{n-2}$ for all $n \geq 7$.

By comparing the numerical values of $v_5 = 8\pi^2/15 \approx 5.264$ and $v_6 = \pi^3/6 \approx 5.168$, we therefore see that the volume of the unit hypersphere is largest in the dimension $n = 5$. In very high dimension, the volume decays to zero as the dimension becomes larger.

The Unit Hypercube in High Dimension Is Mostly Corners, Very "Endy"

Let us inscribe the n-dimensional sphere inside the n-dimensional hypercube. If we are using the unit sphere, with radius $r = 1$, then the corresponding cube will have side length 2 and hence n-dimensional volume 2^n. The proportion of the cube taken up by the inscribed sphere is therefore the ratio of the hypervolume of the sphere to the hypervolume of the hypercube, for a proportion of $v_n/2^n$.

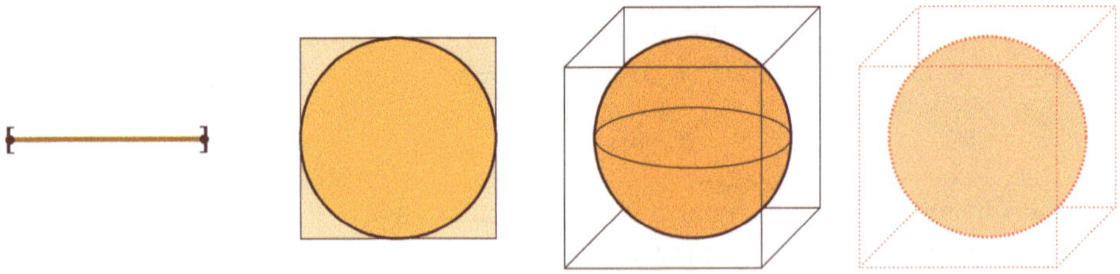

In dimension one, for example, the unit sphere was a line segment, the same as the one-dimensional "cube" in which it is inscribed. In dimension two, the circle forms proportion $\pi/4$ of the square in which it is inscribed, more than three-quarters. In dimension three, the sphere forms the proportion

$$\frac{\frac{4}{3}\pi}{8} = \frac{\pi}{6}$$

of the cube, which is a little over half. When going from dimension $n-2$ to n, the sphere volume is multiplied by $2\pi/n$, but the volume of the cube changes from 2^{n-2} to 2^n, a multiplicative factor of 4. Since $\frac{2\pi}{n} < 4$ for all dimensions $n > 1$, this means that as n increases, the proportion of the hypercube that is inside the inscribed hypersphere becomes less and less, going to zero as n becomes large.

The consequence is that in a large dimension, most of the volume of the hypercube is not near the center of the cube, but rather outside the unit hypersphere. In short, in very high dimension, almost all the volume of the hypercube consists of the regions in the corners of the cube, far from the center. High-dimensional hypercubes are thus very "endy" in a way that is difficult to imagine in lower dimensions. In two dimensions, there are 4 broad corners; in three dimensions, we have 8 corners; and we have 16 corners in four dimensions. It doubles every time. In very high dimensions, a point chosen uniformly at random from the hypercube is most likely stuck into some deep, dark corner of the hypercube—most of the hypercube is like that.

Packing Hyperspheres

Let me close this chapter with another mind-bending phenomenon of high dimension. Let us begin simply with the two-dimensional case. Place four unit circles into the corners of a square, like so:

In the slender spandrel region between these circles, we may fit a smaller blue circle, just touching, like this:

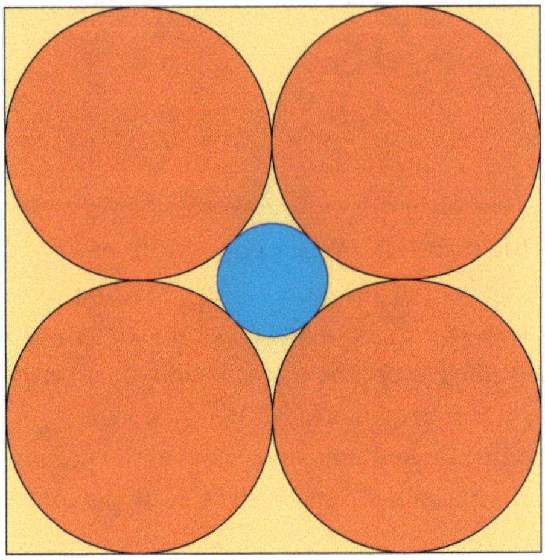

How big is the blue circle? If the original circles have radius 1, then they have diameter 2, making this a 4×4 square. Place the center at the origin, which makes the upper right circle have its center at $(1, 1)$. Draw the line from that center to the origin like this:

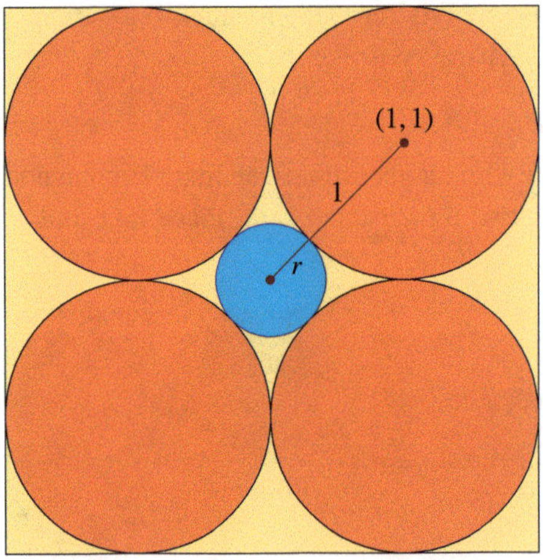

The line segment has length $\sqrt{1^2 + 1^2} = \sqrt{2}$, and since the red circle has radius 1, the leftover portion is the blue radius:

$$r = \sqrt{2} - 1.$$

Very good.

Let us now try the same thing in higher dimensions. In dimension three, we have a $4 \times 4 \times 4$ cube, with eight unit spheres placed into the corners:

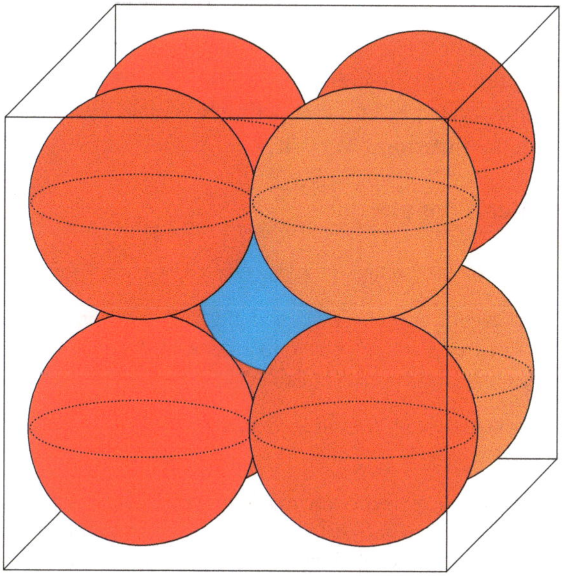

Centering the cube at the origin, the center of the sphere in the principal positive octant is at $(1, 1, 1)$, making its distance to the origin $\sqrt{3}$. Of that length, 1 is for the radius of the red sphere itself, leaving $\sqrt{3} - 1$ as the radius of the blue sphere nestled between them.

In dimension n, we can see similarly that the unit hyperspheres will be centered at $(\pm 1, \pm 1, \ldots, \pm 1)$, which have distance $\sqrt{n}$ from the origin, but length 1 of that is for the unit sphere itself, so the remaining radius for the blue hypersphere is $\sqrt{n} - 1$. Thus, we seem to have solved the problem of determining: How big is the blue sphere that fits just between the orthogonally packed unit spheres in dimension n? It has radius $\sqrt{n} - 1$.

But wait a minute! As n becomes large, this value $\sqrt{n} - 1$ grows without bound! The radius of the blue sphere becomes larger and larger—as large as desired! But how can that be, if the blue sphere fits between the red unit spheres?

For a large enough dimension n, the blue sphere radius $\sqrt{n} - 1$ will exceed 1, which means that in that dimension, the blue sphere will be larger than the unit spheres between which it is nestled. This occurs when $\sqrt{n} - 1 \geq 1$, which happens when $\sqrt{n} \geq 2$, or $n \geq 4$. In dimension 4 itself, the blue sphere is itself a unit sphere, and in dimensions 5 and above, it is strictly larger than a unit sphere. Very weird.

But stranger still, the blue sphere radius $\sqrt{n} - 1$ will eventually exceed 2. This is strange because 2 is the distance from the center of the blue sphere to the walls of the hypercube. To calculate when this occurs, we simply solve $\sqrt{n} - 1 \geq 2$, which is equivalent to $\sqrt{n} \geq 3$, or $n \geq 9$. In dimension 9, the radius of the blue sphere is exactly 2, which means that in that dimension, the blue sphere is tangent not only to the red unit spheres inside the hypercube, but also to the walls of the hypercube itself.

In dimensions higher than 9, the blue sphere has a radius strictly larger than 2, which means that it is not even contained within the hypercube—the blue sphere nestled between the unit spheres actually sticks out from the sides of the hypercube containing those unit spheres. It is amazing!

The intuition one might have had, based on the situation in small dimensions, that the blue sphere was comparatively tiny and definitely contained within the hypercube, is simply false in very high dimensions. Very strange.

Questions for Further Thought

7.1 Is Galileo's argument on the paradox of giants refuted by the existence of dinosaurs?

7.2 In light of Galileo's arguments, what particular physical challenges would be faced by a giant in human form, but twice as large as a normal human? What challenges (or opportunities) would be faced by a miniature human at half-scale? Be specific and realistic.

7.3 What physical reasons might there be that humans are generally the size that they are, rather than twice or half as large?

7.4 Given that modern humans are generally much larger than the fossil remains of early human ancestors (which, after all, might be the result of modern nutrition and health care rather than genes), will this cause an evolutionary pressure in one direction or another for the size-expressing genes?

7.5 Can you provide a sharply tapered version of Gabriel's horn, with both finite volume and finite area? Or a less sharply tapered form, with infinite volume and infinite surface area? Please calculate the volume and area using integrals.

7.6 Can we imagine the entire universe in a nutshell, with all its details intact, but at much smaller scale? Would this be possible with the same physical laws as currently exist? Or would it require some fundamental changes to the laws of physics?

7.7 In which dimension does the unit sphere have the largest surface hyperarea?

7.8 What is the size of the biggest hypercube that fits entirely within the unit sphere in dimension n? Is most of the volume of the unit sphere inside that inscribed hypercube or outside it?

Further Reading

Galileo Galilei (1914 [1638], p. 131). This is Galileo's account of the paradox of giants, including his drawing of what a giant thigh bone would have to look like to support a comparable load.

Credits

Thanks to Leonard Cohen for the secret chord.

8 The Paradox of the Largest Tweetable Number

Abstract. What is the largest tweetable number?

What is the largest number that you can tweet? (Please forgive me for using the older terminology for posting on Twitter, now known, of course, as X.com.) Constrained by the traditional 280-character limit, using only the basic text functionality, what is the largest number that you can describe in a tweet? The question will lead us on a cosmic journey to some truly astounding numbers and beyond, eventually into the evaporating mist of confusions of logical complexity and paradox, while gradually revealing a fundamental underlying issue of mathematical determinacy: Does every mathematical question have a determinate answer?

We can clearly tweet some quite large numbers using ordinary decimal notation—simply fill the text field with digits:

9874578456989843587459839493579375989594
3547393959674535343447583272524262738484
1254132631723230218391738176124365243123
2934876927834928747365462157838473824739
1328647982436529347829364982736482713247
9874586348726186349379487583682479137912
2138783217493485932748732648186927369161

Of course, we would do better to use all 9s, since this would be a larger number:

9999999999999999999999999999999999999999
9999999999999999999999999999999999999999
9999999999999999999999999999999999999999
9999999999999999999999999999999999999999
9999999999999999999999999999999999999999
9999999999999999999999999999999999999999
9999999999999999999999999999999999999999

Numerically, this entry is $10^{280} - 1$, since if we add 1 and perform the sum, we will get 280 carries, and all those 9 digits will turn over to 0, making exactly 10^{280}.

We would have done better simply to write:

One centillion

One centillion (on the short scale discussed in chapter 1) is precisely 10^{303}, which is larger than both of our earlier entries. It has 304 decimal digits altogether—too many to tweet in decimal notation, but easily tweeted as here by using the name of the number.

I hear someone shouting from the back that we should use a *googol*. This is the famously huge number that can be written in decimal as a 1 followed by 100 zeros, like this:

10000000000000000000000000000000000000000000000000000000000000000000000000000000000000000000000000000.

A googol is a truly enormous number, which fits between the margins here only with a very small font. But wait—we have actually already greatly exceeded this number with all three of our previous proposals. A googol has only a 101 digits, after all, but our previous numbers had 280 digits, making them much larger. Since one centillion is 10^{303}, it is a 1000 googol googols times bigger than a googol. A googol is small fry.

I am sure my readers will quickly realize that we can make much larger numbers by using some mathematical operations. Using the factorial, for example, we can enter centillion! or even iterated factorials such as centillion!!! or more. Why not fill the tweet with factorials?

One centillion!!!!!!!!!!!!!!!!!!!!!!!!!!
!!
!!
!!
!!
!!
!!

That is rather sizable. What is the largest tweetable number?

Only Finitely Many Possible Tweets

There are, thankfully, only finitely many possible tweets altogether that can in principle be tweeted, including all possible tweets about cats, about your breakfast—about anything.

If there is a permissible character set of N allowed characters—which would include spaces, lowercase, uppercase, punctuation, and so on—and if any sequence of 280 of these characters is a tweet, then the total number of possible 280-character tweets would be N^{280}. Well, tweets can be shorter than 280 characters, so it is really $\sum_{k \leq 280} N^k$.

Reality, and especially the creations of human culture, are never as simple and clear as mathematics, however, so let me mention that the actual rules regarding the character type and count for tweeting are a tangle of complications. Basically, one can tweet Unicode characters, but not all of them, and furthermore some combinations of Unicode characters are in certain circumstances treated as a single character, such as when control characters used to indicate that accents should be placed on certain letters—in this situation, the control characters do not count against the 280-character limit when used in this way, but they needn't be used in that way, so the calculation of the character counts becomes somewhat muddled. David Madore has mounted a detailed analysis in a Twitter thread, whose details I shall regrettably leave aside here.

The Largest-Tweetable-Number Argument

Most tweets, of course, do not describe any number at all—perhaps you have tweeted a philosophically pithy remark about your breakfast. But some tweets describe numbers. In light of the fact that there are only finitely many possible tweets, there must be only finitely many numbers that can be uniquely described in a tweet—let us call them the *tweetable* numbers—so there must be a largest one.

To summarize, the argument at hand goes like this:

1. Some numbers are tweetable.
2. Every tweetable number is uniquely determined by a tweet of it.
3. There are only finitely many possible tweets.
4. Therefore, there are only finitely many tweetable numbers.
5. Consequently, there is a largest tweetable number.

What is the largest tweetable number?

8.1 Running the Contest

When first writing this chapter, I ran an instance of the largest-number contest on Twitter by posting the following tweet, asking participants to reply with a description of a specific natural number, the bigger the better, and the contest entries began to flow in.

With prudent urgency, I also immediately tweeted the first entry submission.

This was a matter of financial self-protection, you see, in light of the million-dollar prize, which is more than I can easily afford on the weekend. Because the prize was shrewdly to be divided by the winning number, however, this submission seemed to eliminate the risk.

In fact, I needn't have worried about it, since (as I expected) the contest immediately drew some enormous submissions. The million dollar prize will ultimately be divided into dust. Ultimately, the contest is not about winning prize money, but about the simple pride to be earned by winning the contest with the largest number.

The Syntax/Semantics Distinction

Nevertheless, I was initially alarmed by the following submission:

The worry was that this entry indeed contained the most sizable numeral yet submitted—a giant number 1. But if it were to be regarded as the largest submission, then I would seem to be liable again for the full amount of the million dollar prize; dividing a million by 1 is still a million. Have I made a ruinous mistake?

The situation reminds me of my childhood, reading the delightful children's novel, *The Phantom Tollbooth* by Norton Juster (1961), in which numbers come from the number mine under Digitopolis. They found there *the largest number*, a gigantic number 3—over 4 meters tall—made of stone. Broken numbers from the mine were used for fractions, like 5/3, when the number 5 has broken into three pieces.

Both *The Phantom Tollbooth* and this giant red number 1 tweet, of course, conflate numbers with numerals, partaking in the dichotomy between syntax and semantics, also known as the *use/mention distinction*: the distinction between using a word or symbol and mentioning it.

At high table in Oxford a few years ago (a regular gathering of college fellows and scholars at an enjoyable formal meal), my colleague, Alex Moran, citing "customary practice" in the north of England, had said:

> I generally use pants as trousers.

Or perhaps he had said:

> I generally use “pants” as “trousers.”

For the Americans here, let me mention the common British usage of the word “pants” for what would be called “underwear,” “underpants,” or “panties” in the US. I made a discreet glance under the table and can report that he didn’t seem to be using pants as trousers while dining. Since quotation marks are not usually pronounced, not even silently, with two curled fingers in the air, I couldn’t be sure which sentence he had said. But as we had been discussing regional variations in language use, he probably had mentioned the words “pants” and “trousers” rather than used them.

The word “dog” has three letters (a mention of this word), but after eating the homework, my student’s dog may have hundreds of letters inside him (a use of the word).

My daughter Hypatia inquired:

> Does everything rhyme with itself?

I had replied that it did not, having wrongly (and perhaps teasingly) heard her to ask:

> Does “everything” rhyme with “itself”?

But then again, she might have meant:

> Does “everything” rhyme with itself?

or possibly

> Does everything rhyme with “itself”?

So the largest-number contest is about the largest number that you can describe in a tweet, not the largest numeral—we shall be solidly on the semantics side of the dichotomy for the assessment of the winner. Yet syntax is playing a critical role in the contest, for the 280-character limitation is a purely syntactic limitation. The contest is ultimately about the interplay of syntax and semantics, asking which large numbers can be described within the allowed space.

Another Alarming Entry

Perhaps even more alarming was the following submission:

> Zero

After all, in alphabetical order, as we discussed in chapter 1, the number zero is the final number, making it the largest number in the alphabetical order. So if by the “largest” number, we had meant the largest number alphabetically, then zero would indeed be the winning entry. And if indeed it were regarded as winning, then since the prize is to be divided by the winning number, some people might expect the prize to be infinite, since perhaps it is reasonable to think that $\frac{1}{0} = \infty$. But luckily for me, the standard definition of division in the extended real number system stipulates that dividing by zero is undefined rather than infinite. Thus I can escape financial ruin on a technicality since the prize amount would be undefined, thereby invalidating (my lawyers would argue) any obligation I would have to pay anything.

8.2 How to Describe Some Very Large Numbers

The challenge of the contest, of course, is to find succinct descriptions of large numbers. In chapter 1, we saw a list of various large numbers—million, billion, trillion, quadrillion, quintillion, and so on, including googol and centillion. Let us discuss several methods of describing much much larger numbers.

Iterated Exponentials

Consider first the possibility of an *iterated exponential*, which means an exponential expression in which the exponent itself involves an exponential expression, like this:

$$2^{2^{64}}$$

There is a possible ambiguity of notation here, for an expression like

$$a^{b^c}$$

reasonably admits two possible interpretations:

$$\text{Associated downward:} \quad (a^b)^c$$
$$\text{Associated upward:} \quad a^{(b^c)}$$

These two interpretations, however, can be vastly different.

For example, $(10^{10})^{10} = 10^{100}$ is 1 followed by 100 zeros—a googol—which is a big number. But $10^{(10^{10})}$ in decimal is 1 followed by 10^{10} zeros, a vastly bigger number!

Since the downward-associated form simplifies to a^{bc}, we do not actually need iterated exponentials for the downward-associated reading. Furthermore, if $bc < b^c$, which is generally true when these numbers are large, then the upward-associated iterated exponential is larger, often considerably larger. For this reason, iterated exponentials are usually given the upward-associated reading by default. In light of this, there isn't an ambiguity in the notation after all.

Although exponential growth is usually thought of as quite rapid, the iterated exponentials are far more rapid. Compare the following table of the first few values of the exponential 2^n versus the double exponential 2^{2^n}:

n	2^n	2^{2^n}
0	1	2
1	2	4
2	4	16
3	8	256
4	16	65536
5	32	4294967296
6	64	18446744073709551616

So we can use iterated exponentials to make truly big numbers.

Googolplex

As we have mentioned, a googol is 10^{100}. A *googolplex*, in contrast, is 10^{googol}, which is the same as $10^{10^{100}}$. In comparison with numbers that we ordinarily encounter, a googolplex is astoundingly huge. I should like to think of *plex* as an operator that can be applied to any number, defining x plex to mean 10^x, so henceforth I shall write googol plex and not googolplex.

Let us try to compare a googol plex with a googol. We can easily write down the decimal representation of a googol—we just did it a few pages back, sure, but let's do it again—it is just 1 with a hundred zeros, like this:

10000000000000000000000000000000000000000000000000000000000000000000000000000000000000000000000000000.

The decimal representation of a googol plex, in contrast, has a googol zeros, not just 100. How long would it take to write them down? Well, suppose that we could write a million zeros every second. How long would it take to write a googol of them? Physicists tell us that the age of the universe is about 13.8 billion years, which is less than 10^{18} seconds. If we wrote a million zeros every second since the beginning of time, then this would be fewer than $10^{18} \cdot 10^6 = 10^{24}$ zeros, which is still an inconsequential fraction of the 10^{100} digits required. The physical universe is simply much too cramped in both space and time to accommodate the decimal representation of a googol plex.

Let me highlight a certain interesting feature here. Although we were able easily to describe the number googol plex, nevertheless not every number less than a googol plex will be so easy to describe. The typical number less than a googol plex is a number with a googol digits. If they are random enough, then the most efficient way to describe the number in general will be essentially to recite the digits. But as we have just said, you couldn't do this even if you recited a million digits every second since the beginning of time. And yet we seem able to hold this number googol plex as an object of thought in our minds, even though most of the smaller numbers are fundamentally beyond our comprehension.

A googol plex is thus a kind of milestone, a very large number with a comparatively short description. These are precisely the kinds of numbers that the largest-number contest is about. Incidentally, the numbers googol and googol plex reportedly inspired the name of the company Google and its headquarters, the GooglePlex, although the spelling is different.

Googol Bang

A *googol bang* is simply the factorial of a googol:

$$\text{googol bang} \quad = \quad \text{googol!}$$

In general, *x bang* is the whimsical mathematical term for the factorial $x!$.

Question. Which is bigger, a googol plex or a googol bang?

A googol plex is 10^{googol}, and we can write out the factors like this:

$$\text{googol plex} \quad = \quad 10^{\text{googol}} \quad = \quad 10 \cdot 10 \cdot 10 \cdots 10,$$

with a googol factors of 10. A googol bang is

$$\text{googol bang} \quad = \quad \text{googol} \cdot (\text{googol} - 1) \cdot (\text{googol} - 2) \cdots 3 \cdot 2 \cdot 1.$$

This also has a googol factors, but most of them are far larger than 10, and the few that are smaller than 10 are more than made up for by the larger factors. Therefore,

$$\text{googol plex} \quad \ll \quad \text{googol bang}.$$

One can naturally iterate the adjectives that we have introduced, and consider such numbers as googol plex bang and googol bang plex. A googol plex bang, for example, is the factorial of googol plex:

$$\left(10^{10^{100}}\right)!,$$

whereas a googol bang plex is the power of 10 to the googol bang:

$$10^{\left(10^{100}\right)!}$$

Question. Which is bigger, a googol plex bang or a googol bang plex?

Googol Stack

A *googol stack* is the number obtained by the operation of *tetration*, or iterated exponentiation, as follows:

$$\text{googol stack} \quad = \quad 10^{10^{10^{\cdot^{\cdot^{\cdot^{10}}}}}} \Big\} \text{googol}.$$

This number is far larger than a googol plex or a googol bang.

With this terminology that we have introduced, we can now begin to form some truly enormous numbers, such as the following:

> googol plex stack bang bang plex
> stack plex bang bang stack stack

The numbers that arise in this vocabulary form the googol plex bang stack hierarchy.

Knuth Notation

Let me introduce the up-arrow notation of Donald Knuth (1938–), a compact method for expressing the highly iterative processes beyond exponentiation and tetration. A single up-arrow denotes simple exponentiation, or iterated multiplication:

$$2 \uparrow 4 \quad = \quad 2 \cdot 2 \cdot 2 \cdot 2 \quad = \quad 2^4$$

More generally,

$$a \uparrow b \quad = \quad \overbrace{a \cdot a \cdots a}^{b} \quad = \quad a^b.$$

Two arrows indicate iterated exponentiation, which is known as *tetration*:

$$2 \uparrow\uparrow 4 \quad = \quad 2 \uparrow (2 \uparrow (2 \uparrow 2)) \quad = \quad 2^{2^{2^2}}.$$

More generally,

$$a \uparrow\uparrow b \quad = \quad \left. a^{a^{\cdot^{\cdot^{\cdot^{a}}}}} \right\} b.$$

Three arrows indicate iterated tetration, known as *pentration*

$$2 \uparrow\uparrow\uparrow 4 \quad = \quad 2 \uparrow\uparrow (2 \uparrow\uparrow (2 \uparrow\uparrow 2)) \quad = \quad \left. 2^{2^{\cdot^{\cdot^{\cdot^{2}}}}} \right\} \left. 2^{2^{2^2}} \right\} \left. 2^2 \right\} 2$$

More generally,

$$a \uparrow\uparrow\uparrow b \quad = \quad \overbrace{a \uparrow\uparrow (a \uparrow\uparrow \cdots \uparrow\uparrow a)}^{b \text{ terms}} \quad = \quad \overbrace{\left. a^{a^{\cdot^{\cdot^{\cdot^{a}}}}} \right\} \left. a^{\cdot^{\cdot^{\cdot^{a}}}} \right\} \cdots \Big\}}^{b} a\,.$$

In general, each extra arrow indicates that the previous operation is iterated that number of times. If we denote n up-arrows $\uparrow\uparrow \cdots \uparrow$ by $\uparrow^n$, then we can express the recursion like this:

$$a \uparrow^{n+1} b \quad = \quad \overbrace{a \uparrow^n (a \uparrow^n (a \uparrow^n \cdots \uparrow^n a))}^{b \text{ terms}}.$$

And for the anchor case, we take no arrows or $\uparrow^0$ to be multiplication:

$$a \uparrow^0 b = ab \qquad \text{and} \qquad a \uparrow^{n+1} 0 = 1.$$

The functions $k \uparrow^n k$ are closely related to the levels of the Ackermann function, defined by Wilhelm Ackermann in 1928, which is an early instance of a hyperoperation fast-growing function.

Beyond Knuth

We can easily generalize and transcend the Knuth notation simply by defining

$$a \Uparrow b \;=\; a \uparrow^b a \;=\; a \overbrace{\uparrow\uparrow \cdots \uparrow}^{b} a.$$

Note that $\Uparrow$ is very different and generally far larger than $\uparrow\uparrow$ since

$$3 \uparrow\uparrow 3 \;=\; 3^{3^3} \;=\; 3^{27} \;=\; 7625597484987,$$

whereas

$$3 \Uparrow 3 \;=\; 3 \uparrow\uparrow\uparrow 3 \;=\; 3 \uparrow\uparrow (3 \uparrow\uparrow 3) \;=\; 3^{3^{\cdot^{\cdot^{\cdot^{3}}}}} \Big\}\, 7625597484987,$$

which is considerably larger.

Having defined $\Uparrow$, we can proceed with $a \Uparrow\Uparrow b$, $a \Uparrow\Uparrow\Uparrow b$, and indeed more generally $a \Uparrow^n b$, just as we did with the regular up-arrow notation. And after this, we shall of course continue by defining

$$a \text{⤊} b \;=\; a \Uparrow^b a \;=\; a \overbrace{\Uparrow\Uparrow \cdots \Uparrow}^{b} a,$$

which will lead in turn to concepts of $a \text{⤊}\text{⤊} b$ and $a \text{⤊}^n b$ and so on, defined in analogy again with the recursion for $a \uparrow^n b$.

But of course, this is only to invite the generalization

$$a \text{⟰} b \;=\; a \text{⤊}^b a \;=\; a \overbrace{\text{⤊}\text{⤊} \cdots \text{⤊}}^{b} a.$$

One can carry on the generalizations quite a long way, and the numbers grow rather large. Using this new capability for describing enormous numbers, let me make a modest submission entry for the largest-number contest:

$$3 \text{⟰} 3$$

The reader is invited in the questions at the end of the chapter to elucidate just how large this entry is. Of course, this is not to mention the immediate next level:

$$a \text{⟰} b \quad = \quad a \text{⟰}^{b} a \quad = \quad a \overbrace{\text{⤊}\text{⤊}\cdots\text{⤊}}^{b} a,$$

which makes $3 \text{⟰} 3$ unimaginably larger still than $3 \text{⤊} 3$, in light of

$$3 \text{⟰} 3 \quad = \quad 3 \text{⤊}^{3} 3 \quad = \quad 3 \text{⤊}\text{⤊}\text{⤊} 3 \quad = \quad 3 \text{⤊}\text{⤊}(3 \text{⤊}\text{⤊} 3).$$

Graham's Number

Graham's number is a famously huge number, introduced by Ronald Graham in his research work in Ramsey theory as an upper bound for a certain mathematical result that he had proved with Bruce Lee Rothschild. The number appeared in the *Guinness Book of World Records* for the largest number ever used in a serious mathematical proof.

We can describe Graham's number using the Knuth up-arrow notation as the result of a computation with 64 steps as follows:

$$G \quad = \quad \left.\begin{matrix} 3 \underbrace{\uparrow\uparrow\cdots\cdots\cdots\uparrow}\ 3 \\ 3 \underbrace{\uparrow\uparrow\cdots\cdots\uparrow}\ 3 \\ \underbrace{\quad\vdots\quad} \\ 3 \underbrace{\uparrow\uparrow\cdots\uparrow}\ 3 \\ 3 \uparrow\uparrow\uparrow\uparrow 3 \end{matrix}\right\} 64$$

The bottom layer is exactly $3 \Uparrow 4$, using our operation $\Uparrow$. The next layer is therefore $3 \Uparrow (3 \Uparrow 4)$, and then $3 \Uparrow (3 \Uparrow (3 \Uparrow 4))$, and so on. Graham's number is therefore precisely

$$G = \underbrace{3 \Uparrow (3 \Uparrow (\cdots \Uparrow (3 \Uparrow (3 \Uparrow 4))))}_{\text{64 arrows}}.$$

Remember that $\Uparrow$ is far stronger than $\uparrow\uparrow$.

If that final 4 were a 3, this would be precisely $3 \Uparrow\Uparrow 65$, since there are 64 arrows in the expression, and hence 65 terms. But this difference between 3 and 4 is more than made up for by another term, since $4 < 3 \Uparrow 3$ (to put it mildly), so we may deduce

$$3 \Uparrow\Uparrow 65 \quad < \quad G \quad < \quad 3 \Uparrow\Uparrow 66.$$

Consequently, although it is very famous for being extremely large, nevertheless Graham's number is tiny in comparison with the contest entry that I made on the previous page. After all, my contest entry was $3 \text{⟰} 3$, which is the same as $3 \text{⤊}\text{⤊}\text{⤊} 3$, which is $3 \text{⤊}\text{⤊}(3 \text{⤊}\text{⤊} 3)$, whereas Graham's number is less than $3 \Uparrow\Uparrow 66$, an infinitesimal trifle in comparison, a mosquito's whisker.

8.3 Kolmogorov Complexity

Kolmogorov complexity, introduced by Andrey Kolmogorov (1903–1987), is the idea of measuring the complexity of a number not by its size but by how difficult it is to describe how to compute it. We measure the complexity of a number, or actually any string of symbols, by the size of the smallest computer program that would produce it. So a googol plex $10^{10^{100}}$ has rather low Kolmogorov complexity since we can easily write a program that would in principle produce this number (even though running the program would take a long time).

The observation made earlier in this chapter was that most of the numbers less than a googol plex do not have such low Kolmogorov complexity, since the typical random number less than a googol plex has about a googol random digits, which in the typical case would need to be hard-coded into the program for enumeraton, and this would make the program very large.

In any case, we can prove that the number of programs less than a given size is essentially an exponential of the size. So there simply aren't anywhere near enough programs of a small size to compute all the numbers up to a googol plex, and consequently most of these numbers have much higher Kolmogorov complexity.

Kolmogorov Complexity Is Not Computable

There is a deep observation to be made about Kolmogorov complexity—namely, that in general we cannot compute it exactly. To be sure, we can often provide an upper bound for the Kolmogorov complexity of a number simply by providing a specific program that will produce it. The Kolmogorov complexity of the number, by definition, cannot be larger than the size of that program, precisely because it produces that number. But it is impossible in general to compute, or perhaps even to know, the exact Kolmogorov complexity of a number or string.

To see this, suppose that we have an algorithm that for any given number n will compute its exact Kolmogorov complexity. That algorithm has some fixed size k. Now design a modified algorithm that searches for a number n whose Kolmogorov complexity is larger than $10k+1000$ and outputs it when found. The program that does this will involve a simple loop that tries numbers in turn and subjects each of them to the algorithm for computing complexity and testing this against $10k + 1000$. It will eventually succeed in finding such a number since there are only finitely many numbers with complexity bounded by that particular number. Furthermore, the program we described will be only a little larger than its main subroutine, which has size k. Thus we can arrange that our program has size less than $10k + 1000$, but it produces a number that has a Kolmogorov complexity larger than $10k + 1000$. This is impossible in light of the meaning of Kolmogorov complexity. The conclusion is that there can be no algorithm that computes the exact Kolmogorov complexity.

8.4 Winning the Contest

Let us finally strategize more seriously about winning the contest.

The Winning Entry

What I claim is that we can seem to identify a perfect winning entry! Here is the argument. We have already pointed out in the largest-tweetable-number argument on page 95 that there are only finitely many possible tweets, and therefore only finitely many tweetable numbers, and consequently a largest tweetable number.

The strategic observation here is that we can simply tweet that description:

> The largest tweetable number

After all, if our largest-tweetable-number argument is right, then this tweet describes a perfectly definite number, the largest tweetable number, which is therefore also the winning entry. It will win, because it is not possible to tweet a larger number than the largest tweetable number. Right?

Better Than Winning

But now the troublesome problems arrive. Let us do better than winning. Consider the following entry to the contest:

> The largest tweetable number +1

If indeed the concept of *tweetable number* is meaningful to tweet about, then because of the largest-tweetable-number argument, there is a largest tweetable number, and we can in that case certainly add 1 to it. Thus this tweet would seem to describe a definite number, and furthermore, it seems to describe a number that is larger than any number that can be described in a tweet. This tweet is better than winning, since not only is it at least as large as the largest possible entry to the contest, but strictly larger!

But how can that be? After all, we have described it in a tweet. Something is very wrong in our understanding. But what exactly is wrong?

The Paradox of the Smallest Untweetable Number

Similar troubles arrive with the following tweet:

> The smallest untweetable number

We said that there are only finitely many tweetable numbers, but since there are infinitely many numbers, there must be some numbers that are not tweetable, and consequently there must be a smallest untweetable number. In short, if the concept of tweetable numbers is legitimate, then the tweet here exactly describes the number that is the smallest untweetable number.

But we just tweeted it. This is a contradiction. What is wrong with our analysis? There seems to be a serious problem with the concept of tweetable numbers.

Gaps of Untweetability

Sometimes people expect that the smallest untweetable number might be a winning entry.

Question. If indeed there is a smallest untweetable number, would it necessarily win against more ordinary entries?

No! The smallest untweetable number, if indeed the concept makes sense at all, would be tiny in comparison to some of the other large numbers that we have already described as entries. The smallest untweetable number is definitely less than a googol plex, for example, simply because there are not this many possible tweets, so not all those numbers can be tweetable. The basic situation is that the tweetable numbers do not form an initial segment of the numbers—the fact that a large number is tweetable does not mean that every smaller number is tweetable. The tweetable numbers must admit large gaps of untweetability. For example, we can tweet the following:

> The number of possible tweets

This is a perfectly reasonable number and tweetable, but there must be many untweetable numbers below it since most tweets are not tweets of numbers at all, but limericks about your cat and so on, so the number of tweetable numbers is far less than the total number of possible tweets.

Meanwhile, every number up to a googol is tweetable since such numbers have at most 100 digits, and we can simply tweet the decimal representation of any such number. So the smallest untweetable number must be strictly larger than a googol.

Berry's Paradox

The paradox of the smallest untweetable number is an instance of *Berry's paradox*, named for Oxford librarian George Godfrey Berry (1867–1928), whom Bertrand Russell (1872–1970) described as "the only person in Oxford who understood mathematical logic" and reportedly one of the few people with whom Russell corresponded about mathematical logic before publishing his *Principia Mathematica*.

Berry's paradox is simply this. Consider the following number b, which we might call *Berry's number*:

> The smallest number not definable in fewer than a dozen words.

But kindly notice, dear reader, that this very description defines the number b in only 11 words! This is a contradiction—we have entered a logical labyrinth. The paradox of the smallest untweetable number is at heart identical to Berry's paradox, updated from the early twentieth century to the age of Twitter.

8.5 Foundational Issues—A More Sophisticated Perspective

Let me conclude this chapter by moving the discussion briefly to a more advanced level using ideas from mathematical logic, including computable undecidability and logical incompleteness, for those readers who may have some familiarity with these topics. If you have never studied mathematical logic seriously, do not worry about it—just peruse this section or skip it and go to the "Questions for Further Thought" at the end of the chapter.

Computable Undecidability in the Largest-Number Contest

All the largest-number contests, in their most general forms, engage with computable undecidability. There can be no uniform computable procedure for adjudicating winners, even in the special case where submissions are required to provide not merely a description of a number, but rather a specific computable process for producing the number. In the general case, the contest judge would have to decide of two given Turing machine programs p and q, whether indeed they produce a numerical output, and if so, which output is larger.

But this is a computably undecidable decision problem since it would require the judge in effect to solve the halting problem. After all, given any program e, we could design a program p that would simply run e with the plan of giving output 5 if e should ever halt. The program p is a valid entry (in contests allowing program submissions of this size) if and only if e halts, so there can be no uniform computable procedure to determine this.

This argument depends on the "uniform" qualifier, since it shows that there is no single algorithm that will work for all the contests, where the bounds on the allowed program submission size grow ever larger. Meanwhile, for any particular contest with a particular bound on the allowed submission size, such as 280 characters as in the tweetable number contest, the valid-entry and entry-comparison decision problems are both finite problems and thus computably decidable.

Nevertheless, the lack of a uniform judging procedure means that if we shall be running contests of various sizes, we generally shall be unable to judge them computably. That is, there is no computable procedure that on input n produces one of these computable judging procedures for the contest of particular size n, since this would amount to a uniform solution, which is impossible. Furthermore, even for comparatively small values of n, including the tweetable number case and much smaller instances, the fact of the matter is that we simply do not know the answer to the halting problem on all programs of that size. So in fact we have no way of actually judging even small instances of the contest, including the tweetability instance. We simply won't be able to determine the winner.

The Busy Beaver Function

The *busy beaver function* $\mathrm{BB}(n)$ is defined on input n to be the largest number that can be output by a Turing machine program on empty input using at most n states with the two-character alphabet $\{0, 1\}$. The busy beaver function—essentially a computable analogue of the largest-tweetable-number—is not computable because if it were, we could solve the halting problem. Namely, given any program e, we can design a program p that runs e and waits for it to halt; if it does, then p outputs the number of steps that it took. Thus, if e does not halt before $\mathrm{BB}(n)$, where n is the number of states used for p, then it won't ever halt—contradiction. Therefore, the busy beaver function is not computable.

But furthermore, the particular values of the busy beaver function $\mathrm{BB}(n)$ are currently unknown even for very modest values of n. For machines with only six states, for example, the result is not known. Such programs would easily fit within a tweet, and thus we currently have no method of judging entries to the largest tweetable number contest even with submissions consisting of these small Turing machine programs.

Logical Undecidability in the Largest-Number Contest

Not only is there systemic computable undecidability in the largest-number contest, but there is *logical* undecidability as well, regardless of which foundational theory we might choose to adopt. If we work in the theory of Peano arithmetic (PA) or Zermelo-Fraenkel set theory (ZFC), for example, or any other computably axiomatizable consistent theory T extending these (or much less), there must be comparatively small descriptions of numbers σ and τ such that the theories do not settle the question of whether σ or τ defines the larger number—see my proof of this in Hamkins (2025b). Indeed, for any computably axiomatizable theory T, there will be specific Turing machine programs p such that theory T does not settle the question of whether p halts or how long it takes to halt. For such a program, it will be consistent with the theory that the halting time of p is larger than any specific concrete number that we could specify (it will be a nonstandard natural number). For this reason, there can be no provable upper bound on the size of $BB(n)$ for any n above the number of states in this program p.

In this event, if the program fits into a tweet (and certainly it would if we take advantage of the attached-image capability for image files up to a certain number of megabytes), then the door is open wide to a fundamental logical indeterminacy in the nature of the contest. After all, if it is logically independent of our best and strongest mathematical theories whether one entry is larger than another, then who has won the contest? On what basis would we give the prize to one entry over another if the question as to who had the bigger number was not settled, even in principle, by our best understanding of the axioms of mathematics? The facts of the matter begin to seem indeterminate.

Nondefinability of Truth

Alfred Tarski (1901–1983) famously proved that we cannot provide a uniform arithmetic criterion for the truth of arithmetic assertions, and similarly there can be no set-theoretically expressible criterion for the truth of set-theoretic assertions, and the same holds for any foundational theory. A variation of his argument shows that it is not possible to provide a truth criterion for tweetable assertions in such a manner that it can be used in tweets. If we could do this, after all, then we could by the fixed-point methods formulate the tweet asserting its own nontruth by that criterion, leading to contradiction.

Similarly, even if one might believe that there is a fact of the matter about whether a given tweet description succeeds in defining a number, it is not possible to fit the criterion underlying those facts into a tweet in such a way that we can use it. We have already mentioned that different fundamental theories will necessarily disagree about those facts, and it is not possible to provide a full account in a tweet of which particular facts of the matter concerning tweetable numbers that one might have in mind about it. This is precisely because if you could, then you could make the problematic contradictory tweets that we have been discussing.

But now the further observation is that without specifying those further facts of the matter, one is not entitled to refer to tweetable numbers as a singularly definite concept. Rather, these are a disparate collection of indefinite possible facts, whose exact nature depends on the further details of the unspecified underlying theory. It is like someone referring to "the captain of the airplane" at the airport, when there are many airplanes and many captains, none yet specified exactly.

Implicit Commitment to Monism

In light of that analysis, it is evident that much of the confusion surrounding the largest-number contest is the result of an implicit commitment to monism in the foundations of mathematics—the expectation that every mathematical assertion has a definite truth value, without any expectation of doing work to specify what those values may be. Specifically, the confusion arises from the presumption that the questions of whether a given description defines a number or whether one such description is larger than another have singularly definite answers, to which we can refer uniformly without ambiguity. But this is exactly what Tarski's theorem rules out. Thus, the largest-tweetable-number contest begins to engage deeply with the question of pluralism in the foundations of mathematics.

Questions for Further Thought

8.1 Which is bigger, a googol plex bang or a googol bang plex?

8.2 Can you calculate the number 3 ⤊ 3, describing it in more familiar terms? How does it compare with Graham's number? And what about 3 ⟰ 3?

8.3 Is the largest tweetable number argument on page 95 correct? If so, is the tweet given in the discussion starting on page 105 the winning contest entry? If not, in either case, what exactly is the flaw with them?

8.4 Is every number up to a googol tweetable? And what of googol plex? Do the tweetable numbers admit gaps of untweetability?

8.5 Suppose that someone makes the contest entry, "The largest of the other valid entries, plus 1." Is this is a valid entry? What if two or more people submit this entry? And what if another person submits, "The largest of all the other valid entries, plus 2"?

Further Reading

Aaronson (1999). A nice discussion of the largest-number contest by Scott Aaronson, reportedly written when he was 17 years old.

Credits

I first heard about the idea of the largest-number contest in Douglas Hofstadter's *Scientific American* article in the early 1980s, when I was in high school, reprinted in Hofstadter (1985, pp. 737–755). In 2013, I was glad to have the opportunity to hold my own such contest at Fudan University in Shanghai, and I recounted the results on my blog, Hamkins (2013), with the many humorous student submissions. You can find my recent Twitter version of the contest, mentioned in this chapter, at https://twitter.com/JDHamkins/status/1613715098635309057. Some of the discussion on the syntax/semantics dichotomy was adapted from Hamkins (2021b). You can see David Madore's careful analysis of the character count for tweets in https://twitter.com/gro_tsen/status/1563583510430228480. Graham's number was introduced by Ronald Graham in discussion with Martin Gardner in the 1970s as a simplified upper bound of the number appearing in Graham and Rothschild (1971). The remarks about Russell and Berry are from Russell and Moore (2015). Berry's paradox is sometimes referred to as *Richard's paradox*, after Jules Richard, although one can distinguish several forms of paradox, with Richard's diagonalizing against the definable real numbers, whereas Berry's concerns definable natural numbers or ordinals.

9 Potential Versus Actual Infinity

Abstract. Potentialism and actualism offer radically different conceptions of the infinite—the dispute will lead us to ultrafinitism and the modal conceptions of potentialism, revealing a rich variety of potentialist views.

There are infinitely many numbers, to be sure, but mathematicians and philosophers of mathematics sometimes disagree about exactly what this means. What is the nature of existence of infinite collections or infinite objects? According to the philosophy of *potentialism*, the natural numbers are potentially infinite—you can have more and more, as many as you like, but you will never have all of them as a completed infinite totality; according to potentialism, you will never have an *actual* infinity. The philosophy of *actualism*, in contrast, allows for the existence of completed infinities—actualists claim to form actually infinite completed collections and proceed further with them to construct other infinite mathematical objects.

Aristotle emphasizes in his *Physics* the finiteness of everything that is—apparent instances of infinity are seen through a potentialist lens of becoming:

> For generally the infinite has this mode of existence: one thing is always being taken after another, and each thing that is taken is always finite, but always different. Again, "being" has more than one sense, so that we must not regard the infinite as a "this," such as a man or a horse, but must suppose it to exist in the sense in which we speak of the day or the games as existing things whose being has not come to them like that of a substance, but consists in a process of coming to be or passing away; definite if you like at each stage, yet always different. (Physics, 206a27-29)

This potentialist understanding of infinity has been dominant for 2000 years—mathematicians from classical times until the dawn of the twentieth century have almost universally (but not quite) subscribed to a potentialist account of infinity. And yet, over the past century and half, there has been a sea change. The actualist perspective has become dominant in mathematical practice nowadays. Let us explore the differences in these two philosophical perspectives. What is at stake in the choice between them?

Infinitely Divisible Segments

Take a line segment. A potentialist will agree that a segment is infinitely divisible, but this is a potential infinity of divisibility—you may divide the segment into two, and then four, and then eight; or you may divide it by some other proportion, and you can subdivide it again and again as finely as you would like.

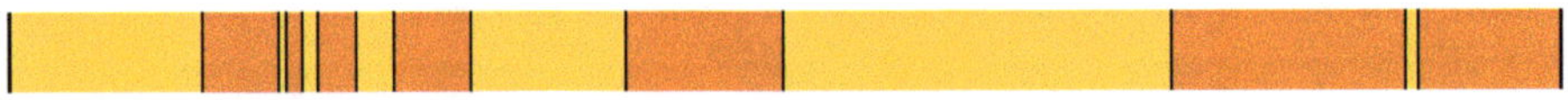

But according to the philosophy of potentialism, we never achieve an actually infinite subdivision. Indeed, according to the potentialist perspective, there is something downright incoherent in the idea of having an actually infinite subdivision of the segment.

After all, if every division that is made is further subdivided, then every segment produced during the process is itself destroyed, chopped finely, and every constituent part of the dividing process thus evaporates into the mist of the limit of an actual infinity of subdivisions. What remains of the infinite subdivision in the end?

Infinitely Extensible Lines

Or, take not a segment, but a line itself. According to the ancient geometers, a line is not a completed linear infinity, but rather a finite fragment of such a line, which might be extended further, as much as desired, by the use of the straightedge:

Euclid's second postulate, for example, asserts that any line can be extended indefinitely at either end. This is a potentialist manner of considering lines, since the view is that one has only finitely much of it at a time, so it may need to be extended.

The actualist, in contrast, considers a line as a completed infinite totality, stretching out with infinite extent. For the actualist, it makes no sense to extend the line further since the line already consists of its entire infinite extension.

Infinite Exhaustion

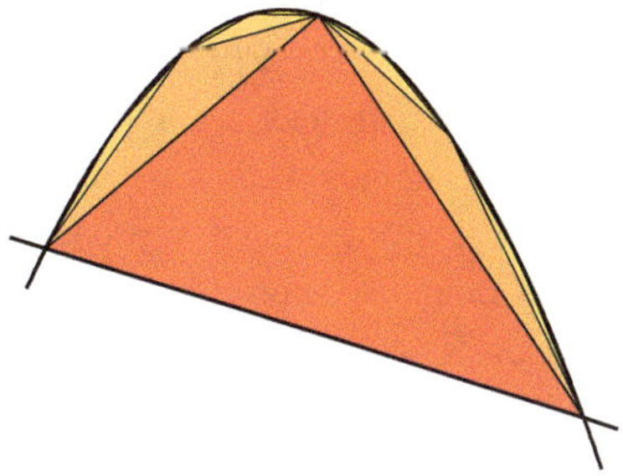

Archimedes had used the method of exhaustion to determine the quadrature of the parabola, filling a parabolic segment with the inscribed triangles, whose total area he analyzed, showing that the parabolic segment was $\frac{4}{3}$ the area of the principal inscribed triangle.

On the actualist account of the method of exhaustion, one considers the infinite exhaustion of the parabolic segment as a completed totality, arguing that the area of the triangles is the actually infinite summation

$$T + \frac{T}{4} + \frac{T}{16} + \cdots,$$

which can be proved to be equal to $\frac{4}{3}T$. And since the parabolic segment is completely exhausted by the inscribed triangles, this is seen as the area of the parabolic segment.

But in his treatise, Archimedes undertakes a potentialist approach to exhaustion without ever referring to a completed process of exhaustion. Rather, he presents a double reductio argument, which requires only arbitrarily large finite instances of the dissection by triangles. Namely, for the lower bound, he argues that the area of the parabolic segment could not be anything less than $\frac{4}{3}$ of the inscribed triangle, because then some finitely many of the inscribed triangles would already exceed it. And conversely, for the upper bound, the difference between the area of the parabolic segment and the finite triangle sums can be made as small as desired—it is reduced by at least half at each stage.

Thus Archimedes mounts his analysis of the quadrature of the parabola as a potentialist construction without ever requiring the infinite exhaustion of the parabolic segment as a completed totality.

Infinitely Many Primes

Euclid proved that there are infinitely many prime numbers, but his presentation of this theorem has an explicitly potentialist character, for the proof shows precisely that any given finite list of primes can be extended with an additional prime.

Let us give the argument.[2] Suppose that $p_1, \ldots, p_n$ are finitely many primes. Let n be the number obtained by multiplying them together and adding 1:

$$n = p_1 p_2 \cdots p_n + 1.$$

[2] You may also find a musical presentation of the proof on YouTube, *Plenitudinous Primes*, arranged and performed by Hannah Hoffman (2021), with lyrics written by me.

Every number above 1 has some prime factor, but any prime factor p of this number n must be different from every prime p_i on the original list since those primes all give a remainder of 1, but p, being a factor of n, has a remainder of zero. Thus, we have found a prime number p that does not appear on the original list. So there is a prime that the list has missed, and thus it can be extended.

This is a potentialist approach to the infinitude of primes, for it shows explicitly that every finite list of primes can be extended to a larger list of primes. You can have as many primes as you like.

9.1 Galileo Argues for Actualism

Remarkably, Galileo resists the potentialist orthodoxy of his time, arguing that actual infinities are possible. He considers how one might bend a segment at angles to form a square or an octagon or a polygon of 40 sides or more. This is not fundamentally different, he argues, from bending the segment into an infinite polygon, forming a circle:

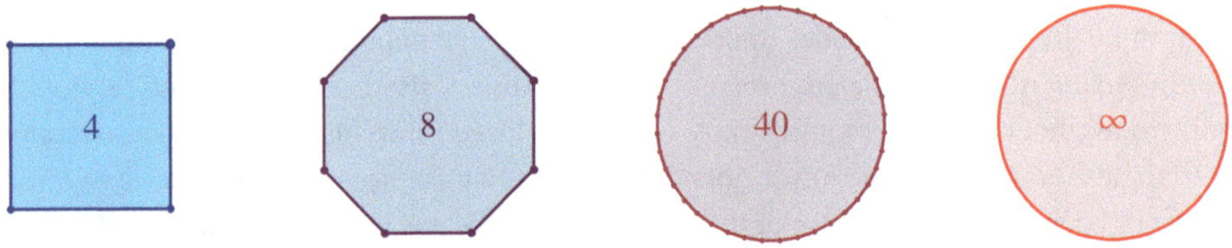

> **Salviati.** If now the change which takes place when you bend a line at angles so as to form now a square, now an octagon, now a polygon of forty, a hundred or a thousand angles, is sufficient to bring into actuality the four, eight, forty, hundred, and thousand parts which, according to you, existed at first only potentially in the straight line, may I not say, with equal right, that, when I have bent the straight line into a polygon having an infinite number of sides, i. e., into a circle, I have reduced to actuality that infinite number of parts which you claimed, while it was straight, were contained in it only potentially? Nor can one deny that the division into an infinite number of points is just as truly accomplished as the one into four parts when the square is formed or into a thousand parts when the millagon is formed; for in such a division the same conditions are satisfied as in the case of a polygon of a thousand or a hundred thousand sides. Such a polygon laid upon a straight line touches it with one of its sides, i.e., with one of its hundred thousand parts; while the circle which is a polygon of an infinite number of sides touches the same straight line with one of its sides which is a single point different from all its neighbors and therefore separate and distinct in no less degree than is one side of a polygon from the other sides. And just as a polygon, when rolled along a plane, marks out upon this plane, by the successive contacts of its sides, a straight line equal to its perimeter, so the circle rolled upon such a plane also traces by its infinite succession of contacts a straight line equal in length to its own circumference. Galileo Galilei (1914 [1638], pp.47–48)

For Galileo, the points of the circle are just as well specified and definite and distinct from one another as are the folding points leading to the large regular polygons, and we might thus take ourselves to be in possession of the actual infinity of the sides of the circle just as much as we were in possession of the 1000-gon or the 100,000-gon.

Is Potentialism Committed to an Actual Infinity of Possibilities?

There is a further subtle point to be made. The potentialist admits that whatever finite subdivision one currently has of a segment, one can always find a further finer subdivision. And thus, Galileo argues, even the potentialist must admit the actual infinity of the *possibilities* of refinement. If a potentialist truly believes that *you can always have more, as many as you like*, then there is an actual infinity of possible things that you can have—that is, the possibilities are actual, as possibilities, even if the things themselves are not yet actual. In this way, potentialism might be seen to admit actual infinities, the infinities of possibility.

9.2 Ultrafinitism

Ultrafinitism is the philosophical view that only comparatively small or accessible numbers exist. According to ultrafinitism, the various extremely large numbers that mathematicians conventionally take themselves to describe, such as 10^{100} or $10^{10^{10}}$, do not actually exist. Ultrafinitism sees it as a kind of illusion to speak of such numbers, and indeed we often find ourselves unable to answer basic questions about them. For example, we have no tools for determining in general even whether an expression such as $e^{e^{e^{e}}}$ represents an integer as opposed to a non-integer real number.[3]

According to ultrafinitism, such expressions and others, such as $5 \Uparrow 5$ or $3 \Uparrow\!\!\Uparrow 3$, are taken merely as formal terms, not necessarily representing any actual number in a meaningful way. The expressions themselves have a manageable size, to be sure, even if their supposed values are outsized, but when using the expressions, ultrafinitists needn't speak of or conceive of the number itself, rather than only of the expressions and the rules for manipulating them. Using the expressions in this way doesn't commit the ultrafinist to the view that there is actually a number being represented by them.

Must Ultrafinitism Commit to an Absolutely Largest Number?

Since ultrafinitism denies the existence of very large numbers, one might naturally wonder where exactly the existence stops—is there a sharp cut-off where suddenly the numbers stop existing? Is ultrafinitism committed to the existence of an absolutely largest number?

The conclusion that there is a largest number strikes many as absurd, since for any given number, it seems possible to imagine a mathematical system with one number more, so the numbers needn't have stopped just there but could have gone on a bit longer. The claim that any particular number is largest thus might seem at best contingent. Perhaps a critic hopes to refute ultrafinitism by burdening the position with what might seem a ridiculous commitment: the existence of a largest number.

[3] It follows from Schanuel's conjecture in number theory that e, e^e, e^{e^e}, and so on are all transcendental and indeed algebraically independent, and hence none are integers; but meanwhile, the truth of Schanuel's conjecture is an open and perhaps intractable question.

But this criticism is frankly too quick; the philosophical position of ultrafinitism is far more subtle. Most accounts of ultrafinitism do not take themselves to be committed to the view that there is an absolutely largest number. Rather, the view is that the arithmetic operation of adding 1 is innocent, whereas constructions such as exponentiation and superexponentiation are seen as misguided, not necessarily leading to actual numbers. The question is whether one can coherently view addition and multiplication as unproblematic, while keeping exponentiation at bay like this, taking it as only partially defined.

No Sharp Cutoff

Harvey Friedman (2002, pp. 4–5) describes an exchange that he had concerning the question of whether ultrafinitism is committed to a sharp end of the numbers:

> I have seen some ultrafinitists go so far as to challenge the existence of 2^{100} as a natural number, in the sense of there being a series of "points" of that length. There is the obvious "draw the line" objection, asking where in
>
> $$2^1, 2^2, 2^3, \ldots, 2^{100}$$
>
> do we stop having "Platonistic reality"? Here this ... is totally innocent, in that it can be easily be replaced by 100 items (names) separated by commas.
>
> I raised just this objection with the (extreme) ultrafinitist Yessenin Volpin during a lecture of his. He asked me to be more specific. I then proceeded to start with 2^1 and asked him whether this is "real" or something to that effect. He virtually immediately said yes. Then I asked about 2^2, and he again said yes, but with a perceptible delay. Then 2^3, and yes, but with more delay. This continued for a couple of more times, till it was obvious how he was handling this objection. Sure, he was prepared to always answer yes, but he was going to take 2^{100} times as long to answer yes to 2^{100} then he would to answering 2^1. There is no way that I could get very far with this.

The picture of ultrafinitism here is perhaps one for which the numbers get a little blurrier and less definite as one ascends to higher realms, without there necessarily being any definite point where one transitions from existence to nonexistence.

But what is the theory of ultrafinitism exactly? Is there an axiomatic account of fundamental ultrafinitist principles? Many philosophical critics of ultrafinitism despair at the lack of a precise account of the position. It is difficult, after all, to criticize a philosophical view that has not been clearly articulated. Can one state precisely what are the mathematical and philosophical commitments of ultrafinitism?

Nontotality of Exponentiation

One of the core commitments of ultrafinitism seems to be the nontotality of exponentiation; that is, the exponentiation function is not defined on all numbers—many ultrafinitists deny the existence of exponentials even with comparatively small exponents, such as 2^{100}. The question is whether for every number n, there is a number fulfilling the definition of what it would mean to be the exponential 2^n, and similarly a^n with other bases a.

In the usual arithmetic theory of Peano, we can prove that exponentiation is a total function on the basis of the standard axioms. Specifically, for every number n, there is a sequence of values 2^k for $k \leq n$ that fulfills the exponential recursion:

$$2^0 = 1$$
$$2^{n+1} = 2^n \cdot 2.$$

One proves by induction that for every n there is such a sequence. This is clear for the anchor case $n = 0$, and if it is true for n, then we can make the sequence for $n + 1$ simply by taking the sequence up to 2^n and then multiplying by 2 to form 2^{n+1}, adding it to the end of the sequence. This argument involves an account of finite sequences of numbers, which requires some arithmetic resources, and also uses the induction scheme for existential assertions.

The main point here is that to deny the totality of exponentiation, an ultrafinitist will have to deny some of the axioms of Peano arithmetic.

Feasible Numbers

One approach to ultrafinitism is based on the concept of *feasible* numbers. We would like somehow to express a realm of feasibility, which is closed under successor and perhaps addition and multiplication, but not under exponentiation. That is, we want a notion of feasible number such that 0 and 1 are definitely feasible, and if n is feasible, then so is $n + 1$, and perhaps if n and m are feasible, then so are $n + m$ and $n \cdot m$, but this closure property does not extend to exponentiation—there should be some feasible numbers n for which 2^n is not feasible.

Of course, the feasible-number idea is challenged directly by the principle of mathematical induction. After all, if 0 is feasible and whenever n is feasible then also $n + 1$ is feasible, then it follows by induction that every number will be feasible.

In order to preserve ultrafinitism and develop a concept of feasible number, therefore, ultrafinitists will have to deny the principle of mathematical induction, at least in the instance relevant for the notion of feasibility itself. One might hope to weaken the induction principle in such a way that a concept of feasible number can be formulated with the desired properties.

A Formal Theory of Ultrafinitism

These observations about the conflict between ultrafinitism and induction point one way toward formulating a precise ultrafinitist theory. Namely, we can simply weaken the axioms of Peano arithmetic to allow only primitive instances of induction. For example, the theory known as $I\Delta_0$ allows only instances of bounded induction:

$$[\varphi(0) \text{ and } \forall x\,(\varphi(x) \to \varphi(x+1))] \to \forall x\,\varphi(x),$$

that is, instances where the statement φ has an especially primitive form, involving no unbounded quantifiers. The quantifier symbol $\forall$ here means "for all," whereas $\exists$ means "there exists."

This provides an ultrafinitist theory because it is known that the theory $I\Delta_0$ is consistent with exponentiation being nontotal—there are models of $I\Delta_0$ in which we can add and multiply numbers as usual, but not every number n has an exponential 2^n. These models thus form a kind of arithmetic universe of numbers that accords with fundamental ultrafinitist ideas—one in which addition and multiplication are total functions, but not exponentiation, thereby establishing the basic coherence of the ultrafinitist position.

In these models, note that while exponentiation is not totally defined, nevertheless there will be no smallest number n that has no exponential, since if 2^k exists, then so does 2^{k+1}, since this is just $2^k \cdot 2$. Because the induction scheme has been limited, however, one cannot prove from this that every number n has an exponential 2^n since the assertion "Number n has an exponential 2^n" is too complex to fall under the bounded induction axiom scheme. Denying the totality of exponentiation seems to require the denial of induction.

In this way, we have found a precisely stated theory that seems to accord with many ultrafinitist ideas. And indeed there are many other such weakened arithmetic theories, generally weakening the induction principle to greater or lesser extents, which give rise to a wide spectrum of alternative accounts of ultrafinitism.

Meanwhile, none of these approaches to a formal theory of ultrafinitism are satisfactory for the most radical ultrafinitist claims, if one works in a non-ultrafinitist metatheory. This is because if one denies the existence of a largest number, then it follows by metatheoretic induction that the theory must admit the existence of 2^{100} and $2^{2^{2^{10}}}$ and $3 \Uparrow 3$ and so on, simply because these numbers are represented by terms in the formal language, even if those terms are extremely large. The numbers n for which the exponential 2^n does not exist in a model of these theories will be *nonstandard* numbers, bigger than any standard number represented by an explicit term, and because all nonstandard numbers are necessarily very large, in this sense the models of the ultrafinitist theories do not feel very ultrafinitist.

Indeed it seems difficult (or perhaps impossible) to formulate a precise arithmetic theory that denies the existence of a largest number, while also denying the existence of 2^{100}.

For these reasons, ultrafinitism poses a special challenge in formalization since we seem to require ultrafinitism also in the metatheory in which the formalization is carried out. Thus, we seem to have to know already what ultrafinitism is before saying what it is. Perhaps this explains why the more radical ultrafinitist positions often lack a fully satisfactory formal foundation, and it can be seen as an open question to provide a precise formal account of what the commitments of ultrafinitism are exactly.

Epistemological Versus Ontological Potentialism

Some philosophers distinguish between *epistemological* potentialism and *ontological* potentialism. According to epistemological potentialism, it is our knowledge that is incomplete—we are learning more all the time—even if the actual universe stands behind appearances as a fully completed reality. According to ontological potentialism, in contrast, the universe itself is incomplete, regardless of what we might believe or know about, or indeed, whether any agents are contemplating it at all.

9.3 A Modal Conception of Potentialism

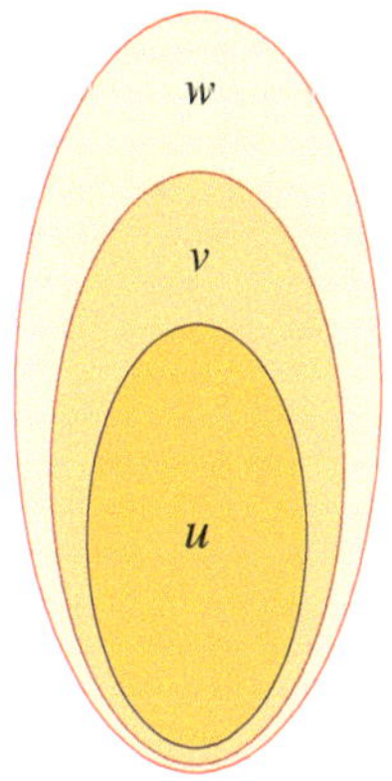

Contemporary philosophical work has emphasized the modal aspect of potentialism by expressing various potentialist conceptions in a modal vocabulary of possibility and necessity. The main idea is to conceive of potentialism via realms of feasibility, in terms of the various possible finite universe fragments that might be actual or possible at a given time. If the natural numbers are potentially infinite, in the sense that you can have more and more of them, as many as you like, but never all of them as a completed totality, then let us consider the possible collections of them that you can have, considering each as a possible "world," which might become the actual world for a time, if only for a moment. A small world u might be extendible to a larger world v, which is itself extendible to a still larger world w. The potentialist outlook consists at bottom of the idea that any given world u might be successively expanded in this way to larger worlds.

This situation enables a natural modal semantics and formal language of potentiality. Namely, we say that an assertion φ is *possible* at world u, denoted as $\Diamond\varphi$, if there is some world v extending u at which φ becomes true; and φ is *necessary* at world u, denoted as $\Box\varphi$, if φ is true at all such worlds v. For assertions made at world u, the existential quantifier $\exists x$ and the universal quantifier $\forall x$ range over the values of x that are actual in world u.

These modal operators provide a formal language capable of expressing fine shades of meaning with regard to potentialist conceptions. The potentialist can express not only what is true of a given finite part of the world already in possession, but also about what might later become true with a larger collection or what will be necessarily true, necessarily possible, or possibly necessary.

Modal Potentialism with Finite Universe Fragments in the Natural Numbers

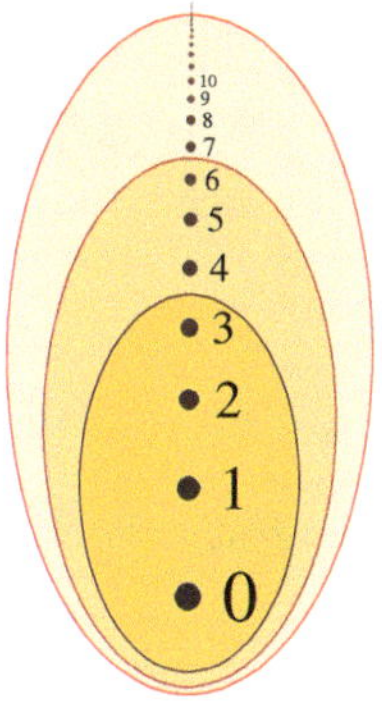

Let us imagine the potentialist system, in which the possible universe fragments or worlds consist of all the numbers up to some arbitrary given number n, which provides a larger universe fragment when n is made larger, as large as desired:

$$u = \{0, 1, 2, \ldots, n\}.$$

What will be the nature of the modal potentialist assertions? Notice that a given world u will not satisfy the assertion

Every number has a successor: $\forall x\, \exists y\; y = x + 1$

because if n is the largest number in u, the currently largest number at hand, then there is no actual $n + 1$ yet available. But it is *possible* that $n + 1$ exists since there is another universe fragment that goes a bit longer. Thus, although the finite worlds of this potentialist system do not fulfill the assertion that every number has a successor, they nevertheless do fulfill the assertion

Every number *possibly* has a successor: $\forall x \Diamond \exists y\, y = x + 1.$

Indeed, since this will be true in every universe fragment, we can say that it holds necessarily:

Necessarily, every number possibly has a successor:

$$\Box \forall x \Diamond \exists y\, y = x + 1.$$

Similarly, we express the potential existence of sums and products:

Necessarily, every two numbers possibly have a sum:

$$\Box \forall x, y \Diamond \exists z\, z = x + y.$$

Necessarily, every two numbers possibly have a product:

$$\Box \forall x, y \Diamond \exists z\, z = x \cdot y.$$

The modal conception of potentialism thus views many ordinary mathematical claims as modal in nature.

How do we express the infinitude of primes? One way to do so is to say that above any given number, there is a prime number. But the potentialist modal context makes this assertion subtle in several respects. First, when saying "above any given number," we do not want to refer only to the currently actual numbers, but rather to any possible number that we might have later with a larger universe fragment. So we should be saying *necessarily*, above any given number there is a prime. But a second issue is that saying it that way exactly wouldn't quite be right, since perhaps we extend to a universe fragment that has a number, but there is yet no actual prime yet above that number. What we want to say is that it is *possible* that above that number there is a prime number. Putting all this together, the assertion that there are infinitely many primes can be expressed in the modal potentialist language as follows:

Necessarily, for every number it is possible that there is a prime above it:

$$\Box \forall x \Diamond \exists p\, (x < p \text{ and } p \text{ is prime}).$$

The reader is invited in the questions at the end of the chapter to translate various other arithmetic assertions into the potentialist modal language.

Nonlinear Potentialism

But that is not the only way to organize the potentialist system. Let us consider an alternative conception of potentialism, for which having a number as actual does not imply that all smaller numbers are also already actual.

This is a sensible idea, in light of the fact that we can often describe or define a very large number more easily than many of the numbers below it. For example, it is relatively easy to describe the number googol plex $10^{10^{100}}$, which in decimal notation has an initial digit 1 followed by a googol many zeros. Thus, a googol plex has about a googol digits, and most of the numbers less than a googol plex also have about a googol digits. But for such a number with essentially random digits, there will generally be no simpler means of describing the number exactly except by pronouncing the digits one after the other. But as discussed in chapter 8, even if we pronounced 1 million digits every second since the beginning of time at the big bang, we will have hardly made progress on the googol many digits of our number. Most of the numbers below a googol plex are thus far more difficult for us to describe or even conceive than is a googol plex itself. In a precise sense, then, a googol plex is more amenable to thought than most of the numbers that are less than a googol plex. In light of this, it seems reasonable to consider a form of potentialism in which the numbers might arrive in the order of the complexity of their descriptions rather than in numerical order.

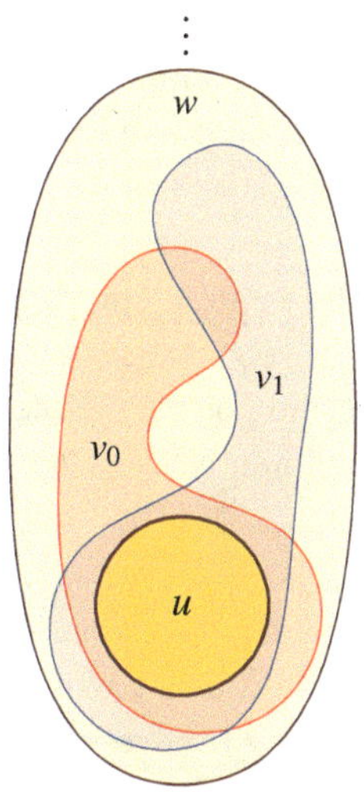

Let us consider more generally the potentialist system, in which the universe fragments consist simply of any finite set of natural numbers—let us take literally the idea that you can have more and more, as many as you like, any finite set. According to this conception of potentialism, we may have any finite set of natural numbers as actual, while any further finite set remains possible. Thus, a finite world u accesses the possibilities in worlds v_0, v_1 containing u, worlds which might be incomparable to each other, although both will be contained in a still larger world w. Let us use the symbols $\diamondsuit$ and $\boxdot$ to represent possibility and necessity for this potentialist conception.

This form of potentiality has a somewhat different structure than the earlier initial-segment formulation of potentialism, since those worlds were linearly ordered by inclusion, but here they are not. Nevertheless, with the arbitrary-finite-set conception, the worlds are *directed*, which means that any two of them are contained in a common extension. This directed-but-nonlinear feature causes some differences in the manner of expressing valid modal potentialist assertions.

For example, to say that number x is *even*, one ordinarily would say that there is some y such that $2y = x$. But if x is actual, with this arbitrary-set conception of potentialism, it doesn't follow that the y for which $2y = x$ will yet be actual, so one should say that *possibly* there is some y for which $2y = x$:

$$x \text{ is even:} \quad \diamondsuit \exists y \, (2y = x).$$

Similarly, to say that a number p is prime, one should refer to the possible factors rather than the actual factors:

$$p \text{ is prime:} \quad 1 < p \text{ and } \neg \Diamond \exists x, y < p\ (xy = p).$$

Those contortions were not required in the initial-segment formulation of potentialism.

9.4 Philosophical Varieties of Potentialism

The modal conception of potentialism, expressed in terms of potentialist systems of possible worlds, enables us to analyze and express certain differing philosophical features that we seek to have in our various potentialist conceptions, as well as to express sweeping general principles of potentialism. So let us discuss several varieties of potentialism that arise naturally in this modal framework and use the modal resources to express those differences.

Common Fundamental Modal Principles

First, let us discuss and verify certain fundamental modal principles that hold in common amongst all the diverse potentialist conceptions. One such principle, for example, is the assertion that every necessary statement is true, expressed by

$$\Box\varphi \to \varphi.$$

This is a valid modal principle in potentialism since if statement φ is true at all the worlds that you can access from world u, then in particular, it is true in world u itself, since we regard potentialist accessibility as a reflexive relation—every world can access itself.

Similarly, every necessary statement is necessarily necessary:

$$\Box\varphi \to \Box\Box\varphi.$$

This is valid because if φ holds in all worlds accessible from u and u accesses a larger world v, then φ will also be true in all worlds accessible from v, as these are also accessible directly from u—potentialist accessibility is a transitive relation.

Finally, if an implication $\varphi \to \psi$ holds in all accessible worlds and the hypothesis φ also holds in them, then the conclusion ψ also holds in all those worlds. This can be expressed by the modal assertion:

$$[\Box(\varphi \to \psi) \wedge \Box\varphi] \to \Box\psi.$$

These three principles constitute the central axioms of the modal theory known as S4.

Linear Inevitability

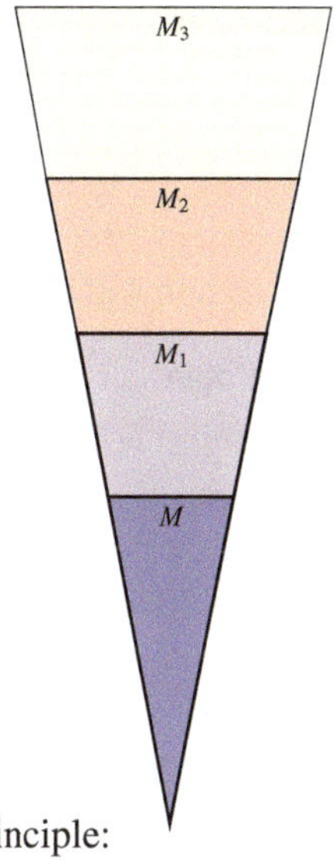

Consider now a form of potentialism, like the initial-segment form of potentialism considered initially, in which the universe fragments of possible worlds are linearly ordered. One moves from smaller universes to larger, and there is a kind of inevitability of the atomic structure of the objects that appear. Linear potentialism automatically validates certain kinds of potentialist assertions. For example, in linear potentialism, every possibly necessary assertion is also necessarily possible:

$$\Diamond\Box\varphi \to \Box\Diamond\varphi.$$

The reason is that if world M thinks that φ is possibly necessary, then there is some larger world M_1 thinking that φ is necessary, true from that point on. But now every world N will access some world larger than M_1, so every world N will think that φ is possible.

Similarly, linear potentialism also validates the following modal principle:

$$\Diamond\varphi \wedge \Diamond\psi \to [\Diamond(\varphi \wedge \Diamond\psi) \vee \Diamond(\psi \wedge \Diamond\varphi)].$$

The reason is that if both φ and ψ are possible over world M, then there are worlds M' and M'' containing M in which φ and ψ are true, respectively. But since the worlds are linearly ordered, either M' appears before M'' or the converse, and this therefore realizes one of the two possibilities in the conclusion of the principle, thereby establishing the validity.

Convergent Potentialism

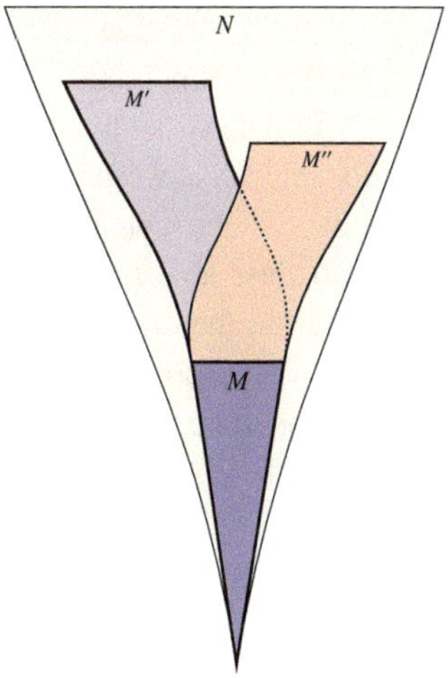

Consider next a form of potentialism in which the universe fragments of possible worlds are not necessarily linear ordered, but any two worlds can be amalgamated into a common extension. Thus, the universe fragments are altogether converging to an ultimate, actualist limit structure. Like linear potentialism, this convergent form of potentialism also validates the assertion that every possibly necessary assertion is also necessarily possible:

$$\Diamond\Box\varphi \to \Box\Diamond\varphi.$$

If φ is possibly necessary in world M, then there is a larger world M' in which φ is necessary, but for any other extension M'' of M, there is a common extension N of M' and M'', in which φ will be true since it extends v, so φ was possible in M. Thus, φ was necessarily possible in M, as claimed.

But in general, convergent potentialism does not generally validate the linearity principle considered earlier:

$$\Diamond\varphi \wedge \Diamond\psi \to [\Diamond(\varphi \wedge \Diamond\psi) \vee \Diamond(\psi \wedge \Diamond\varphi)]\,.$$

The reason is that a world M can access two worlds M' and M'' in which φ and ψ are true, but incomparable and never true again in larger worlds. In this situation, the premise of the implication will hold in M, but the conclusion will not.

The Potentialist Translation

Convergent potentialism, which includes the case of linear potentialism, supports a certain translation from actualism to potentialism. Namely, if the universe fragments have a convergent nature, then from the actualist perspective, we may view them as converging to a certain limit structure.

The main idea of the translation is systematically to replace actualist existential assertions $\exists x$ with possible existence $\Diamond\exists x$, and actualist universal assertions $\forall x$ with necessary universals $\Box\forall x$. We had already performed this translation several times in our earlier account giving the potentialist interpretation of various statements in elementary number theory, such as the infinitude of primes. The point that I should like to make now is that this is a completely general method of translating any statement from the actualist context to potentialism. Actualist assertions about the limit model are translated to potentialist assertions about the universe fragments of which it is constituted. Furthermore, one can prove that any actualist assertion about the limit model is equivalent to the potentialist translation assertion made about the universe fragments.

The conclusion is that for convergent potentialism, the potentialist seems to have all the resources in the purely potentialist ontology to refer to the objects, the structure, and even the truth assertions of the limit model. The potentialist, to be sure, holds that the limit model does not exist. And yet he seems nevertheless to know everything about that actually infinite structure anyway.

Philosophically, we can take this observation two ways. On the one hand, it shows the power of the potentialist perspective since in convergent systems, the potentialist is nevertheless able to express all the truths and understanding that the actualist would want. But on the other hand, we could view the observation critically, taking it to show that convergent forms of potentialism are *implicitly actualist*—in convergent potentialism, there seems to be nothing at stake in the potentialist claim that the limit model does not exist since the potentialist has a complete account of it.

Radical Branching Potentialism

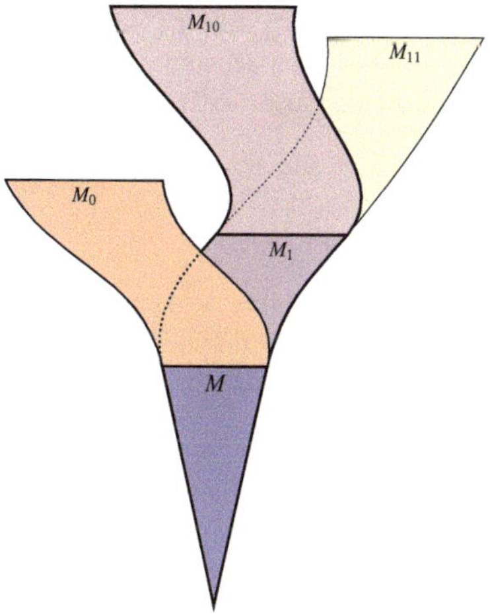

What remains to consider is a more radical form of potentialism, in which the potentialist translation will fail, and indeed the nature of possibility and necessity will depend on the particular manner in which the universe unfolds. As objects become actual, they may do so in such a way that precludes other possibilities from becoming actual, even when they might have been possible before. The issue is one of *branching possibility*, where extending a universe fragment M to M_0 or M_1 might exhibit certain features that cannot afterward be amalgamated. In world M, we may have two potentialities M_0 and M_1, both as possibilities. But once one of them becomes actual, the other is permanently closed off as a possibility. Having moved to world M_1, say, we may find further possibilities M_{10} and M_{11}, which were not possible over M_0. Perhaps number-theoretic phenomena, for example, are revealed in M_0 and M_1 that are incompatible with amalgamation. Perhaps a certain computational process halts with output 0 in M_0, but it halts with output 1 in M_1. There can be no common extension of these worlds since the computational process would have to conform with both of the earlier worlds.

In my article, **Hamkins:The-modal-logic-of-arithmetic-potentialism**, I show that the models of Peano arithmetic under end-extension fulfill the nonamalgamation properties of radical-branching potentialism. Each model of PA is an arithmetic realm of feasibility, with potentiality exhibiting this fundamental branching-possibility nature. Indeed, the branching occurs with computational processes in exactly the manner I just described.

9.5 The Rise of Actualism

Let me conclude this chapter by discussing more explicitly the sea change that has taken place in mathematics from potentialism to actualism. In this chapter, I have discussed the potentialist outlook at length, whereas most of the rest of the book is probably best understood as presented from an actualist point of view.

The change started to occur in earnest with Bernard Bolzano (1781–1848), Richard Dedekind (1831–1916), and Georg Cantor (1845–1918), at first gradually and then all at once. In contemporary mathematics research, the overwhelming majority of mathematicians have an actualist outlook rather than potentialist. The challenging troubles of actualism are viewed as having been straightened out, the paradoxes have been explained, and mathematicians these days find great inspiration and insight from the actualist perspective.

Questions for Further Thought

9.1 Present two detailed accounts of the quadrature of the parabola side by side, one following an actualist account of the exhaustion and one following a potentialist account. How do the arguments differ?

9.2 Assess the argument that potentialism, in asserting that you can always have more, as many as you like, is thereby committed to an actual infinity—namely, the infinite collection of possibilities.

9.3 Is our inability actually to count up to 10^{100} a problem for the existence of this number? What is the connection between the limitations of human abilities in reasoning and computation and the ontological questions about the existence of large numbers?

9.4 Can one formulate a successful notion of feasible numbers, and thus a form of ultrafinitism, by saying that feasible numbers are precisely those that one can count to in a reasonable amount of time? This would include all the smallish numbers, and if we can count to n in a reasonable amount of time, then it might also seem that we can count to $n + 1$ in a reasonable amount of time; and yet we cannot count to 10^{100} in a reasonable amount of time since even if we say a million numbers every second since the beginning of time at the big bang, we will have counted only a tiny portion of it.

9.5 Which of the following modal principles are valid for the finite-initial segment conception of potentialism?

1. $\Diamond\Box\varphi \to \varphi$
2. $\Box\varphi \to \varphi$
3. $\Box\varphi \to \Box\Box\varphi$
4. $\Diamond\Box\varphi \to \Box\Diamond\varphi$
5. $\Box\Diamond\varphi \to \Diamond\Box\varphi$
6. $\Diamond\Box\varphi \vee \Diamond\Box\neg\varphi$

9.6 Give an example of a *switch*, a statement φ whose truth value can be turned on and off again endlessly as one ascends to larger universe fragments. Can you provide an infinite family of *independent* switches φ_n, meaning that any finite on/off pattern is necessarily possible over any given universe fragment?

9.7 Give an example of a unpushed *button*, a statement ψ that is not true in a given universe fragment but is possibly necessary in that world, becoming true from some point on in larger worlds. Can you provide two *independent* buttons ψ_0 and ψ_1, meaning that both are initially false in some world u, but there are extensions v_0 and v_1 such that in v_0, the assertion ψ_0 is necessarily true and ψ_1 is false, and in v_1, the assertion ψ_0 is false and ψ_1 is necessarily true? Does the answer depend on whether one uses the finite-initial-segment conception of potentialism or the arbitrary-finite-collection conception of potentialism? Can you make an infinite independent family of buttons for one of the potentialist conceptions?

9.8 Translate the following arithmetic assertion into the potentialist modal language, using the notion of possibility by which a possible world consists of an initial segment of the numbers $\{0, 1, \ldots, n\}$. Express the assertion in natural language by using the words "necessarily" and "possibly," but also in a more formal expression using the modal operators $\Box$ and $\Diamond$.

1. There is no largest number.
2. Every even number is the sum of two primes.
3. There is always a prime between any number n and $2n$.
4. $7n^2 + 9n < n^3$ for all sufficiently large n.
5. There are infinitely many prime pairs (primes differing by 2).
6. The square root of 2 is irrational.

9.9 Translate the assertions of the previous question into the potentialist conception by which a possible world is any finite set of numbers, not necessarily closed under initial segment. Use the modal operators ◈ and ▣ for this conception of possibility and necessity. What are the differences in formalization for this potentialist conception versus the initial-segment conception?

Further Reading

Galileo Galilei (1914 [1638], pp. 47–52). Read Galileo's discussion of potential and actual infinity, with the example of bending a segment into polygons or a circle.

Hamkins and Wołoszyn (2024). An introduction to modal model theory that implements potentialism in diverse parts of mathematics.

Hamkins and Linnebo (2022). A technical account of various kinds of set-theoretic potentialism.

Hamkins (2026a). A exploration of ultrafinitism, viewed through a potentialist lens.

10 Equinumerosity and Comparison of Size

Abstract. What does it mean for two sets to have the same size? Galileo's paradox identifies a tension between natural notions of equinumerosity and Euclid's principle that the whole is greater than any proper part.

10.1 Equinumerosity

Before the dinner party, young Gottlob happens to pass the dining room prepared for the guests, and he observes at a glance, without counting, that the number of plates on the dining table is the same as the number of knives. How does he do this without counting? He simply notices, of course, that every place setting has one plate and one knife. By thus associating every plate with its counterpart knife at that setting, he has formed a one-to-one correspondence between the plates and knives, which implies that their numbers will be the same without any need to actually count them. It is not necessary to know how many plates or knives there are to know that the numbers are the same.

In the lecture hall, I might similarly tell at a glance that the number of people at the lecture is the same as the number of noses, while counting neither people nor noses. I can do this simply by placing the people in a one-to-one correspondence with the noses, associating every person with their own nose. If indeed everyone has a nose and there are no dreadful extra noses in someone's pocket, then indeed this is a one-to-one correspondence, which shows that the two numbers are the same, whatever they are, with no need to count them.

In general, we say that sets A and B are *equinumerous* if there is a one-to-one correspondence between their elements.

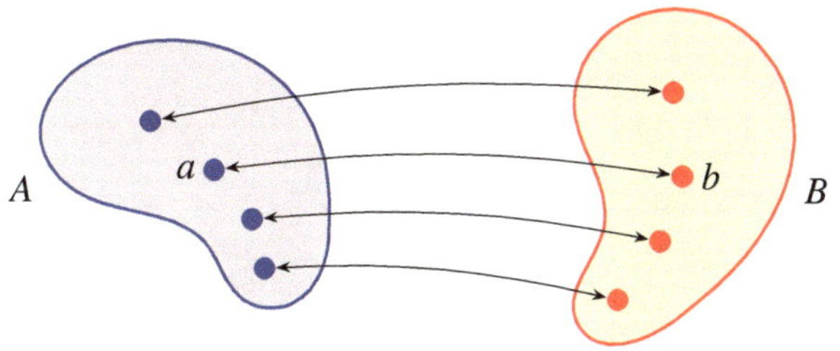

Every individual a of the first set A is matched with a corresponding individual b in the second set B in such a way that all the individuals have exactly one partner in the other set. The shepherd counts off his sheep on his fingers.

The main idea here is that equinumerosity seems to provide us with a criterion for saying that two sets have the same number of elements, in a manner that is conceptually prior to having a suitable concept of number or method of counting. For example, we can speak coherently of equinumerosity with infinite sets, even if we may not yet have figured out an appropriate notion of counting or numbering them. This idea has come to be known as the *Cantor-Hume* principle, expressing our expectations for how numbers should behave, whether in finite or infinitary contexts—the principle expresses a necessary requirement on whatever number concept we may eventually settle on.

Cantor-Hume principle. *Two sets have the same number of elements if and only if those sets can be placed in a one-to-one correspondence.*

The Cantor-Hume principle is also widely known simply as *Hume's principle*, in light of David Hume's brief statement of it in 1739:

> When two numbers are so combin'd, as that the one has always an unite answering to every unite of the other, we pronounce them equal; Hume (1739, p. I.III.I)

The idea has come up many times in the philosophy of numbers and infinity. Long before Hume, Galileo considers equinumerosity in his treatment of several confounding matters of infinity in his final scientific work, *Dialogues Concerning Two New Sciences* (1638). At the end of the nineteenth century, Gottlob Frege (1848–1925) founds his theory of arithmetic directly upon the Cantor-Hume principle. Slightly before that, Georg Cantor places equinumerosity at the very center of his spectacular treatment of infinite and especially uncountable cardinalities. It is evidently a very natural idea.

The Paradox of Aristotle's Wheel

Before discussing Galileo, let us roll Aristotle's wheel across the page. Consider the large blue circle as shown here, centered at A with radius AC, containing a smaller red circle with radius AB within it:

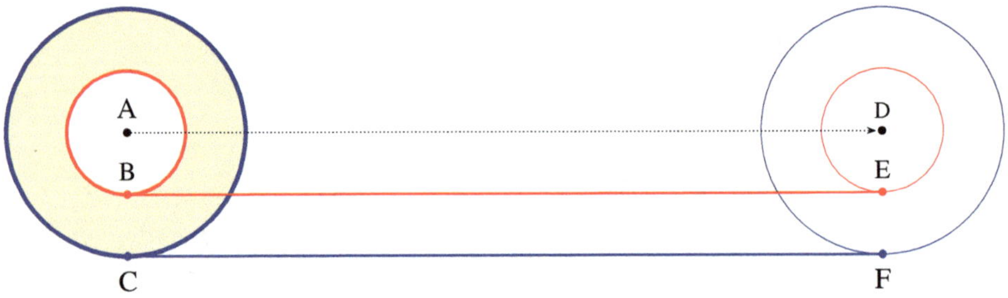

A swift kick will set the wheel rolling from left to right, bringing its center from A to D, smoothly completing exactly one revolution. During the journey, every point on the circumference of the blue circle is therefore matched with its point of contact on the ground, which changes continuously as the wheel rolls. The blue circumference is thus laid out onto the segment CF. In particular, the circumference of the blue circle is exactly the same as the length of segment CF.

A more troublesome observation, however, concerns the smaller red circle. Being rigidly attached to the same wheel, the red circle similarly lays itself out on the line from B to E, also changing its point of contact continuously as the wheel rolls. Every point on the red circumference is thus associated with one and only one point on the segment BE. We might be led to deduce that the circumference of the red circle is the length of segment BE. But the conundrum is that... the segments BE and CF have the same length! Are we actually to conclude that the two circles, both the smaller and the larger, have exactly the same circumference, or at least the same number of points on them? How can that be?

Simplicio and Salviati

Galileo rolls Aristotle's wheel into his treatment of the paradoxical and confusing nature of infinity in *Two New Sciences*, leading his characters Simplicio and Salviati into fits of confusion with several troubling matters of infinity.

To help with his analysis of Aristotle's wheel, Salviati rolls out an accompanying clunky hexagon, as shown here, an outer blue hexagon with an inner red hexagon rigidly attached to it on the same axle:

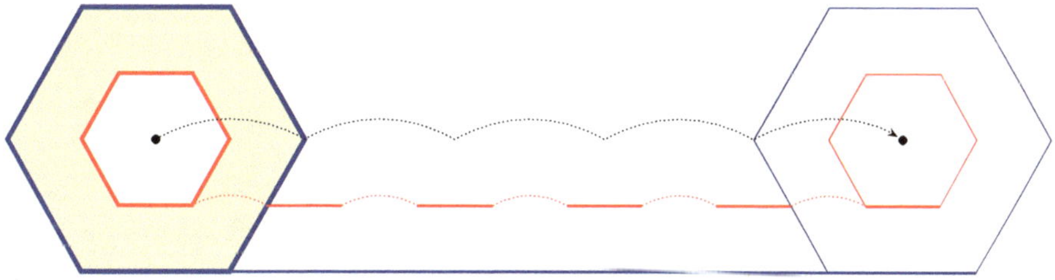

Heaving against the blue hexagon, they tip it up on its vertex and push it over to rest on the next edge, *kerthunk!* and then again, like a cut stone block rolling awkwardly. Finally, a complete revolution is made. Do you see how the red hexagon moves during the traversal? As the blue hexagon is tipped up, the red edge lifts completely off the horizontal line before coming to rest upon it again as the blue hexagon drops into the next place. The points on the circumference of the red hexagon are thus associated only intermittently with the horizontal red line since there are these regular gaps between the resting places.

The same feature would hold if we had used a septagon or an octagon—no matter how many sides we had. Salviati explains:

> Now this exposition which I have given in the case of these hexagons must be understood to be applicable to all other polygons, whatever the number of sides, provided only they are similar, concentric, and rigidly connected, so that when the greater one rotates the lesser will also turn however small it may be. You must also understand that the lines described

> by these two are nearly equal provided we include in the space traversed by the smaller one the intervals which are not touched by any part of the perimeter of this smaller polygon.
>
> Let a large polygon of, say, one thousand sides make one complete rotation and thus lay off a line equal to its perimeter; at the same time the small one will pass over an approximately equal distance, made up of a thousand small portions each equal to one of its sides, but interrupted by a thousand spaces which, in contrast with the portions that coincide with the sides of the polygon, we may call empty. Galileo Galilei (1914 [1638], p. 22)

And yet, as the number of sides increases, we approach the case of the circle, in which there seem to be no gaps at all in the line traced out by the inner red circle. For Galileo the lesson concerns the confusing combination of infinity and indivisibility; Salviati identifies the difficulties like this:

> . . . infinity and indivisibility are in their very nature incomprehensible to us; imagine then what they are when combined. Yet if we wish to build up a line out of indivisible points, we must take an infinite number of them, and are, therefore, bound to understand both the infinite and the indivisible at the same time. Galileo Galilei (1914 [1638], p. 30)

Equinumerosity of Any Two Similar Arcs

In *The Paradoxes of the Infinite* (1851), Bernard Bolzano expands on Galileo's paradoxes, pointing out that any two similar arcs, regardless of size, can be placed in a one-to-one correspondence simply by following the common rays from the center. Each point on the red arc here is thus associated with exactly one blue point and vice versa.

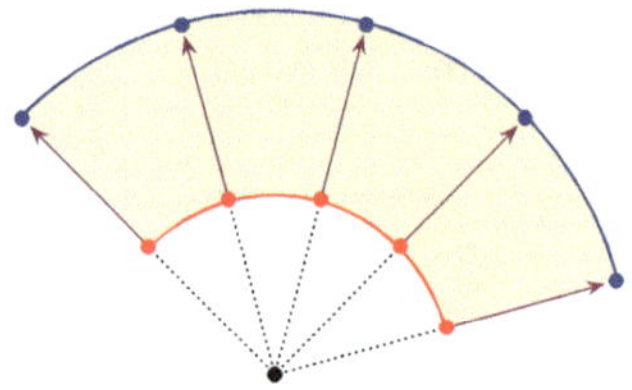

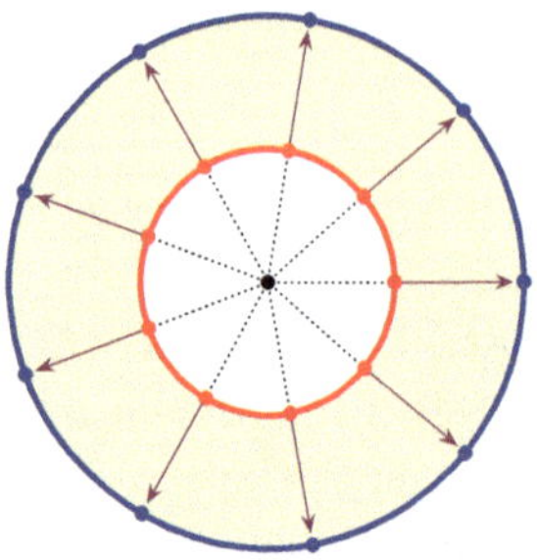

The same idea, which Bolzano traces back to Aristotle, shows that any two circles can be placed in a one-to-one correspondence—simply orient them concentrically and follow the common rays. Every red point is matched with a distinct blue point and vice versa. Bolzano (1851, p. 137) argues that this analysis refutes the opinion that he calls a "mistake,"—namely, "that two infinite sets, as well as two finite ones, are bound to be equal so soon as each member of the one can be coupled off against one member of the other." Thus he seems flatly to contradict the Cantor-Hume principle. Have we made a mistake or has Bolzano?

Equinumerosity of Line Segments and Lines

We may similarly observe that the points on any two line segments, regardless of length, can be placed in a one-to-one correspondence by a fan of similarity, as in the figure that follows. Every point on the short red segment is matched with a distinct point on the longer blue segment in a one-to-one manner. The paradox is that while one would naturally want to say that the shorter segment has fewer points, nevertheless the correspondence provides a sense in which the two segments have exactly the same number of points:

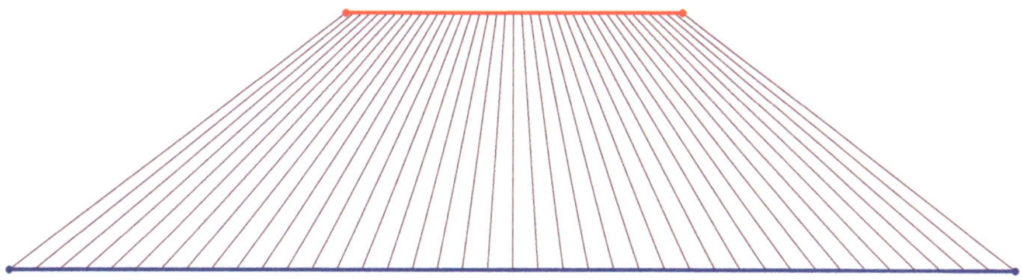

A finite line segment can even be placed in one-to-one correspondence with an entire line. The points of the red line segment here (not including the endpoints) are projected down to the semicircle and then projected radially outward to the infinite line. Every point on the segment is thus associated with a unique point on the line and vice versa:

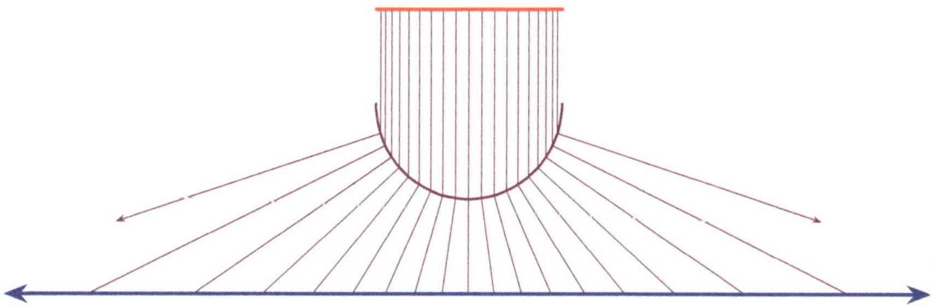

In this way, we see that a finite line segment has the same number of points as an entire line—it seems incredible, but it's true.

10.2 Galileo's Paradox

Galileo's characters continue their conversation by observing that the natural numbers can be placed in a one-to-one correspondence with the perfect squares by associating each number n with its square n^2:

$$\begin{array}{cccccccc} 0 & 1 & 2 & 3 & 4 & 5 & 6 & \cdots \\ 0 & 1 & 4 & 9 & 16 & 25 & 36 & \cdots \end{array}$$

The situation is deeply paradoxical—can you see why? There naturally seem to be many more numbers than there are perfect squares since not every number is a square, and indeed the gaps between the squares are growing increasingly large. And yet, the correspondence suggests that there are *exactly the same* number of numbers as squares. Precisely because we can put these two collections into a one-to-one correspondence, it seems that the two collections should be taken as having the same size—they are equinumerous.

The Tension Between the Cantor-Hume Principle and Euclid's Principle

Galileo's paradox thus showcases a fundamental tension between the Cantor-Hume principle and the classical *Euclid's principle*, as follows:

Euclid's principle. *The whole of any object is greater than any proper part.*

The point is that Galileo has placed the whole set of natural numbers into one-to-one correspondence with the set of perfect squares by associating each number n with its square n^2, associating 2 with 4, and so on. On the basis of the Cantor-Hume principle, therefore, the set of natural numbers would have the same size as the set of perfect squares. But on the basis of Euclid's principle, it should be strictly larger since the squares form only a proper part of the numbers—there are so many other numbers between the squares. So these two principles lead to contradictory conclusions.

Galileo recognizes, through his charismatic Salviati, that we must give one of them up. Sadly, Salviati concludes that comparisons of size simply do not apply in the realm of the infinite:

> But if I inquire how many roots there are, it cannot be denied that there are as many as there are numbers because every number is a root of some square. This being granted we must say that there are as many squares as there are numbers because they are just as numerous as their roots, and all the numbers are roots. Yet at the outset we said there are many more numbers than squares, since the larger portion of them are not squares. Not only so, but the proportionate number of squares diminishes as we pass to larger numbers. Thus up to 100 we have 10 squares, that is, the squares constitute 1/10 part of all the numbers; up to 10000, we find only 1/100 part to be squares; and up to a million only 1/1000 part; on the other hand in an infinite number, if one could conceive of such a thing, he would be forced to admit that there are as many squares as there are numbers all taken together.
>
> ...
>
> So far as I see we can only infer that the totality of all numbers is infinite, that the number of squares is infinite, and that the number of their roots is infinite; neither is the number of squares less than the totality of all the numbers, nor the latter greater than the former; and finally the attributes "equal," "greater," and "less," are not applicable to infinite, but only to finite, quantities. When therefore Simplicio introduces several lines of different lengths and asks me how it is possible that the longer ones do not contain more points than the shorter, I answer him that one line does not contain more or less or just as many points as another, but that each line contains an infinite number.
> Galileo Galilei (1914 [1638], p. 32)

10.3 Toward a Comparative Notion of Size

Despite Galileo's conclusion, let us discuss how we might overcome his paradox and use the ideas underlying the Cantor-Hume principle to provide a successful comparative notion of size, which might work even when the sets in question are infinite. What should it mean exactly to say that one set is larger than another, or that it is at least as large?

The tension of Galileo's paradox would relax completely if we were to weaken Euclid's principle to the following reflexive form:

Reflexive Euclid's principle. *The whole of any object is at least as great as any part.*

The paradox gets no traction at all with this reflexive formulation of Euclid's principle—the contradiction evaporates. The situation of Galileo's example is simply that the set of natural numbers is at least as great as the set of perfect squares, and indeed, they have the same size in light of the one-to-one correspondence.

In place of contradiction, therefore, what Galileo's paradox seems to reveal from this perspective instead is a curious feature of infinity—an infinite set can have the same size as a proper part. Although very different from what we are used to with finite sets, this by itself is not actually a contradiction. In this way, the reflexive weakening of Euclid's principle seems to dissolve the paradox.

At Least as Large

Although the Cantor-Hume principle provides us with means for expressing that two sets or classes have the same size—they are equinumerous—a robust concept of number and size would enable us also to make *comparative* size judgements, to say that one set or class is *at least as large* as another. For this, we would seem to need a principle going strictly beyond the Cantor-Hume principle.

Comparative size principle. *One set is at least as large as another if and only if it has a subset that is equinumerous with it.*

In other words, a set Y is *at least as large as* X, written as $X \leq Y$, if there is a one-to-one correspondence between X and a subset of Y. Equivalently, $X \leq Y$ if there is a one-to-one mapping of X into Y; it needn't map X onto all of Y.

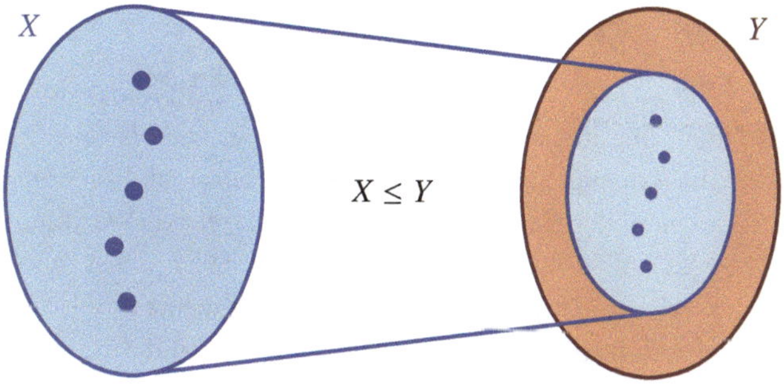

It follows immediately from this definition that every set is at least as large as all its parts, so the definition expresses the reflexive formulation of Euclid's principle. But it does so in a way that is compatible with the Cantor-Hume principle, by allowing the situation that indeed an infinite set might also have the same size as a proper part.

The notion of comparative size here is *reflexive*, meaning that every set is at least as large as itself, or $X \leq X$. It is also *transitive*, meaning that $X \leq Y \leq Z$ implies $X \leq Z$. This makes the comparative size relation a *preorder* relation.

Every preorder relation gives rise to an accompanying strict order relation. In our case, we may define that one set Y is *strictly larger than* another set X if it is at least as large, but not conversely. That is,

$$X < Y \qquad \text{if and only if} \qquad X \leq Y \quad \text{and} \quad Y \nleq X.$$

Notice that the definition of $X < Y$ here is not saying merely that X is equinumerous with a proper part of Y, since as we have seen with Galileo's paradox, that can occur when X and Y are equinumerous or even identical—an infinite set can be equinumerous with a proper part of itself. Rather, the definition says that the strict size comparison holds $X < Y$ exactly when Y is at least as large as X but not conversely.

We may notice that the strict order relation is *irreflexive*, meaning that no set is strictly larger than itself, and also that it is transitive, meaning that $X < Y < Z$ implies $X < Z$. Despite Galileo's reservations, therefore, we seem to provide in this way a sensible and robust comparative theory of size, applicable even in the context of infinite sets.

A Missing Piece of the Analysis

I should like to emphasize a certain subtle issue of the comparative size issue that is unfortunately often overlooked in philosophical discussion. Consider the situation where Y is at least as large as X, and also X is at least as large as Y. In light of the ordinary meaning of these words, we should want this statement to imply, of course, that X and Y are the same size. But does it? After all, we have given these phrases particular technical meanings in terms of the existence of certain kinds of one-to-one correspondences, and it is at bottom a purely mathematical question whether this conclusion follows from those premises.

Question. If $X \leq Y$ and $Y \leq X$, must X and Y be equinumerous?

It is tempting to think that this conclusion is both trivial and obvious, but to my way of thinking, this would be a mistake. This is truly a subtle issue, and for someone to take the conclusion as obvious is perhaps to miss what is at stake here. We want it to be true, of course, that $X \leq Y \leq X$ implies that X and Y have the same size, since this would be a basic feature of any successful account of comparative size. But precisely for this reason, we had better ensure that our notions of comparative size indeed exhibit this expected feature.

So let us consider the matter carefully. To say that both $X \leq Y$ and $Y \leq X$ means that we have an injective correspondence of X with a part of Y and simultaneously an injective correspondence of Y with a part of X:

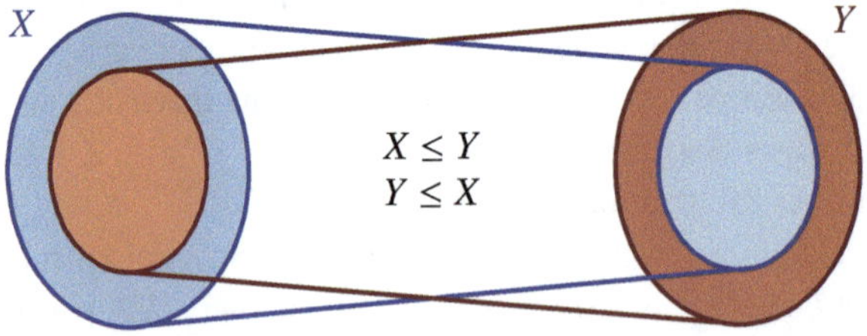

What needs to be done, in order to conclude that X and Y have the same size, is to construct from those associations a one-to-one correspondence between X and Y themselves, with no parts missing. Is this possible?

Yes, indeed it is, and this conclusion is precisely the content of what is known as the Cantor-Schröder-Bernstein theorem.

Cantor-Schröder-Bernstein Theorem. *If X is equinumerous with a subset of Y and Y is equinumerous with a subset of X, then X and Y are themselves equinumerous.*

The theorem expresses the *antisymmetry* property for the comparative-size and same-size relations. The main idea of the proof is to find suitable restrictions of the two correspondences in each direction in such a way that they no longer conflict, and in this way assemble a one-to-one correspondence of the entire set X with the entire set Y.

The theorem has a convoluted history. It was stated by Cantor in 1887 without proof; proved but not published by Richard Dedekind in 1888; proved by Cantor in 1895, but only by assuming the linearity of cardinals; proved by Ernst Schröder (1841–1902) in 1896 as a corollary of a theorem of William Stanley Jevons (1835–1882) and then again in 1897 independently (but this proof was later found to be faulty); and then proved independently by Felix Bernstein (1878–1956) in 1897. As Cantor had observed, the theorem is an immediate consequence of the assumption that every set can be well ordered, an assumption equivalent to the axiom of choice, but since one does not need this assumption to prove the theorem, it is better to do so without it.

Although I have stressed that this theorem is subtler than a naive outlook would lead one to expect, nevertheless the theorem is not so terribly difficult to prove, and one can find a proof of it in any good set-theory textbook. I provide an elementary proof, for example, in my proof-writing book, Hamkins (2020, theorem 117), based on the (1906) proof of Julius Kőnig (1849–1913).

To my way of thinking, the fact that the Cantor-Schröder-Bernstein theorem is true can be taken as evidence that we have indeed found a robust comparative conception of size—one set is at least as large as another when it is equinumerous with a subset of it. And the theorem shows that two sets are at least as large as each other if and only if they have the same size by the criterion of the Cantor-Hume principle. Further, I say that whenever someone proposes a philosophical conception of size applicable to infinite sets, then it is incumbent upon them to engage with the antisymmetry property, which states that whenever two sets are at least as great in size as the other, then they are equinumerous.

Meanwhile, the philosophical literature is full of discussion and analysis of the Cantor-Hume principle, expressing the concept of same-size—philosophers have written about it at length—but there is usually little accompanying treatment of the comparative notion of at-least-as-large or the concept of larger-than. I view this as a regrettable gap in the philosophical literature. To my way of thinking, one has not fully engaged with the number-assignment problem without also looking at the comparative-size problem, and any serious engagement with that will lead eventually to the Cantor-Schröder-Bernstein theorem.

So let me recommend to you, dear reader, that if you should ever find yourself writing in any philosophical work about Hume's principle or the Cantor-Hume principle, you should immediately seek to augment your discussion with a treatment of comparative size and the issue of antisymmetry.

Questions for Further Thought

10.1 What tension is there, if any, between the Cantor-Hume principle and Euclid's principle, and then how might it be resolved? Is one of the principles somehow more fundamental?

10.2 Does the Cantor-Hume principle express our best conception of the size of a set?

10.3 Does a line segment consist of the set of points incident upon it? If so, are we correct to speak of the number of points on the line segment in terms of equinumerosity? Do segments of different lengths have the same number of points?

10.4 Does Bolzano's observation that any two circles, regardless of size, are equinumerous refute the Cantor-Hume principle?

10.5 Which is the more fundamental philosophical conception, the Cantor-Hume principle or the Comparative size principle?

Further Reading

Galileo Galilei (1914 [1638]). Galileo's dialogues—the discussion on pages 20–33 is particularly apt for the topic of this chapter.

Hamkins (2020). Chapter 13 contains an elementary proof of the Cantor-Schröder-Bernstein theorem.

Credits

The story of young Gottlob at the dinner party is a reference to Gottlob Frege's *Grundlagen der Arithmetik* (1884), in which he describes equinumerosity of plates and knives. Parts of this chapter were adapted from Hamkins (2021b).

11 What Is the Infinite?

Abstract. In this chapter, we shall ponder the question of what it means, exactly, to say that a set is infinite. Can we provide precise mathematical definitions of the finite and the infinite?

11.1 What Is the Infinite?

We have gotten ahead of ourselves, talking so much about the infinite but without having a clear definition of what it means to be infinite. What actually is the infinite? What exactly does it mean to say that a given set is infinite? How are we to describe this sublime, elusive notion? We should want a crisp, clear account—one suitable for mathematics.

Suppose that I turn the tables and ask the dual question: What is the finite? You might think, *How absurd*—obviously we know all about the finite. Right? To my way of thinking, however, the two questions are equally hard, for they carve out the same boundary between the finite and the infinite. To define the finite is to define the infinite by complement; to have a clear criterion for one is to have a clear criterion for the other.

Meanwhile, it is not so easy and clear after all how to define either the finite or the infinite. In mathematical history, there have been many proposals, so let us explore a little the spectrum of concepts on offer.

Aristotle

Aristotle (384–322 BC) defines the infinite as that which proceeds without end, but as discussed in chapter 9, he emphasizes that we can never achieve actual infinity. We always find ourselves rather with only a finite quantity at hand, even while the infinite potentiality looms before us, tempting us. That perspective persisted in mathematics for millennia, although there was a sea change at the dawn of the twentieth century and views have been completely overturned—we're almost all actualists now.

Galileo

Galileo was a rare, early critic of potentialism, and he undertook to investigate the true nature of actual infinity. We discussed in chapter 10 his observation in *Two New Sciences* (1638) that the natural numbers can be placed in one-to-one correspondence with the perfect squares by associating each number n with its square n^2 like this:

n	1	2	3	4	5	6	$\cdots$
n^2	1	4	9	16	25	36	$\cdots$

The situation is deeply paradoxical, we said, for the gaps between the perfect squares grow increasingly large, so most natural numbers are not perfect squares. And yet, the correspondence suggests that there should be exactly the same number of numbers as squares—they are equinumerous. Thus we observe the tension between Euclid's principle, according to which the whole is strictly greater than any proper part, and the Cantor-Hume principle, which asserts that sets placed in a one-to-one correspondence will have the same size. The characters in Galileo's dialogue ultimately throw up their hands in confusion.

Dedekind Finite

The mathematician Richard Dedekind (1831–1916) proposes that we consider this feature as a definition. Namely, a set is finite—we call it *Dedekind finite*—when it is not equinumerous with any proper subset, and *Dedekind infinite* when it is. So the natural numbers come out as Dedekind infinite, of course, just as Galileo observes, and this is what we want. But will the criterion give the right result in all cases? After all, do we really expect that *every* infinite set will have this paradoxical property of being equinumerous with a proper part? Perhaps there could be some strange, inhomogeneous infinite set that did not admit such kinds of self-directed correspondences, but would we want to say that just because of this, the set should count actually as finite?

Order Infinite

Meanwhile, other approaches to defining the dichotomy between the finite and the infinite are grounded in elementary concepts of ordering. One naively attractive idea may be that a finite set is one for which we can place the elements in a discrete, linear order, with a first point, a next point, and a next, up to the very last point. Every intermediate point should have a next point and a previous point, like this:

The idea, of course, is that there should be only finitely many points between, but it would be circular to make that requirement part of the definition because we are attempting to define in the first place what it means to be finite.

But there is a regrettable problem here, you see: Merely having a discrete order with a first and last element, unfortunately, does not prevent an infinity sneaking into the middle like this:

Although infinite, this red-yellow set is nevertheless placed in a linear order with a first and last point, and every point in between has a definite next point and previous point, so it is a discrete order. The naive approach doesn't quite work.

To improve it, one thing to notice about this red-yellow order is that we can rearrange the two halves and produce an order with no first or last element, like this:

Harnessing this observation, let us define that a set is *order infinite* if it admits a linear ordering without any largest element. This proposal would seem to fit well with Aristotle's definition, after all, since if the order has no largest element, then indeed it proceeds without end. But what if a set had no linear ordering at all? If such a strange set is possible—namely, a set that cannot be placed in a linear order—then it would not be realized as order infinite, but we wouldn't want therefore to regard it as finite, since don't we expect that every finite set can be placed in a linear order? Simply lay down the elements in a line until you run out.

Linearly Finite

In light of this, let us therefore define instead that a set is *linearly finite* if it admits a linear order and all linear orders of the set have a final element. It follows that all linear orders of the set will have both a first and last element since if there were no first element, then the inverse order would have no last element. The red-yellow set shown above is not linearly finite since the rearranged order has no final element.

Discretely Finite

Another observation would be that the original red-yellow set above admits a rearrangement that is not discrete, by moving the rightmost yellow ball to the center, like this:

The middle point now has neither an immediate successor nor an immediate predecessor, so this order is not discrete. This suggests that we say that a set is *discretely finite* if it admits a linear order, but all linear orders of it are discrete. It follows from this that the orders will also have a first and last element since if not, we could place one of the points from the middle to the extreme, resulting in a nondiscrete order. So every discretely finite set is also linearly finite. The converse also holds since if there is a nondiscrete linear order, then we can rearrange it to place the failure of discreteness at the beginning or end and thereby violate the first/last element condition. So linearly finite and discretely finite are actually the same concept. This is welcome news since it shows that our finiteness concepts are coming into alignment.

Stäckel Finite

Perhaps a little simpler than rearrangements is to consider whether there might be a bad subset of the order, a nonempty subset that has no first or no last element. For example, we rule out the order discussed previously by finding a disqualifying subset like this:

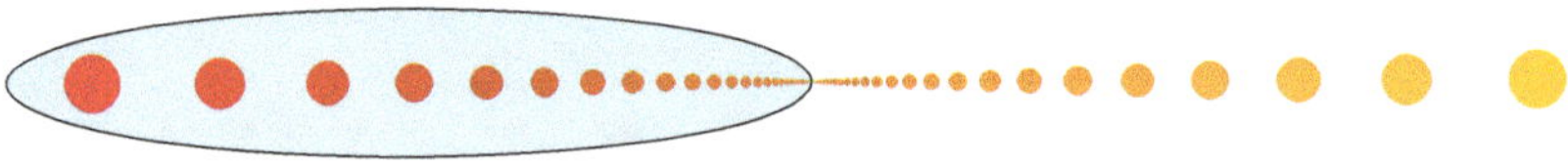

The mathematician Paul Stäckel (1862–1919) similarly defines that a set is finite—let us call it *Stäckel finite*—if it can be placed in a linear order, such that every nonempty subset of the set admits a least and greatest element. (This is equivalent to saying that the set has a well-order that is also conversely well ordered, to use a concept that will reappear several times later in this book.) This implies that the set is discretely ordered—every intermediate point has a next point and a previous point, since if not, there would be a disqualifying subset, and the converse also is true.

Tarski Finite

The logician Alfred Tarski (1901–1983) proposes several additional concepts of finite in the early twentieth century, several of which have to do with collections of subsets of the set. For example, a set is *Tarski finite* if every nonempty family of subsets of the set has a minimal such set, that is, a subset in the family that does not admit any proper subset in the family. So in the end, we have many finiteness concepts.

Numerically Finite

Apart from these order-theoretic notions of finiteness, another natural approach to the problem is to try to use the finite numbers themselves to define which sets are finite (and thus to define the infinite by complement). That is, we say that a set is *numerically finite* if it can be counted off with the finite numbers from 1 up to some finite number n, like so:

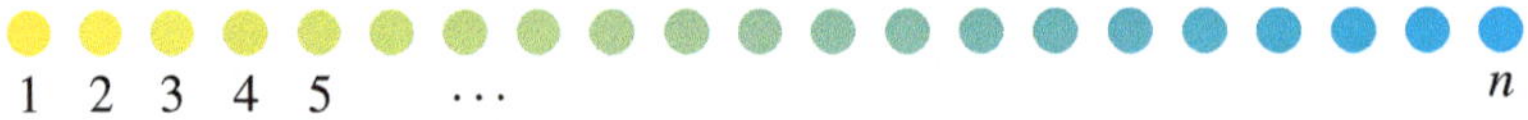

This certainly agrees with ordinary usage—the shepherd counts the sheep in his flock: one, two, three, and so on. But one worries whether it might be hopelessly circular. After all, we would seem to need to know exactly which are the finite numbers already in order to know which sets are finite, and vice versa. So how can this idea help us to define the finite?

Frege's Concept of Finite

Nevertheless, the logician Gottlob Frege attempts in the late nineteenth century to solve this circularity problem by providing a direct independent account of the finite numbers. Frege mounts an elaborate investigation of the nature of abstract objects and concepts, the *logicism* program, by which he aims to reduce all of mathematics to logic. In Frege's account, we abstract from any set of objects to its number in such a way that fulfills the Cantor-Hume principle that sets placed in a one-to-one correspondence have the same size. Frege then defines that the successor of a number is simply the number arising from a set with one additional element. Using this, he provides a definition of finiteness: The *finite* numbers are precisely those numbers that have every property of zero that is also transferred from every number to its successor.

Do you see how it works? If a property holds of zero and is transferred from every number to its successor, then it also must hold of 1, and therefore also of 2, of 3, and so on successively by iterating the transfer property. Furthermore, the idea is that finiteness itself will be one of these magical properties since zero fulfills the definition, and if a number is finite according to the definition, then its successor will also be finite. The finite numbers will therefore be the smallest collection of numbers with these properties. In a sense, he is defining the finite numbers and the finite sets simultaneously, but not in a circular manner. Rather, it is a clever kind of recursive ramping-up—ultimately, it is a form of mathematical induction. In this way, Frege provides us with a definition of the finite numbers, based in logic, which perhaps is a core success of his logicist project.

Dedekind Arithmetic

Dedekind mounts a similar idea before Frege, when he identifies the axioms that characterize the natural numbers up to isomorphism. Namely, the theory of Dedekind arithmetic posits a set of numbers, including the number zero, with an operation S called the *successor* operation—the operation of adding 1 to the number—with the properties that (1) zero is not the successor of any number; (2) the successor operation is one-to-one; and (3) every number is generated by S from zero, in the sense that every set of numbers containing 0 and closed under S has all the numbers. This latter axiom is known as the *induction* axiom.

Dedekind proves that these axioms are *categorical*—they determine the structure up to isomorphism, meaning that if you have your number system $\langle \mathbb{N}, 0, S \rangle$ with your zero 0 and your successor operation S, and I have my alternative system $\langle \bar{\mathbb{N}}, \bar{0}, \bar{S} \rangle$ with my alternative zero $\bar{0}$ and alternative successor operation $\bar{S}$, but both of our systems fulfill the Dedekind theory, then in fact there is an isomorphism $n \mapsto \bar{n}$, a translation from your system to mine revealing them to be copies of one another. Dedekind furthermore proved that in his theory, one can undertake definitions by *recursion*, defining addition and multiplication and indeed developing all the key ideas of elementary number theory.

This was the beginning of the philosophy of structuralism in mathematics, by which we do not say nor care to say what the mathematical objects *are* in themselves, but rather only how they relate to each other in a mathematical system. All that matters are the structural relations rather than their individual identity or essence, since other objects playing the same structural roles would serve just as well for mathematics.

The fact that the structure of natural numbers is uniquely determined by Dedekind's theory shows that our concept of finite number is well grounded in set theory. We thus obtain a rigidly definite concept of finite number, from which we can define our concept of numerically finite and consequently also the accompanying conception of the infinite.

11.2 Mathematical Induction

Let me dwell on the topic of mathematical induction a little longer, for I should like to emphasize its truly fundamental nature. Mathematical induction features as a core principle in Dedekind's categorical theory of the successor; it was also the key idea underlying Frege's definition of natural number. Mathematical induction underlies all known accounts of the foundations of arithmetic—essentially every fundamental fact of arithmetic is proved using induction. Let me illustrate with a few examples.

Lowest Terms

We use induction to prove the familiar fact that every fraction can be placed in lowest terms. The claim is that every fraction has a representation p/q, where the numerator and denominator have no nontrivial common factor. To prove this, consider any fraction. Let p be the smallest natural number arising as the numerator of a representation p/q of the given fraction. If p and q had a nontrivial common factor, we could factor it out, thereby obtaining a representation with a smaller numerator. But p was the smallest, so this is impossible. Therefore, p/q must already be in lowest terms.

Fundamental Theorem of Arithmetic

Another example would be the fundamental theorem of arithmetic, asserting that every positive integer can be factored uniquely into primes (we include 1 as having the empty prime factorization). This is really two claims, an existence claim and a uniqueness claim, both of which can be proved by induction. Let us prove the existence claim, that every number admits a prime factorization. To see this, we observe that there cannot be a smallest violation, a smallest number n that cannot be factored into primes. The reason is that if $n > 1$, then either the number n is itself prime, which counts as its factorization, or it can be factored as the product ab of smaller numbers, which are not necessarily prime. But because a and b are smaller than n, they can be expressed as a product of primes, so we can express n itself as the product of those factorizations. So we have found a prime factorization of n after all, contradicting our assumption that n had none.

Many Forms of Induction

These arguments both use induction in the form of the *least number principle*, asserting that every nonempty set of natural numbers has a least element. This form of induction is, at bottom, about ruling out *minimal counterexamples*, for if there can be no minimal counterexample, then by the least number principle, there can be no counterexample at all. Some elementary accounts of induction emphasize the principle of *common induction*, asserting that if 0 has a certain property, and if that property is transferred from every number n to $n + 1$, then every natural number has that property. The principle of *strong induction*, in contrast, asserts that if we have a property P of natural numbers, with the feature that a number has that property whenever all smaller numbers have it, then indeed every number has that property.

In fact, all these induction principles are equivalent to one another—one can prove strong induction from common induction or the least number principle and vice versa, for all the induction principles that I have mentioned. Induction is truly robust, and there are many equivalent ways to express it. It is fundamental to our conception of the finite.

The Pigeonhole Principle

The *pigeonhole principle* is a core principle of finite combinatorics, asserting that if you have finitely many pigeons, but more pigeons than pigeonholes, and you place every pigeon into a pigeonhole, then at least two pigeons will be sharing a pigeonhole. In purely mathematical language, it asserts that there can be no one-to-one association of a set of finite size n with a set of smaller size $k < n$.

Perhaps the principle seems obvious, but let me point out that we actually seem to have abundant empirical evidence *against* it. Imagine that two people are faced with a mountainous heap of pennies, and they both start counting them. Are you confident they will get exactly the same answer? If they get different answers, is this not evidence against the pigeonhole principle? The two ways of counting off the pennies would provide a one-to-one correspondence of the two numbers. In close elections, there are sometimes recounts of the votes, and it is rare that the count comes out exactly the same. Is this not evidence against the pigeonhole principle?

Meanwhile, we can indeed prove the theorem, and on that basis, we usually attribute that kind of empirical evidence to human error in the counting process rather than some weirdness in the foundations of arithmetic.

To prove the pigeonhole principle, we argue that there can be no minimal counterexample to it. Suppose that there were a one-to-one association of a set of size n with a proper subset of itself. We may take the smallest value of n for which this is possible. But now by simply removing an element from the set of size n—one should remove an element not in the proper subset—then what remains is a one-to-one association between a set of size $n - 1$ with a proper subset of itself, contradicting the minimality of n. So there can be no counterexample at all, and thus the pigeonhole principle is true.

11.3 Dedekind Finite Revisited

Let us also consider a little more closely the notion of Dedekind finiteness. We defined that a set is Dedekind finite if it is not equinumerous with any proper subset, and otherwise it is Dedekind infinite. This definition in effect takes the pigeonhole principle as the defining property of finiteness. Namely, finite sets should not be equinumerous with any proper subset, but all infinite sets, as in Galileo's paradox with the perfect squares and the other infinite sets that we have seen, should be equinumerous with a proper subset.

Question. Is Dedekind finite the same as numerically finite?

All the numerically finite sets that we have encountered so far are also Dedekind finite, and all the numerically infinite sets that we have encountered are Dedekind infinite. The natural numbers $\mathbb{N} = \{0, 1, 2, \ldots\}$ are equinumerous with the positive integers $\mathbb{N}^+ = \{1, 2, 3, \ldots\}$, simply by associating n with $n + 1$, so $\mathbb{N}$ is Dedekind infinite. It follows that whenever you can find in a set a copy of $\mathbb{N}$ as a subset, then the set will be Dedekind infinite, and this will include essentially all the familiar infinite sets, the set of real numbers $\mathbb{R}$, the set of infinite binary sequences, the plane $\mathbb{R}^2$, space $\mathbb{R}^3$, and so on. Each of these infinite sets is also Dedekind infinite. If the finiteness properties are not equivalent, then the counterexample will be a rather strange set.

Every Numerically Finite Set Is Dedekind Finite

The fact that every numerically finite set is also Dedekind finite is precisely what the pigeonhole principle asserts. To be Dedekind finite, at bottom, is to fulfill the pigeonhole principle, and since every numerically finite set fulfills the pigeonhole principle, every numerically finite set is Dedekind finite.

Must Every Dedekind-Finite Set Be Finite?

The converse question, however, is extremely subtle. Is it conceivable that there is an infinite set that is not equinumerous with any proper subset? We have mentioned already that if a set contains a copy of the natural numbers as a subset, then since that set already is equinumerous with a proper subset, the whole set will be equinumerous with a proper subset. Thus, any infinite set that contains a copy of $\mathbb{N}$ will also be Dedekind infinite. But does every numerically infinite set contain a countably infinite subset?

On the one hand, this might be seen as obvious. If a set X is infinite, then we can simply pick an element a_0 from it—it doesn't matter which element. Since X is infinite, we can pick another, distinct element a_1, and then a_2, a_3, a_4, and so on—all different. At no stage will we have exhausted all the elements of X since it is numerically infinite, and therefore we have found a countably infinite subset:

$$\{a_0, a_1, a_2, a_3, a_4, \ldots\} \quad \subseteq \quad X.$$

This argument relies on a principle known as the *axiom of choice*, which asserts that there is indeed a way of making such choices in a mathematical construction of this kind (we shall discuss the axiom of choice at length in chapter 21). In the standard axiomatization of set theory, which includes the axiom of choice, therefore, we may conclude that every infinite set is Dedekind infinite. In short, the axiom of choice implies that Dedekind finite is the same as numerically finite.

Some mathematicians, however, prefer not to assume the axiom of choice, and in this case, we can no longer prove that every Dedekind-finite set is numerically finite. Indeed, we can prove that without the axiom of choice, the concepts split apart—there can be a Dedekind-finite set that is nevertheless numerically infinite. In a mathematical universe with such a set, the pigeonhole principle simply does not characterize finiteness.

In such a mathematical world, furthermore, it is not true that every numerically infinite set contains a copy of the countably infinite set $\mathbb{N}$. In such a world, the infinity of the natural numbers $\mathbb{N}$ is consequently not the smallest infinity, in the sense of being less than or equal to all the others, since the infinity of the natural numbers would be *incomparable* to the infinity of a numerically infinite Dedekind-finite set, with neither set being equinumerous with a subset of the other. Such a strange mathematical world that would be . . .

11.4 On Defining Finite and Infinite

Altogether, we have thus seen in this chapter several distinct concepts of the finite—but are they equivalent? The answer turns out to be subtle and interesting. The first thing to say is that in the standard Zermelo-Fraenkel (ZFC) set theory, yes, indeed, we can prove that all the finiteness definitions mentioned here are equivalent. In this sense, we have got it right—we know what it means to be finite and what it means to be infinite in a way that is fruitfully clear and precise for mathematics.

Meanwhile, however, some of the equivalences are not provable in the weaker theory ZF—that is, set theory without the axiom of choice. In particular, I mentioned already that we know that it is relatively consistent with ZF that there can be a numerically infinite set that is Dedekind finite. Such a set would be infinite, in that one cannot count off its members with the predecessors of a natural number, but still it would not happen to be equinumerous with any proper subset, and consequently, it would be Dedekind finite. It is also relatively consistent with ZF that there is a set that admits of no linear order at all, and such a set would be numerically infinite set but not order infinite, thereby pulling these finiteness concepts apart.

Nevertheless, even without the axiom of choice, we can prove that the concepts of numerically finite, linearly finite, discretely finite, Stäckel finite, and Tarski finite are all equivalent, and any set that is finite in these senses is also finite in all the other proposed senses of finiteness. So we might take these notions collectively as the most primitive finiteness concept. Perhaps it is the concept of numerical finiteness that most easily and directly implies these and all the other notions, and I take this as evidence that this is what we should mean by finite—we should define that a set is finite if and only if it is numerically finite. And indeed this is the definition of finiteness that is almost universally accepted today as the right definition of finiteness.

In conclusion, we define that a set is *finite* if and only if it is numerically finite, which means that it is equinumerous with a natural number—it has n elements for some natural number n. This definition is grounded in the prior and independent concept of natural number provided by Dedekind's categorical characterization of the natural numbers, which does not rely on any prior understanding of finiteness. Numerical finiteness expresses directly our most natural, intuitive ideas about finiteness—that we can count out the elements of the set up to some natural number—while also implying the properties expressed by all the other finiteness conceptions that have been offered. It is ideal.

Consequently, having defined our concept of finiteness, we also arrive at the corresponding notion of the infinite. Namely, a set is *infinite*, by definition, if it is not finite—that is, it is not numerically finite. So the infinite sets are those that do not have size n for any natural number n. This is what we shall mean when saying that a set is infinite, and it has proved a truly robust, fruitful notion.

Questions for Further Thought

11.1 Prove directly (without using the axiom of choice) that linearly finite, discretely finite, Stäckel finite, and numerically finite are all equivalent.

11.2 Prove that every subset of a numerically finite set is numerically finite.

11.3 If a set has no linear order, then neither does any superset of it. So if these kinds of sets exist, then there would be sets that are not order infinite, and furthermore, no superset of them is order infinite, even after adding a definitely order-infinite set to them. For example, $A \sqcup \mathbb{N}$ would not be order infinite if A had no linear order. Is this a decisive refutation of order-infinite as a suitable definition of the infinite?

11.4 What does it mean for a set to be finite? What grounds, if any, do you find in support of the various concepts of finiteness, such as a set being Dedekind finite, linearly finite, discretely finite, Stäckel finite, Tarski finite, or numerically finite? Which definition ultimately is best?

11.5 Can one coherently take the Dedekind-finite conception as defining finiteness if one does not also accept the axiom of choice as expressing a fundamental truth of set theory?

Further Reading

Bolzano (1950 [1851]). The remarkable treatise on the infinite by Bernard Bolzano.

Hamkins (2021b, chapter 1). My discussion of the concept of number and the roles played by Frege, Dedekind, Giuseppe Peano, and others.

Hamkins (2020, chapter 4). My elementary introduction to mathematical induction.

Credits

This chapter was adapted from my essay, "What Is the Infinite?" Hamkins (2024c), which was included in the *Infinity Essays* collection.

12 Hilbert's Grand Hotel

Abstract. The hotel is full, but it nevertheless can accommodate new guests, even infinitely many new guests. The parable of Hilbert's Grand Hotel thus illuminates the nature of countable infinity.

12.1 Hilbert's Grand Hotel

Let us hear the parable of Hilbert's Grand Hotel. The hotel has infinitely many rooms, each a luxurious full-floor suite, numbered like the natural numbers: room 0, room 1, room 2, and so on, without end. Every room is occupied—the hotel is completely full, with infinitely many guests.

A new guest arrives seeking a room. She inquires at the reception desk—is there any possibility of accommodating her? The manager replies coolly, "No problem, just a minute," while he sends an announcement up to all the current guests: Everyone must change rooms to the next higher room:

$$\text{Room } n \quad \mapsto \quad \text{Room } n+1$$

There was fine print in the agreement that everyone had signed at check-in, you see, stipulating that guests might have to change rooms under the management's direction. Everyone complies, with the result that room 0 has now become vacant, available for the new guest.

Over the next several days, more guests arrive, and although the hotel is completely full, the manager is able to accommodate them each time by directing the current guests to move up again, thereby freeing up an additional room when needed.

On the weekend, a crowd shows up all at once—1000 guests. What to do? The manager expertly accommodates them in one migration by having all current guests move up 1000 rooms:

$$\text{Room } n \quad \mapsto \quad \text{Room } n + 1000.$$

This makes available all the rooms from 0 to 999, which accommodates the crowd.

Hilbert's Bus

The following week, a considerably larger crowd arrives—Hilbert's bus has arrived. The bus has infinitely many seats, numbered like the natural numbers, with seat 0, seat 1, seat 2, and so on, and sitting in every seat is a new guest wanting to check into the hotel. Can the manager accommodate them?

The manager cannot do what he did before, by placing the bus passengers at the bottom of the hotel and the previous guests somehow above them, for the bus passengers would already use up the entire hotel that way. It would make no sense, after all, to ask the current guests to move up infinitely many rooms—there are no such rooms, as every individual room has a finite room number. But is there some other kind of rearrangement that will enable the manager to accommodate all the guests and the bus passengers in the hotel? Please try to figure out a method on your own before reading further.

Think about it...

Did you find a solution? The answer is yes—the guests can be accommodated. For one solution, the manager can direct that the current guests should all double their room numbers, moving from n to room $2n$, which will free up all the odd-numbered rooms, and then the passenger in seat s should take room $2s + 1$. This is certainly an odd number, and people in different seats will get different odd-numbered rooms:

$$\begin{aligned} \text{Room } n \quad &\mapsto \quad \text{Room } 2n \\ \text{Seat } s \quad &\mapsto \quad \text{Room } 2s + 1. \end{aligned}$$

Thus, everyone is accommodated in a room of their own, with the previous guests in even-numbered rooms and the bus passengers in odd-numbered rooms.

Hilbert's Train

The next day, Hilbert's train arrives in the station just under the hotel:

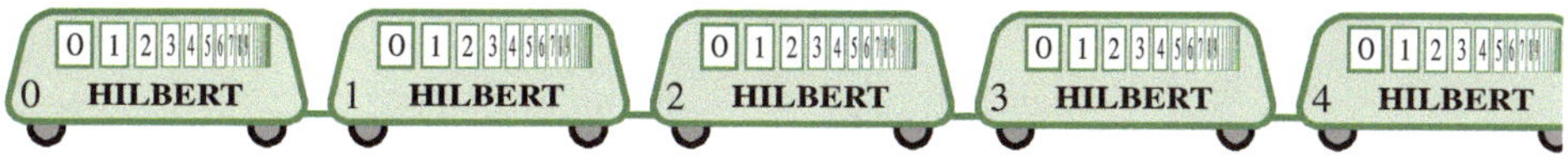

The train has infinitely many train cars, each with infinitely many seats, and every seat is occupied. Each train car is a little like Hilbert's bus. Every passenger is identifiable by two pieces of information (their car number c and their seat number s), and they are all eager to check into the hotel. Can the manager accommodate them all?

Think about it...

Yes, indeed. There are many solutions, but here is one way to do it. The manager again directs the current residents to double their room numbers, moving from room n to room $2n$, which frees up the odd-numbered rooms as before. The manager then directs the train passenger in car c, seat s to check in to room 3^c5^s:

$$\begin{aligned} \text{Room } n \quad &\mapsto \quad \text{Room } 2n \\ \text{Car } c,\ \text{Seat } s \quad &\mapsto \quad \text{Room } 3^c5^s. \end{aligned}$$

Notice that 3^c5^s is certainly an odd number, being the product of odd numbers, so it is available for a new guest. Different train passengers will land in different rooms precisely because of the uniqueness of the prime factorization—every room number of the form $3^c \cdot 5^s$ determines the exponents c and s, which correspond to a unique train passenger. So again everyone is accommodated in the hotel.

Hilbert's Half-Marathon

The big race is this weekend: Hilbert's half-marathon—this crowd likes fractions—and the runners are densely packed in amongst themselves. Each runner has a race number, one for every nonnegative rational number. All the runners want to check into the hotel.

Can the hotel manager accommodate all the runners?

Yes. After freeing up the odd-numbered rooms as before, he can place runner p/q in room 3^p5^q, where p/q is in lowest terms:

$$\begin{array}{rcl} \text{Room } n & \mapsto & \text{Room } 2n \\ \text{Runner } \dfrac{p}{q} & \mapsto & \text{Room } 3^p5^q. \end{array}$$

Thus the entire half-marathon fits into Hilbert's hotel. Notice how similar the manager's arrangements were for the half-marathon as for the train. This is simply because every runner is determined by a pair of numbers—the numerator and the denominator—just as the train passengers were determined by their car and seat numbers.

12.2 Countability

The main concept at play in the parable of Hilbert's hotel is the notion of *countability*, for a countable set is one that fits into Hilbert's hotel.

Definition 4. A set is *countable* if it can be placed in one-to-one correspondence with a set of natural numbers.

Thus, a set A is countable if there is a one-to-one function $r : A \to \mathbb{N}$, a method of associating with every element $a \in A$ a corresponding number $r(a)$. Such a function r amounts to a room assignment for Hilbert's hotel, placing individual a into room $r(a)$. For example, the function

$$r(\text{train passenger in car } c \text{ seat } s) \quad = \quad 3^c \cdot 5^s$$

is the function that we used to assign the train passengers to their hotel rooms.

Another way to think about countability is that a nonempty set A is countable if and only if one can enumerate the elements of A by a list, indexed with the natural numbers

$$a_0,\ a_1,\ a_2,\ a_3,\ \ldots$$

possibly with repetition (in case the set is finite).

Notice that according to this terminology, the finite sets are included as countable. After all, if a set is finite, then we can certainly accommodate it in Hilbert's hotel. Even the empty set counts as a countable set. Meanwhile, a set is *countably infinite* if it is infinite and countable. These sets are exactly those with an enumeration indexed by the natural numbers

$$a_0,\ a_1,\ a_2,\ a_3,\ \ldots$$

using distinct elements a_i—that is, without repetition.

A Countable Set Plus One More Element Is Countable

The main initial observation of Hilbert's hotel shows that if you have a countably infinite set and add a new element, then the result is a countably infinite set. The current hotel occupants, even though they fill the hotel completely, can nevertheless be rearranged by moving up one room to accommodate a new guest.

This shows a sense in which adding an element to a countably infinite set does not make the set any larger—the new set of elements can be placed in one-to-one correspondence with the previous set, so we may regard them by the Cantor-Hume principle as having the same size, Galileo's paradox notwithstanding.

The Union of Two Countable Sets Is Countable

The Hilbert's bus scenario shows that the union of two countable sets is countable. In other words, if A and B are countable sets, so is the set $A \cup B$, which includes both of them. To see this, assume that A and B are countable sets. Perhaps we imagine A as the current guests of the hotel and B as the bus passengers. Since the set A can be put into a one-to-one correspondence with a set of natural numbers, by doubling all those numbers, we can place A in one-to-one correspondence with a set of even numbers. Similarly, since B can be put into one-to-one correspondence with a set of natural numbers, by composing with the map $n \mapsto 2n + 1$, we may place B into one-to-one correspondence with a set of odd natural numbers. By taking these two correspondences simultaneously—they do not interfere with each other—we have found a one-to-one correspondence of the union set $A \cup B$ with a set of natural numbers.

Here is an alternative argument. If A is countable and nonempty, then we may enumerate the elements of A as a_0, a_1, a_2, and so on. And similarly, if B is countable and nonempty, we may enumerate the elements as b_0, b_1, b_2, and so on. Thus, we may enumerate the elements of the union set $A \cup B$ simply by interleaving these two enumerations:

$$a_0, b_0, a_1, b_1, a_2, b_2, \ldots$$

Therefore, the union set is countable.

As an application of this, let me argue that the set of integers $\mathbb{Z}$ is countable. The integers include the natural numbers and their negatives:

$$\mathbb{Z} = \{\ldots, -2, -1, 0, 1, 2, \ldots\}.$$

This is countable since it is the union of two countable sets—namely, the natural numbers $\{0, 1, 2, \ldots\}$ and the negative integers $\{-1, -2, -3, \ldots\}$. So it is countable by the arguments we just gave in this discussion. We can also see that $\mathbb{Z}$ is countable simply by enumerating it like this:

$$0,\ 1,\ -1,\ 2,\ -2,\ 3,\ -3,\ \ldots$$

The Set of Pairs of Natural Numbers Is Countable

Consider the set of pairs of natural numbers, which can be viewed as the points arising in the natural-number lattice as pictured here, with natural-number coordinates (n, m). What I claim is that the set of such pairs, denoted by $\mathbb{N} \times \mathbb{N}$, is a countable set. The points of the natural-number lattice can be placed in one-to-one correspondence with the natural numbers themselves.

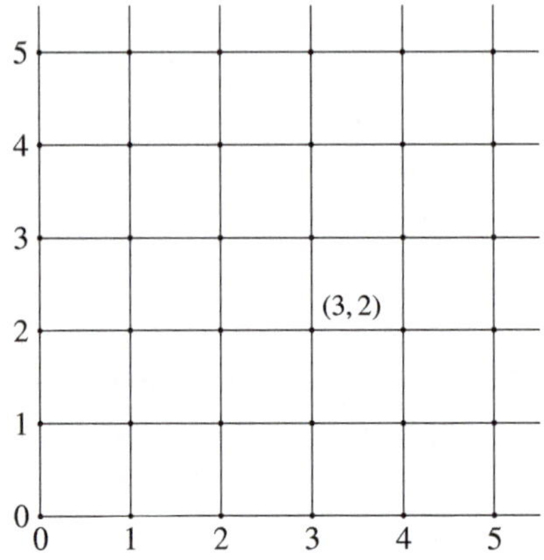

Let us see several proofs of this. The first is to use the idea underlying Hilbert's train. Namely, we can associate any pair of natural numbers (n, m) with the number $3^n 5^m$. This provides a function $d : \mathbb{N} \times \mathbb{N} \to \mathbb{N}$, which is one-to-one because of the uniqueness of prime factorizations. Thus, $\mathbb{N} \times \mathbb{N}$ is in one-to-one correspondence with a set of natural numbers, so it is countable.

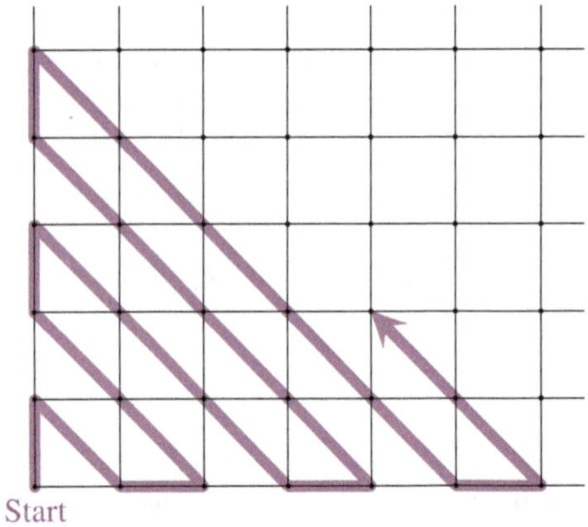

For a second proof, let us borrow an idea we had used back in our analysis of the Chocolatier's Game in chapter 5. Namely, let us view $\mathbb{N} \times \mathbb{N}$ as the set of grid points of the natural number lattice and then trace a winding path as in the figure here. We shall enumerate the grid points as we encounter them on this winding path, starting at the origin and then proceeding to zig and zag progressively up and down the diagonals. The main point is that we shall eventually encounter any given lattice point on this path if we should simply walk long enough. We may associate the initial point on the path with 0, the next with 1, then with 2, and so on; we associate every natural number n with the nth point (x_n, y_n) encountered on this path. In this way, the function

$$n \mapsto (x_n, y_n)$$

is a one-to-one correspondence between $\mathbb{N}$ and $\mathbb{N} \times \mathbb{N}$. So the set of pairs of natural numbers is countable.

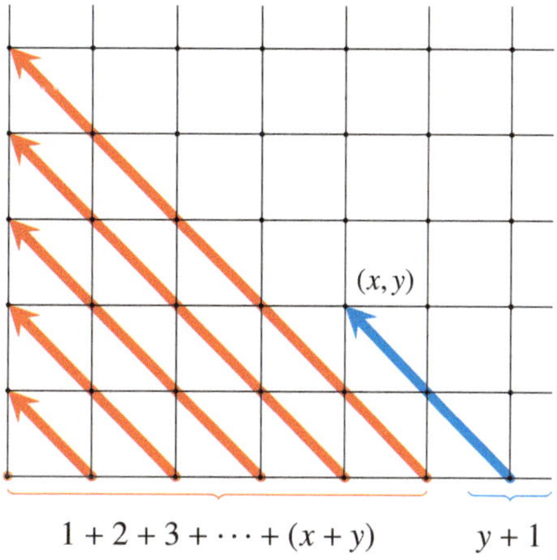

A variation of the previous argument leads to a stronger result, yielding not just a one-to-one correspondence between $\mathbb{N} \times \mathbb{N}$ and $\mathbb{N}$, but a specific polynomial function providing the correspondence. For this argument, instead of the winding purple path of the previous argument, we shall instead follow the upward diagonals as pictured in the figure here. We start at the origin and then systematically follow each upward diagonal in turn. Let us calculate how many points we have encountered by the time we reach point (x, y). First, we will have completed a certain number of earlier diagonals, pictured in red. The number of points on these diagonals is 1, 2, 3, and so on. Since each diagonal fulfills the equation $x + y =$ constant, with the constant increasing by 1 each time (and 1 less than the number of points on it), it follows that the final full red diagonal has $x + y$ points. So the total number of points on the red diagonals is the triangular number $1+2+\cdots+(x+y)$, which is equal to $(x+y)(x+y+1)/2$. Next, the number of points on the final blue line leading up to (x, y) is $y+1$. Thus, the total number of points up to and including (x, y) is $(x + y)(x + y + 1)/2 + y + 1$. But we want to associate each point with the number of *preceding* points, in order that we will start with 0, then continue with 1, 2, and so on, as we walk the path. So the polynomial

$$p(x, y) = \tfrac{1}{2}(x + y)(x + y + 1) + y$$

is therefore a polynomial bijection between $\mathbb{N} \times \mathbb{N}$ and $\mathbb{N}$, as desired. This polynomial function is known as the *Cantor pairing function*, which assigns to each pair of numbers a single number that in effect codes the pair, because each output value of the function determines the input pair that gave rise to it.

It follows that the full integer lattice $\mathbb{Z} \times \mathbb{Z}$ also is countable since this is the union of four quadrants, each of which looks just like $\mathbb{N} \times \mathbb{N}$.

The Set of Rational Numbers Is Countable

Let us argue next that the set of rational numbers, the fractions, is a countable set—a fact that people sometimes find surprising. The fact of the matter is that the set of rational numbers $\mathbb{Q}$, despite its complexity and density within the real numbers, nevertheless can be placed in a one-to-one correspondence with the natural numbers.

We can prove this simply by using our previous observation that $\mathbb{Z} \times \mathbb{Z}$ is countable. Namely, we associate each rational number (p, q), in lowest terms and with a positive denominator, with the point (p, q) in the integer lattice $\mathbb{Z} \times \mathbb{Z}$. Since $\mathbb{Z} \times \mathbb{Z}$ is countable, we have associated the fractions with a subset of a countable set, so it is countable.

Alternatively, we may argue as we did with Hilbert's half-marathon. Namely, let us associate every nonnegative fraction p/q in lowest terms with the number $3^p \cdot 5^q$, which is a one-to-one correspondence with a set of natural numbers. And let us simultaneously associate the negative fractions $-p/q$ with $2 \cdot 3^p \cdot 5^q$. Since these associations do not interfere with one another, we have therefore constructed a one-to-one association of the set of fractions with a set of natural numbers, thus $\mathbb{Q}$ is countable.

The Union of Countably Many Countable Sets Is Countable

The arguments that we have given here generalize to a sharper conclusion that people sometimes find surprising. Namely, suppose that we have a countable list of countable sets:

$$A_0 \quad A_1 \quad A_2 \quad A_3 \quad A_4 \quad A_5 \quad \cdots$$

The claim is that the union of all these sets together is also a countable set:

$$A_0 \cup A_1 \cup A_2 \cup A_3 \cup \cdots$$

It is truly remarkable.

Here is the argument. Since A_0 is countable, we may enumerate the elements of it in a manner indexed by the natural numbers. Let us imagine that we have laid off the elements of A_0 on the nodes of the natural number lattice on the column above 0. Similarly, we may place the elements of A_1 on nodes of the column above 1, and we may choose a manner of placing the elements of A_n on the column above n—lay them out on that column however you like. Thus, the entire union set has been associated with the nodes of the lattice, which we have already proved is countable. So the union set $\bigcup_n A_n$ is countable.

There is an extremely subtle point about this proof, which I should like to discuss briefly. Namely, the argument that we gave used a certain uniform selection process, when we moved from knowing that each set A_n is countable to having a particular way of enumerating it on the nth column of the natural-number lattice. In this step, we gave no rule or construction method for selecting a specific such method of enumeration, but rather just said that you should choose a way of doing it, however you like. The general principle asserting that such kind of choices can be made is known as the *axiom of choice*, and deep results in set theory show that there is no way to avoid this kind of choice when proving this theorem. Meanwhile, there is a tradition in mathematics to highlight uses of the axiom of choice in mathematical arguments, as I am doing here, and some mathematicians study what set theory can be like without the axiom of choice. In those theories, one cannot say as a general principle that the countable union of countable sets is countable. Meanwhile, in ordinary mathematical arguments, it is perfectly acceptable to use the axiom of choice whenever it is convenient, and this axiom is considered one of the standard fundamental axioms of set theory.

The Product of Two Countable Sets Is Countable

Let us argue that if A and B are both countable sets, then so is the product $A \times B$, the set of pairs (a, b) where a is an element of A and b is an element of B. If A can be placed in one-to-one correspondence with a set of natural numbers, and the same holds for B, then by simply combining those correspondences, we can place $A \times B$ into one-to-one correspondence with a subset of $\mathbb{N} \times \mathbb{N}$. But this set as a whole is countable, so every subset of it is as well. Therefore, $A \times B$ is countable.

The Set of Words Over a Countable Alphabet Is Countable

Let us now make a further remarkable observation as a consequence of all this. Consider the set of all possible words that one can form using the symbols of a given alphabet A, a set of letters or symbols. Here, I am referring to an abstract conception of the term, by which a *word* is simply any finite sequence of elements of the alphabet A. We denote the set of all such words by A^*.

These abstract kinds of words over the familiar alphabet $\{a, b, c, \ldots, z\}$, for example, would include such sequences as *aaba*, *azzzbcz*, *bowwowyow*, and so on, as well as extremely long words with billions of letters and also the *empty* word, often denoted by ε or $\langle\rangle$, which is the sequence of length zero. The set $\mathbb{N}^*$ is the set of all words over the natural numbers—all the possible finite sequences of natural numbers.

The remarkable fact is that the set of words in a countable alphabet is also countable. In other words, if A is countable, so is the set of words A^*.

This can be seen as follows. We may view the alphabet A itself as essentially the set of one-letter words. And then $A \times A$ is the set of pairs or two-letter words, and $A \times A \times A$ is the set of three-letter words:

$$
\begin{aligned}
A &= \text{one-letter words} \\
A \times A &= \text{two-letter words} \\
A \times A \times A &= \text{three-letter words} \\
&\vdots
\end{aligned}
$$

The n-fold product A^n consists of the n-letter words (note that A^0 is the set of words of length 0, which is to say the empty word). All these sets are countable by induction, since A itself is countable and each arises as the product of A with the previous set. But now just observe that the set of all words is simply the union of these sets:

$$A^* \quad = \quad \text{all words} \quad = \quad \bigcup_n A^n.$$

This is a countable union of countable sets, and hence countable. So the set of words over a countable alphabet is countable.

We can give a direct argument in the case of finite sequences of positive integers. Namely, for each finite sequence of positive integers

$$s = s_0 s_1 s_2 \cdots s_n,$$

let us associate with it the number

$$2^{s_0} 3^{s_1} 5^{s_2} \cdots p_n^{s_n},$$

which is obtained by placing s_i as the exponent of the ith prime p_i. This is a one-to-one association precisely because of the uniqueness of prime factorizations—from any number we can extract the sequence of exponents and thus identify the unique word with which it was associated (if any). So the set of all finite sequences of positive integers is countable. From this, it follows that the set $\mathbb{N}^*$ of all finite sequences of natural numbers is countable, and from that, one can establish directly that the set of words A^* over any countable alphabet A is countable.

12.3 Cantor's Cruise Ship

Returning to the parable of Hilbert's Grand Hotel, let me set the stage for the fascinating events that unfold there next. Specifically, the morning after the last of the frenetic rearrangements discussed earlier, you see, with all the bus and train passengers, the runners, and the earlier guests ensconced sleeping in their suites, a long, ominous shadow falls across the hotel. It is Cantor's cruise ship entering the harbor, carrying a passenger for every real number. Every passenger on the ship has a ticket, stamped with a distinct serial number—a real number—and every real number arises as the stamp on some ticket. Thus there is passenger $\sqrt{2}$, passenger e, passenger π, and so on, all of them eager to have finally arrived at the famously accommodating hotel.

Will the passengers of Cantor's cruise ship fit into Hilbert's Grand Hotel? Check in with chapter 13 to hear more about that ...

Questions for Further Thought

12.1 Suppose that guests arrived at Hilbert's Grand Hotel in the manner described in this chapter: First one guest arrives, and then 1000 all at once, and then Hilbert's bus, followed by Hilbert's train, and finally Hilbert's half-marathon. If the manager followed the procedure mentioned in the chapter, describe who are the occupants of rooms 0 through 100. How did they arrive, and with which party? In which car or seat were they when they arrived if they arrived by train or bus? Which fraction did they wear in the marathon? If you were the very first guest to arrive, where do you end up in the end? And what is the first room above you that is occupied? How did that guest arrive?

12.2 Down the street from Hilbert's Grand Hotel is Hilbert's Cubical Co-op apartment complex, which is an infinite cubical building, like $\mathbb{N} \times \mathbb{N} \times \mathbb{N}$, where every occupant's residence can be described by a floor number n, a hallway number h, and a corridor r. Because the interior rooms have very little light, the entire cooperative wants to move to Hilbert's Grand Hotel. How can the manager accommodate them?

12.3 Consider the set of all possible words you can form from the ordinary letters of the alphabet A to Z, allowing any arbitrary finite sequence of letters. Describe a procedure for placing them into Hilbert't Grand Hotel.

12.4 A real number is *algebraic* when it is a root of a nontrivial polyomial over the integers, such as $3x^3 + 5x - 7$. Argue that the set of algebraic numbers is countable. (Hint: Argue that there are only countably many polynomials over the integers since every polynomial is determined by the finite sequence of coefficients; next, appeal to the fact that every nontrivial polynomial has only finitely many roots.)

Further Reading

Hamkins (2021b, chapter 3). The chapter on infinity in my book on the philosophy of mathematics.

Credits

Much of this chapter is adapted from Hamkins (2020, chapter 13) and Hamkins (2021b, chapter 3). David Hilbert (1862–1943) introduced his Grand Hotel in a 1924 lecture, "Über das Unendliche" ("On the Infinite")—see Hilbert (2013, p. 730). The presentation with Hilbert's bus, Hilbert's train, Hilbert's half-marathon, and Cantor's cruise ship is my own invention from the early 1990s. The image of Cantor's ship is by Cunard ("Postcards from the Early 1920s"), available in the public domain via Wikimedia Commons.

13 Uncountable Infinity

Abstract. The set of real numbers is an uncountable infinity, larger than the countable infinity of the natural numbers. Thus, there are different sizes of infinity—indeed, infinitely many infinities—but no largest infinity. To be sure, there are more infinities than any one of them.

Cantor's cruise ship has pulled into the harbor in front of Hilbert's Grand Hotel, carrying a passenger for every real number. Every passenger has a ticket with a distinct real serial number, and every real number is realized—passenger $\sqrt{2}$, passenger e, passenger π, and so on, all eager to check in to the famously accommodating hotel. Can the hotel manager fit them in?

The fascinating answer, shocking and profound, is the passengers on Cantor's cruise ship will not fit into Hilbert's hotel—there is no system of room assignments that will assign every passenger their own private room. In plain mathematical language, the set of real numbers cannot be placed in a one-to-one correspondence with a set of natural numbers. Thus Cantor proved that the set of real numbers is *uncountable*.

The fantastic conclusion is inescapable: There are different sizes of infinity. The infinity of the real numbers is a larger infinity than the infinity of the natural numbers:

$$\mathbb{N} < \mathbb{R}.$$

Indeed, there are infinitely many different sizes of infinity, but no largest infinity. To be sure, there are more infinities than any particular one of them.

Although Georg Cantor's ideas were controversial in his day, sometimes meeting with stubborn rejection, mathematicians eventually recognized his enormous accomplishment. In 1925, David Hilbert extolled Cantor's theory of sets and infinite cardinals, announcing, "No one shall cast us from the paradise that Cantor has created for us." Let us begin to see what Hilbert was talking about.

13.1 Uncountability of the Real Numbers

The central claim of Cantor's theorem is that the set of real numbers $\mathbb{R}$ is uncountable—it cannot be placed in one-to-one correspondence with the natural numbers $\mathbb{N}$, and therefore the size of $\mathbb{R}$ is a higher-order infinity than that of $\mathbb{N}$.

The proof is truly ingenious—let us go through it. To begin, we suppose toward contradiction that $\mathbb{R}$ is a countable set, so we can enumerate all the elements of $\mathbb{R}$ as r_1, r_2, r_3, and so on, with r_n for every natural number $n \geq 1$. The enumeration is in effect a way of fitting the real numbers into Hilbert's hotel. We have assumed that every real number appears as some r_n on this list. Using the list, we shall now describe a particular real number z, which will be useful in our argument, by specifying its decimal digits:

$$z = 0.d_1d_2d_3\ldots$$

The main idea is that we shall choose the nth digit d_n specifically to be different from the nth digit of the decimal representations of r_n. Let us specifically take $d_n = 1$ unless the nth digit of a decimal representation of r_n is 1, in which case we use $d_n = 7$ instead. In the construction, we ensure that $z \neq r_n$ already by our decision made at stage n, precisely because whatever else happens, these two numbers will differ in their nth decimal digits (and note that z, having all digits 1 or 7, has a unique decimal expansion). Since this is true for every n, it follows that the real number z does not appear on the list. But this contradicts our assumption that every real number appears on the list. Therefore, our assumption that we could make such a list in the first place must be wrong. Since that it precisely what it would mean for the set of real numbers to be countable, we have thus established that the real numbers form an uncountable set, completing the proof.

To deepen our understanding of the argument, let me illustrate the construction with an explicit example. Perhaps the initial list of numbers happened to begin like this:

$$\begin{aligned} r_1 &= \pi \\ r_2 &= e \\ r_3 &= \sqrt{2} \\ r_4 &= \ln(2) \\ r_5 &= \sqrt{29} \\ &\vdots \end{aligned}$$

We assume that the enumeration continues, eventually listing every single real number exactly once.

We can write out the decimal expansions of these numbers like this:

$$
\begin{aligned}
r_1 &= 3.\,\mathbf{1}\,4\,1\,5\,9\,2\,6\,5\,3\,5\,8\ldots\\
r_2 &= 2.\,7\,\mathbf{1}\,8\,2\,8\,1\,8\,2\,8\,4\,5\ldots\\
r_3 &= 1.\,4\,1\,\mathbf{4}\,2\,1\,3\,5\,6\,2\,3\,7\ldots\\
r_4 &= 0.\,6\,9\,3\,\mathbf{1}\,4\,7\,1\,8\,0\,5\,5\ldots\\
r_5 &= 5.\,3\,8\,5\,1\,\mathbf{6}\,4\,8\,0\,7\,1\,3\ldots\\
&\vdots \qquad\qquad \ddots
\end{aligned}
$$

By inspecting those red diagonal digits, we may follow the defining rule for the digits of the diagonal real z to produce

$$z = 0.\,7\,7\,1\,7\,1\ \ldots$$

The rule tells us to use digit 1 or 7 in each place, chosen to be different from the corresponding red digit on the diagonal. We use digit 1, unless the corresponding red digit is 1, in which case we use digit 7. In this way, the real z that we construct will be different from the first real r_1 in the first digit, different from the second real r_2 in the second digit, and so on—it is different from r_n in the nth digit. Thus the real number z is different from every r_n, so z is not on the list, which contradicts our assumption that every real number was on the list.

The real number z is called the *diagonal real number* for the given enumeration because the digits of z are defined by referring to the highlighted diagonal digits on the original list of numbers, the nth digit of the nth number. For this reason, Cantor's argument method is known as a *diagonal argument*, an extremely robust method of argument that appears now in perhaps thousands of mathematical arguments, many of them quite abstract. Indeed we shall see several more diagonal arguments in this chapter. The general feature of a diagonal argument is that one fulfills an infinite list of requirements in an infinitely long construction by satisfying the nth requirement on the nth step.

A Minor Issue About Nonunique Decimal Representations

Let me briefly discuss a subtle aspect of the proof that is sometimes confusing but is ultimately a minor matter. The issue is that not all real numbers have unique decimal expansions. For example, in an earlier chapter we discussed "the most contested equation in middle school," referring to

$$0.9999999999\ldots \quad = \quad 1 \quad = \quad 1.0000000000\ldots$$

The point here is that this identity shows that some real numbers have more than one decimal representation. The same phenomenon occurs whenever a decimal expansion ends in all 9s or all 0s; for example:

$$1.2467999999\ldots \quad = \quad 1.2468000000\ldots$$

Because of this, it is unfortunately not always meaningful to speak of *the* decimal expansion of a real number since some numbers have two such representations, although the only instances of this are exactly these eventual 9 or eventual 0 situations. Since our diagonal real z used only digits 1 and 7, however, it does have a unique decimal expansion. We are therefore entitled to refer to *the* decimal expansion of z. But furthermore, we need not have done anything special with any particular digit to ensure that $z \neq r_n$, when r_n does not have a unique representation, since this very fact itself already ensures that $z \neq r_n$. For this reason, the nonuniqueness of decimal representations becomes a moot point in the argument.

Cantor's Original Argument

As a historical matter, Cantor's original proof did not refer to the digits of the numbers in the manner of argument that we have just given, but rather proceeded by constructing a certain nested sequence of intervals—the diagonal-digits argument was a later presentation. So let us go through his original argument.

Suppose again toward contradiction that the set of real numbers is countable. In this case, we can enumerate the real numbers as r_1, r_2, r_3, and so on, as a list containing every real number. We shall now proceed to construct a certain nested sequence of closed intervals:

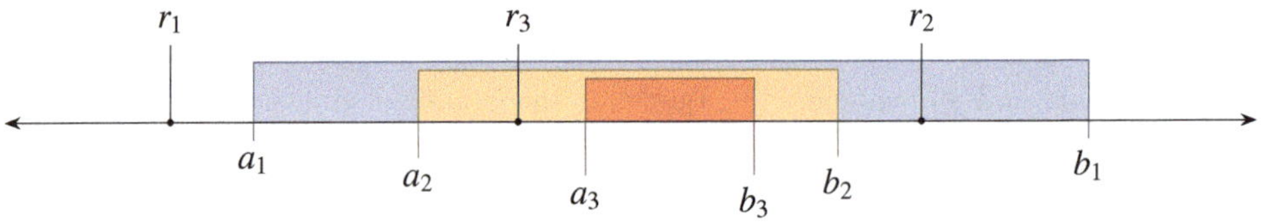

Namely, we shall describe a sequence of numbers $a_n < b_n$, which will be nested like this:

$$a_1 < a_2 < a_3 < \cdots < b_3 < b_2 < b_1.$$

The main idea is that we should choose the nth interval $[a_n, b_n]$ specifically to exclude the number r_n. That is, we choose the first interval $[a_1, b_1]$ specifically to exclude the number r_1, we choose the next interval $[a_2, b_2]$ to exclude r_2, and so on, progressively through the natural numbers. Let $z = \sup_n a_n$ be the supremum of the left endpoints of the intervals. This is a real number that is therefore inside all the intervals. It is consequently different from every number r_n because those numbers were each systematically excluded from the intervals. So z is a real number that is not on the orginal list of numbers. But this contradicts the assumption that the list had included every real number. So the set of real numbers is uncountable.

The Two Proofs Are Fundamentally the Same

Some authors have emphasized what they see as an enormous distinction between the two arguments, claiming that Cantor's original argument did not have the diagonalization idea, which, they say, came only much later (and indeed, one can find polemical discussions online about this). My perspective, however, is that although the two proofs are superficially different, the first using diagonalization and the second using a nested sequence of closed intervals, nevertheless the two arguments are fundamentally the same construction—I view it as merely two manners of presenting the same underlying idea.

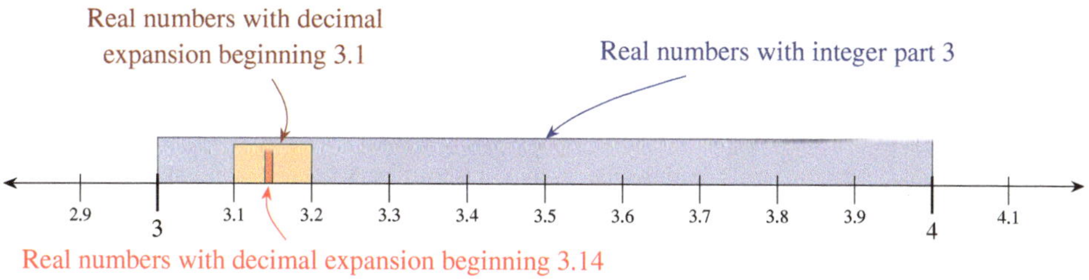

The reason is that to specify the first several digits of a real number in the decimal expansion is exactly to restrict to a certain interval of real numbers, the numbers whose expansion begins with those digits. Thus, to specify an additional digit of the diagonal real number to be different from the corresponding digit of r_n amounts to restricting to an interval of numbers excluding r_n. So the two arguments are describing the same underlying idea; the digit diagonalization argument is simply a more concrete, and perhaps more attractive way to describe the particular intervals that are being chosen.

Mathematical Cranks

Cantor's theorem and proof are sometimes misunderstood in certain characteristic ways, and they have attracted the phenomenon of the mathematical "crank," an unfortunate person who takes himself or herself as an unrecognized genius who has refuted the established dogma of mathematics but who actually has done nothing of the sort. I regularly receive emailed manuscripts from people claiming that Cantor's proof is all wrong, or that all of mathematics is wrong, a conspiracy of delusion. Sadly, the objections they raise have no force.

Let me try to give a taste of it. In some manuscripts, the proposal is essentially that after having constructed Cantor's diagonal real number z, if it does not appear on the original list $r_1, r_2, r_3, \ldots$, then one should simply *add it to the list*! Put the diagonal real number z in front, they say, and everything will be fine.

But will it?

The proposal, of course, is without merit. It simply does not matter that the diagonal real number can appear on another list, different from the original list. The point of the diagonal real number is that because it is not on the original list, it refutes the supposition that the original list already contained all real numbers. This contradiction shows that it is impossible to make a list containing all real numbers, and that is exactly what it means to say that the set of real numbers $\mathbb{R}$ is uncountable.

Another strategy that one sees is that a person will simply ignore Cantor's argument and instead try to mount a direct proof that the set of real numbers is countable. For example, the person might start by proving that the set of real numbers with finite terminating decimal expansions is countable, and next, point out that every real number is a limit of such numbers; so they conclude that there are only countably many real numbers.

This argument starts out fine, since indeed, the numbers with a finite terminating decimal expansion are precisely the rational numbers expressible with a denominator that is a power of 10, and indeed there are only countably many such numbers. The second step is also fine since every real number is indeed a limit of a sequence of such rational numbers, which approximate it ever more closely. The last step of the argument, however, is not correct; this part of the argument is usually left with no proof or with an erroneous proof, and the conclusion is simply incorrect. The set of real numbers does have a countable dense set, but this does not mean that the real numbers themselves are countable.

13.2 Cantor on Transcendental Numbers

Cantor uses his theorem to provide a new proof of Joseph Liouville's 1844 theorem that some real numbers are transcendental. A real number r is *algebraic* if it is a root of a nontrivial polynomial in one variable over the integers. For example, $\sqrt{2}$ is one of two roots of the polynomial $x^2 - 2$, the other being $-\sqrt{2}$. Similarly, the number

$$x = \sqrt[3]{2 + \sqrt{5}}$$

is algebraic since $x^3 = 2 + \sqrt{5}$, so $x^3 - 2 = \sqrt{5}$, which means that $(x^3 - 2)^2 = 5$, showing that x is a root of $(x^3 - 2)^2 - 5$. Similar arguments work with essentially any expression by radicals, but not every algebraic number is expressible that way.

A real number is *transcendental*, in contrast, if it is not algebraic—it is not the root of any nontrivial polynomial over the integers. We now know that e and π (and indeed many other famous real numbers) are transcendental, but in earlier days, this was not yet known, and it was not initially known whether there were any transcendental real numbers at all.

Liouville's Lakes

Joseph Liouville (1809–1882) was the first to prove that transcendental numbers exist. To do so, he considers a class of *lacunic* numbers, whose expansions have increasingly large lakes of zeros. He considers, for example, the following specific number:

$$0.110001000000000000000001000\ldots$$

now known as the *Liouville constant*, which has the digit 1 in decimal places given by the factorials 1, 2, 6, 24, and so on.

The lakes of zeros in the Liouville constant have a very interesting consequence that has been investigated by the number theorists—namely, this number has rational approximations with surprising accuracy: 0.11 is correct to 5 decimal places, and 0.110001 is correct to 23 places. The basic situation is that the rational approximation up to the closer shore of each lake extends in accuracy across it to the opposite shore.

The main point of the lacunic numbers, however, consists of the observation that if the lakes grow sufficiently in size, then the isolated 1s between the lakes eventually do not interfere with one another in rational polynomial expressions. In particular, these parts of the number cannot conspire to cancel each other out in algebraic combinations. Precisely for this reason, such numbers cannot solve a nontrivial algebraic equation over the rational numbers, and therefore they are transcendental.

Cantor's Proof of Transcendental Numbers

Cantor, in contrast, proves not only that transcendental real numbers exist, but furthermore, *most* real numbers are transcendental. He proves specifically that while there are only countably many algebraic real numbers, there are uncountably many transcendental real numbers—a strictly larger infinity.

To do this, Cantor observes first that there are only countably many real algebraic numbers. In algebra, one proves that every nontrivial polynomial has only finitely many roots. Because of this, every algebraic real number is uniquely determined by specifying a finite list of integers: the coefficients of a polynomial of which it is a root and its place in order amongst the roots of that polynomial. Every polynomial is determined by the finite sequence of its coefficients (including coefficient zero up to the degree), and since the collection of all finite sequences of integers is countable, there are consequently only countably many algebraic real numbers.

Next, since Cantor had proved that there are uncountably many real numbers, it follows that there must be some real numbers that are not algebraic, which is to say that they are transcendental. Indeed, since the union of two countable sets is countable, there must be uncountably many transcendental real numbers, but only countably many algebraic numbers. In this sense, most real numbers are transcendental.

Constructive Versus Nonconstructive Arguments

Let me highlight a certain subtle issue raised by this argument. The claim that there are transcendental numbers is an existential claim, one saying that a certain kind of mathematical object exists (a transcendental number). Existential claims in mathematics can often be proved with a *constructive* proof, one that explicitly exhibits the mathematical object in question. A nonconstructive proof, in contrast, might prove that there is such an object without providing any specific instance or directing us to any specific example of that kind of object. A nonconstructive proof of an existential claim, for example, might proceed by deriving a contradiction from the assumption that there are no such objects.

Consider the following argument: There must be a cat in this warehouse since we do not see any mice. The argument is nonconstructive, since it does not exhibit any specific cat; a constructive argument would show us the actual cat. Similarly, when we argue that there must be a transcendental real number because otherwise there would be only countably many real numbers, then we do not seem to learn about any specific transcendental number, while a constructive argument like Liouville's would exhibit an actual instance. Mathematicians often prefer constructive proofs, when possible, if the difficulty of the argument is comparable. Are constructive arguments more convincing? Perhaps, because when the object is explicitly provided, then it seems less likely that there has been a mistake in the argument; when you show me an actual cat in the warehouse, it is easy to believe that there is a cat in the warehouse. Nevertheless, constructive proofs can often be fussy and more complicated than the alternatives, precisely because they make a more detailed claim. Another reason to prefer constructive proofs is that they often seem to provide more information about the nature of the object asserted to exist than merely the raw statement of existence. When you show me the actual cat, then I might see its markings and learn whether it is friendly, and so forth.

One must be aware that there is a certain clash in the usage of the *constructive* terminology. Namely, on the one hand, mathematicians commonly use the term in the loose way that I have just described, to refer to a situation where a mathematical argument provides a means of building—in some sense—the mathematical object that the argument is about, even in cases where those means might be infinitary or noncomputable. But meanwhile, on the other hand, mathematicians working in the area known as *constructive mathematics* use the term with a more exact meaning, imposing much stricter requirements on what counts as constructive. For example, in this usage, constructive arguments are not allowed even to use the law of excluded middle in their logic, such as might arise in an argument by cases, even though such a use might still be considered constructive in the more permissive ordinary usage.

Is Cantor's proof of the existence of transcendental numbers constructive? Liouville's earlier 1844 argument certainly was constructive; he provided a specific real number, the Liouville constant, by specifying its digits, and he proved that it was transcendental. Does

Cantor's argument also provide specific transcendental numbers? Or is it instead only a pure existence proof, showing that there must be some transcendental numbers, but we know nothing about them, like the unseen cats in the warehouse.

One can often find mathematicians asserting that indeed, Cantor's argument is not constructive; I once saw an extremely prominent mathematician make this assertion to hundreds of people in a big public lecture. Mathematicians making this claim usually explain that Cantor's proof of the uncountability of the set of real numbers proceeds by contradiction—it is uncountable because any proposed enumeration is inadequate—and one then deduces, also by contradiction, that there must be transcendental numbers because otherwise there would be only countably many real numbers, which is a second contradiction. This way of arguing indeed appears nonconstructive, since it proves that some real numbers must be transcendental, but it does not seem to exhibit any particular transcendental number.

Nevertheless, I claim that Cantor's argument, when undertaken properly, is actually constructive and provides a specific transcendental number. To see this, notice first, that Cantor's proof that the algebraic numbers are countable provides an explicit method for enumerating them; and second, that Cantor's diagonalization method provides for any given enumeration of real numbers a method of constructing a specific number—the diagonal real—which does not appear on that enumeration. By combining these two steps, therefore, first enumerating the algebraic numbers and then producing the corresponding diagonal real number, we thereby construct a specific real number that is not algebraic. So Cantor's proof is constructive after all.

13.3 Uncountably Many Binary Sequences

Consider the space of infinite binary sequences—infinite sequences of zeros and 1s. We have the all-1s sequence:

$$111111111111\ldots$$

as well as the all-zeros sequence:

$$000000000000\ldots$$

and sequences that alternate in various patterns:

$$101010101010\ldots$$

$$010101010101\ldots$$

$$110011001100\ldots$$

and various sequences that are entirely random:

$$110101110010\ldots$$

The space of all such infinite binary sequences is known as *Cantor space*. Note that we are thinking of the binary sequences in Cantor space purely as sequences, not as real numbers obtained via binary expansion or some such interpretation.

The claim is that Cantor space is uncountable. To prove this, suppose that it were countable. In this case, we could place all the infinite binary sequences in a list: s_1, s_2, s_3, s_4, and so on. We may then define the diagonal binary sequence

$$z = d_1 d_2 d_3 \ldots$$

by specifying the nth binary bit d_n simply to be the opposite of the nth bit of s_n. That is, the first bit d_1 is the opposite of the first binary bit of s_1, the second bit d_2 is the opposite of the second binary bit of s_2, and so on. If the nth bit of s_n is 0, then we make it 1 in z and vice versa.

For example, if the list were to begin like this:

$$\begin{aligned}
s_1 &= \mathbf{1}\,1\,1\,1\,1\,1\,1\,1\,1\,1\,1 \ldots \\
s_2 &= 0\,\mathbf{0}\,0\,0\,0\,0\,0\,0\,0\,0\,0\,0 \ldots \\
s_3 &= 1\,0\,\mathbf{1}\,0\,1\,0\,1\,0\,1\,0\,1 \ldots \\
s_4 &= 1\,1\,0\,\mathbf{0}\,1\,1\,0\,0\,1\,1\,0 \ldots \\
s_5 &= 1\,1\,0\,1\,\mathbf{0}\,1\,1\,1\,0\,0\,1 \ldots \\
&\ \vdots \qquad \ddots
\end{aligned}$$

Then the resulting diagonal sequence would begin like this:

$$z = 01011\ldots$$

The main point is that because z differs from s_n on the nth bit, it follows that $z \neq s_n$, and since this is true for every n, it means that z is an infinite binary sequence that does not occur on the list. This contradicts our assumption that the list contains all infinite binary sequences. Therefore, our assumption that it was possible to make such a list in the first place was wrong—we cannot place the infinite binary sequences into a one-to-one correspondence with the natural numbers. In short, Cantor space is uncountable.

13.4 More Subsets than Elements

Cantor generalizes his diagonal argument to show that every set has strictly more subsets than elements. That is, no set X is equinumerous with its *power set* $P(X)$, the set of all subsets of X.

Consider first the case of $P(\mathbb{N})$, the set of all subsets of the natural numbers. If this were countable, then we could assign each number n a subset $X_n \subseteq \mathbb{N}$, in such a way that every subset was represented.

But then we can define the diagonal set

$$D = \{ n \in \mathbb{N} \mid n \notin X_n \}.$$

Notice that D cannot be equal to any particular X_n, since $n \in D$ if and only if $n \notin X_n$. So $D \neq X_n$ for every n, contradicting the assumption that every set was represented.

Power Sets Are Always Larger

The same argument works with any set X to show that the power set $P(X)$ is larger than X. Suppose toward contradiction that a set X is equinumerous with its power set $P(X)$. So there is a one-to-one correspondence between the elements of X and the subsets of X, associating each element $a \in X$ with a subset $X_a \subseteq X$ in such a way that every subset is realized. Given this association, let us define the diagonal set:

$$D = \{ a \in X \mid a \notin X_a \}.$$

This is a perfectly good subset of X. The key thing to notice is that for every element a, we have $a \in D$ exactly when $a \notin X_a$. This immediately implies that $D \neq X_a$ for every $a \in X$. So D is a subset that is not associated with any element $a \in X$, contrary to our assumption that we had a one-to-one association of the elements of X with all the subsets of X. So $P(X)$ is not equinumerous with X. In fact, the argument shows that there can be no map from a set X onto its power set.

This argument amounts to an abstract form of Cantor's essential diagonal idea. We are not drawing a diagonal through the digits as before; rather, we have placed an element a into the diagonal subset D depending on whether a is in X_a. That is, at coordinate a, we work with set X_a, in effect working on the abstract diagonal of the set $X \times X$.

So far, we have proved that no set X is equinumerous with its power set $P(X)$. Meanwhile, it is easy to see that every set X has at least as many subsets as elements, for we can associate each element $x \in X$ with the singleton subset $\{x\}$. Therefore, the power set of a set is strictly larger than the set, $X < P(X)$.

No Largest Infinity

It now follows immediately that there is no largest infinity—we can always find a larger one. For any set X, no matter how large, the power set $P(X)$ will be larger.

The power set of the natural numbers $P(\mathbb{N})$, for example, is larger than $\mathbb{N}$, so this is an uncountable set. By simply iterating the power set operation, we may therefore produce larger and larger sizes of infinity:

$$\mathbb{N} \quad < \quad P(\mathbb{N}) \quad < \quad P(P(\mathbb{N})) \quad < \quad P(P(P(\mathbb{N}))) \quad < \quad \cdots$$

Each successive power set is a strictly larger uncountable infinity, so in particular, there are infinitely many different sizes of infinity.

On the Number of Infinities

How many infinities are there? Well, there are infinitely many infinities, of course, as we have just observed. But which infinity? So far, by iterating the power set, we have only seen countably many different infinities. Are there more? Yes, indeed there are. We can form a set larger than the ones we saw simply by taking the union

$$\mathbb{N} \cup P(\mathbb{N}) \cup P(P(\mathbb{N})) \cup P(P(P(\mathbb{N}))) \cup \cdots$$

This set is at least as big as any of the sets appearing in the sequence, but since that sequence was strictly increasing in size, this set is strictly larger than all of them—a new, larger infinity.

More generally, suppose that we have κ many different infinite cardinals κ_i, indexed by elements i in some index set I of size κ. So each cardinal κ_i is the cardinality of some set X_i. Let us fix particular such sets X_i, and let $X = \bigcup_{i \in I} X_i$ be the union of all these sets, which is therefore at least as large as any of them. The power set $P(X)$, therefore, is strictly larger than every set X_i. So we have found a cardinality larger than every κ_i. What we have proved is that no set of cardinal numbers can have all the cardinal numbers.

The general conclusion, therefore, is the following:

> There are more different infinities than any given infinity.

There are a few set-theoretic subtleties to the argument that we just gave, which I should like briefly to mention. First, in picking the sets X_i, the argument used what is called the *axiom of choice*, which we shall discuss in chapter 21. But this use is eliminable, however, and the result can be proved in Zermelo-Fraenkel set theory without the axiom of choice—that is, in ZF rather than ZFC. Meanwhile, the conclusion cannot be proved in the weaker Zermelo system, which lacks the replacement axiom; indeed, Zermelo's weaker set theory is consistent with the existence of only countably many different cardinalities.

13.5 Allegorical Accounts of Cantorian Logic

Let us see how Cantorian logic plays out in several rather more whimsical instances.

The Number of Committees

Let us recast Cantor's argument in anthropomorphic form by arguing that the number of possible committees that may be formed from a given body of people is strictly greater than the number of people, whether there are finitely or infinitely many people. A committee may be formed from any set of people from that body. So we have the general committee, of which everyone is a member, and the empty committee, of which nobody is a member, and all the one-person committees and the various two-person and three-person committees, and so on. From the one-person committees, we see immediately that there are at least as many committees as people. This by itself does not show that there are more committees than people, however, for in the infinite case, perhaps there could still be a one-to-one correspondence between people and committees.

So let us argue via Cantor diagonalization that there is no such correspondence. Suppose toward contradiction that there is a one-to-one correspondence between people and committees. Such a correspondence can be interpreted simply as a committee-naming scheme, by which we have named each committee after a different person. Let us now form the committee D, consisting of all and only those people who are not members of the committee that is named after them. This is a perfectly good committee, so it must be named after someone—let us say person d. And now we ask the question: Is person d a member of committee D? If yes, then d is a member of the committee named after d, in which case d should *not* be a member of D; but if d is not a member of D, then d is not a member of the committee named after d, so d *should* be a member of D. In either case, we have reached a contradiction, so there can be no such naming system, and the number of committees must therefore be strictly larger than the number of people.

The Number of Fruit Salads

Suppose that you have an assortment of fruits and we wonder how many different possible fruit salads one can make from them, where we think of a fruit salad as determined by the kinds of fruit that are in it. The claim is that there are more fruit salads than fruits. If not, then we could name every fruit salad after some fruit, and then consider the diagonal salad, consisting of all the fruits that are not in the fruit salad that is named after them. This is a perfectly valid salad, but it cannot be named after any fruit since if it were named after apples, say, then apples would be in the diagonal salad if and only if they were not, which is a contradiction.

The Number of Propositions

Russell highlights the puzzling propositional formulation of the paradox. Namely, in light of Cantor, there must be more classes of propositions than propositions. And yet, for every class of propositions, there seems to be the proposition asserting that they are all true. This would seem to place the propositions into one-to-one correspondence with the classes of propositions. For any set of propositions, is it a proposition to assert that they are all true? And is this a different proposition for different sets?

Russell performs the diagonalization as follows:

> Again, we can easily prove that there are more propositional functions than objects. For suppose a correlation of all objects and some propositional functions to have been affected, and let φ_x be the correlate of x. Then "not-$\varphi_x(x)$," i.e. "φ_x does not hold of x," is a propositional function not contained in the correlation; for it is true or false of x according as φ_x is false or true of x, and therefore it differs from φ_x for every value of x.
> Russell (1903, §348)

Thus we find again Cantor's diagonal reasoning.

13.6 The Russell Paradox

Let us consider next the general comprehension principle, asserting that for any property, we may form the set of all objects that have that property.

General comprehension principle. *For any property P, one may form the set of all x with property P(x). That is,* $\{\, x \mid \varphi(x) \,\}$ *is a set.*

According to this principle, we may form the set of all elephants, the set of all prime numbers, and the set of all topological spaces.

Perhaps the principle seems very appealing at first—we might even take it to express a fundamental truth of set theory. Nevertheless, the fact of the matter is that the general comprehension principle is wrong. It is inconsistent; it has self-contradictory instances. Ultimately, we can look on it as a logical fallacy, like denying the antecedent; it is initially very plausible, but in the end, it is easily refuted. Perhaps precisely because the principle combines simple clarity and attractiveness with contradiction, it was the cause of a rude coming-of-age for set theory in its teenage years—a dramatic mathematical catastrophe. Let me relate the story.

Gottlob Frege had for years poured himself, you see, into a monumental work on the logical and philosophical foundations of arithmetic, the *Grundgesetze der Arithmetik*, Frege (1893/1903). The first volume appeared in 1893, and the second was nearly complete in 1901, destined to become the core of the *logicist* program, striving to found all of mathematics, starting with arithmetic, on purely logical principles. Frege aimed in short to reduce mathematics to logic. To do so, he presented a logical system and method of formal reasoning in which he had worked out in laborious detail how to derive all the fundamental principles of arithmetic from his purely logical principles. Unfortunately, amongst those purely logical principles in effect was the general comprehension principle, which was woven deeply and inextricably into the functioning of his system.

Just as the second volume of the *Grundgesetze* was being completed, however, Russell observed that the underlying logical system was fundamentally flawed and inconsistent. The general comprehension principle was simply refutable—some properties cannot define sets at all, upon pain of contradiction.

Russell's argument is this. Let R be the set of all sets x that are not members of themselves; that is,

$$R \quad = \quad \{\, x \mid x \text{ is a set and } x \notin x \,\}.$$

According to the general comprehension principle, this is a set. Let us ask: Is R a member of itself? The members of R are precisely the sets x with $x \notin x$, and therefore we see that

$$R \in R \qquad \iff \qquad R \notin R.$$

But this is a contradiction, so this instance of the general comprehension principle cannot be true.

You might say that the property $x \notin x$ is rather peculiar to consider, since we do not often expect sets to be members of themselves. After all, the set of all elephants is not an elephant (it is a *set* of elephants), and it is therefore not a member of itself; the set of all prime numbers is not itself a prime number, for it is not a number at all. Very well then, the objection shows that the property of being non-self-membered is quite ordinary or even banal—it seems to hold of most, or even all, sets—and therefore the general comprehension principle is false for some quite ordinary properties.

Russell wrote a letter to Frege in 1902 explaining the inconsistency that he had discovered—a devastating objection to the latter's enormous work. And indeed, Russell's letter arrived just as the printing was being undertaken. To his credit, Frege quickly wrote and included an epilogue addressing the objection:

> Hardly anything more unwelcome can befall a scientific writer than to have one of the foundations of his edifice shaken after the work is finished. This is the position into which I was put by a letter from Mr. Bertrand Russell as the printing of this volume was nearing completion. Frege (1893/1903, epilogue), translated in Frege (2013, p. 253)

It must have been heartbreaking.

The Barber

Consider the allegory of the barber who shaves all those in town who do not shave themselves. Does the barber shave himself? Well, if he does not, then he should; and if he does, then he should not. This is precisely the contradictory logic of the Russell paradox, the same as in Cantor's theorem on power sets.

The Barista

Or consider the barista who happily serves up coffee each morning to all and only those who do not make coffee for themselves. Does she make coffee for herself? If so, then she should not; and if not, then she should.

Questions for Further Thought

13.1 By successively taking the power set, we produced sets of increasing infinite size:

$$\mathbb{N} \quad < \quad P(\mathbb{N}) \quad < \quad P(P(\mathbb{N})) \quad < \quad P(P(P(\mathbb{N}))) \quad < \quad \cdots$$

Can you produce infinities larger than these? How many different sizes of infinity can you describe above them?

13.2 Use Cantorian reasoning directly to show that for any set A with at least two elements, there are more functions $f : A \to A$ than there are elements of A.

13.3 Prove that the set of complex numbers $\mathbb{C}$ is equinumerous with the set of real numbers $\mathbb{R}$.

13.4 Argue that the real line $\mathbb{R}$ is equinumerous with three-dimensional space $\mathbb{R}^3$, and indeed with $\mathbb{R}^n$ in any finite dimension. And what about infinite dimensional space $\mathbb{R}^{\mathbb{N}}$?

13.5 Are there more integer-valued functions on the real numbers, $f : \mathbb{R} \to \mathbb{Z}$, or real-valued functions on the integers, $g : \mathbb{Z} \to \mathbb{R}$?

13.6 Describe a set that arises naturally in mathematics in a subject other than logic or set theory, which has very large cardinality. How big is it? What is the largest such set you can find?

Further Reading

Hamkins (2020). Chapter 13 of my book on proof-writing is an elementary introduction to the topics mentioned in this chapter.

Hamkins (2021b). Chapter 3 of my book on the philosophy of mathematics gives an extended account of the central ideas in this chapter, and chapter 8 explores further issues in the philosophy of set theory.

Credits

Major parts of this chapter were adapted from Hamkins (2021b, chapter 3) and Hamkins (2020, chapter 13). The fruit salad example comes from Martha Storey. The barista example comes from Barbara Gail Montero. The quotation of the 1877 letter from Cantor to Dedekind is from Dauben (2004, p. 941). Demian Nahuel Goos (2025) argues, in light of newly discovered 1873 letters from Dedekind to Cantor in their exchange leading up to Cantor's 1874 article, that the proof of the countability of the algebraic numbers should be credited to Dedekind, rather than to Cantor, who does not mention Dedekind's contributions in his article.

14 How to Count

Abstract. Shall we count together in the ordinals? Let us venture into that transfinite realm beyond infinity, counting to ω, to ω^2, to ε_0, and far beyond. Is infinity even or odd? Can we count to an uncountable ordinal?

14.1 The Ordinal Numbers

The ordinal numbers are the natural continuation of the natural numbers into the transfinite realm beyond infinity. Anyone, even a child, can learn to count into the ordinals, so let us explore this realm together.

We start with zero at the bottom and count through the familiar natural numbers:

$$0 \quad 1 \quad 2 \quad 3 \quad 4 \quad 5 \quad \cdots$$

But we have hardly begun. Let us imagine that we have finished counting through *all* the finite numbers. Afterward, we shall encounter the first infinite ordinal number, the first ordinal after all finite numbers—this is the ordinal known as ω, the Greek letter *omega*:

$$0 \quad 1 \quad 2 \quad 3 \quad 4 \quad 5 \quad \cdots \quad \omega$$

What? How can there be any numbers *after* all the finite numbers? After all, if we were counting them, there wouldn't be any time left to pronounce ω since we would use up all our time counting the finite numbers, right?

Some people find infinite ordinals easier to grasp if we imagine counting through the finite numbers faster and ever faster in the supertask manner discussed in earlier chapters—we pronounce the number 0 in the first half-hour, the number 1 in the next quarter-hour, 2 in an eighth-hour, and so on. After one hour, we will have pronounced every finite number, with plenty of time remaining for more.

In my view, however, that supertask process is an unnecessary extravagance. To understand the ordinals, it is not necessary to embed them into what we may perceive as the structure of our temporal reality. Rather, the ordinals have their own order structure, their own integrity. Let us understand the nature of the ordinals directly in terms of their own order structure, such as it is. Let us count in *ordinal time* rather than physical time, giving each ordinal its own pace and dignity.

The ordinal ω, the first infinite ordinal, is a *limit* ordinal, which means that it is approached by smaller ordinals, the finite numbers, but it has no immediately preceding number. After all, the successor of a finite number n is the number $n + 1$, which is still finite. There is no ordinal number immediately prior to ω.

Counting to ω^2

Let us count now in earnest. After any ordinal, **we can always add 1**, so having reached ω, we may form the ordinal $\omega + 1$. Adding 1 again, we form $\omega + 2$, then $\omega + 3$, and so on:

$$0 \quad 1 \quad 2 \quad 3 \quad 4 \quad 5 \quad \cdots \quad \omega \quad \omega+1 \quad \omega+2 \quad \omega+3 \quad \cdots$$

We have thus discovered in effect a new era of numbers, a second copy of ω above the original—the ordinals proceeding from ω to $\omega + 1$ and $\omega + 2$ and so on have just the same order structure as the finite ordinal numbers.

$$\boxed{0 \quad 1 \quad 2 \quad 3 \quad 4 \quad 5 \quad \cdots} \boxed{\omega \quad \omega+1 \quad \omega+2 \quad \omega+3 \quad \omega+4 \quad \cdots}$$

The ordinals indeed look the same proceeding from any given ordinal as though starting anew. There will be a next ordinal, and a next, and so on until the next limit ordinal, and so on, just as when counting from the beginning. Every set of ordinals will have a smallest ordinal above them all.

After forming the finite successors $\omega + k$ in this next era of infinity, we may again find a supremum just as we did before—the ordinal $\omega + \omega$, which can also be written as $\omega \cdot 2$, occurs first after them all. And of course, it starts its own next era of numbers, continuing with $\omega \cdot 2 + 1$, and so on:

$$\boxed{0 \quad 1 \quad 2 \quad 3 \quad \cdots} \boxed{\omega \quad \omega+1 \quad \omega+2 \quad \cdots} \boxed{\omega\cdot 2 \quad \omega\cdot 2+1 \quad \cdots}$$

We shall eventually reach the next limit ordinal $\omega \cdot 3$, starting yet another era of infinity, then $\omega \cdot 4$ after that, and indeed $\omega \cdot n + k$ for every finite n and k:

$$\boxed{0 \ 1 \ 2 \ 3 \ \cdots} \boxed{\omega \ \omega+1 \ \omega+2 \ \cdots} \boxed{\omega\cdot 2 \ \omega\cdot 2+1 \ \cdots} \ \cdots \ \boxed{\omega\cdot n \ \omega\cdot n+k \cdots} \ \cdots\cdots \ \boxed{\omega^2}$$

Thus we have counted to the ordinal ω^2. The ordinals encountered along the way take the form $\omega \cdot n + k$ for finite n and k.

The ordinal ω^2 is the first *compound* limit ordinal—a limit ordinal that is a limit of limit ordinals since ω^2 is the limit of $\omega \cdot n$ as n increases in ω. In other words, ω^2 is a limit ordinal, but there is no largest limit ordinal below it. A *simple* limit ordinal, in contrast, is a limit ordinal that is not a compound limit—all simple limits take the form $\alpha + \omega$ for some ordinal α.

Just Like Counting to 100

Counting to ω^2 is rather like counting to 100. When we count to 100 you might notice that within each decade—the teens, the twenties, the thirties, and so on—it is just like counting to 10 again. In counting to 100, or 10^2, we thus count to 10 altogether 10 times.

Similarly, when we count to ω^2, we encounter ω many eras, each of size ω. We started with the era of finite numbers, the original copy of ω. In the next era, we proceeded from ω to count ω many more numbers, of the form $\omega + k$. Then we did it again from $\omega \cdot 2$, again from $\omega \cdot 3$, and so on. We counted to ω altogether ω many times.

Just as the numbers up to 100 have two digits in base ten, with the form $10 \cdot n + k$, similarly the ordinals up to ω^2 have the form $\omega \cdot n + k$, which is two digits in base ω.

Counting to ω^2 Poster

A few years ago, I made a poster of the process of counting to ω^2, shown in figure 14.1. The PDF is available at a link in the Credits section on page 188 at the chapter's end, along with a short YouTube video that I made walking through the counting process.

14.2 Ordinal Versus Cardinal Numbers

Linguists distinguish between two uses of number words—namely, the *cardinal* number words are *one*, *two*, *three*, and so on, the possible sizes of a set, measuring how big the set is or how many elements it has. The *ordinal* number words, in contrast, such as *first*, *second*, *third*, and so on, have to do with the order of the elements of a set or how one might count off the elements of a set in order.

In the case of finite numbers, these concepts are closely connected since if a set has five elements, then the various ways of placing them in order all have the same underlying structure—first, second, third, fourth, and fifth. The particular elements in those places are not determined, of course, but the underlying order type itself is determined from the size of the set for the finite numbers.

The corresponding fact for infinite sets, however, is no longer true since a countably infinite set such as the integers can be arranged in order type ω if we enumerate them like this:

$$0 \quad 1 \quad -1 \quad 2 \quad -2 \quad 3 \quad -3 \quad \cdots$$

Or it could be arranged in order type $\omega + \omega$ if we first count off the nonnegative integers, and then after that the negative integers:

$$0 \quad 1 \quad 2 \quad 3 \quad \cdots \quad -1 \quad -2 \quad -3 \quad \cdots$$

This second order has a totally different ordinal structure since the number -1 occupies the ωth position, then -2 is in the $(\omega + 1)$th position, and so on, but there are no analogues of these positions in the first ordering.

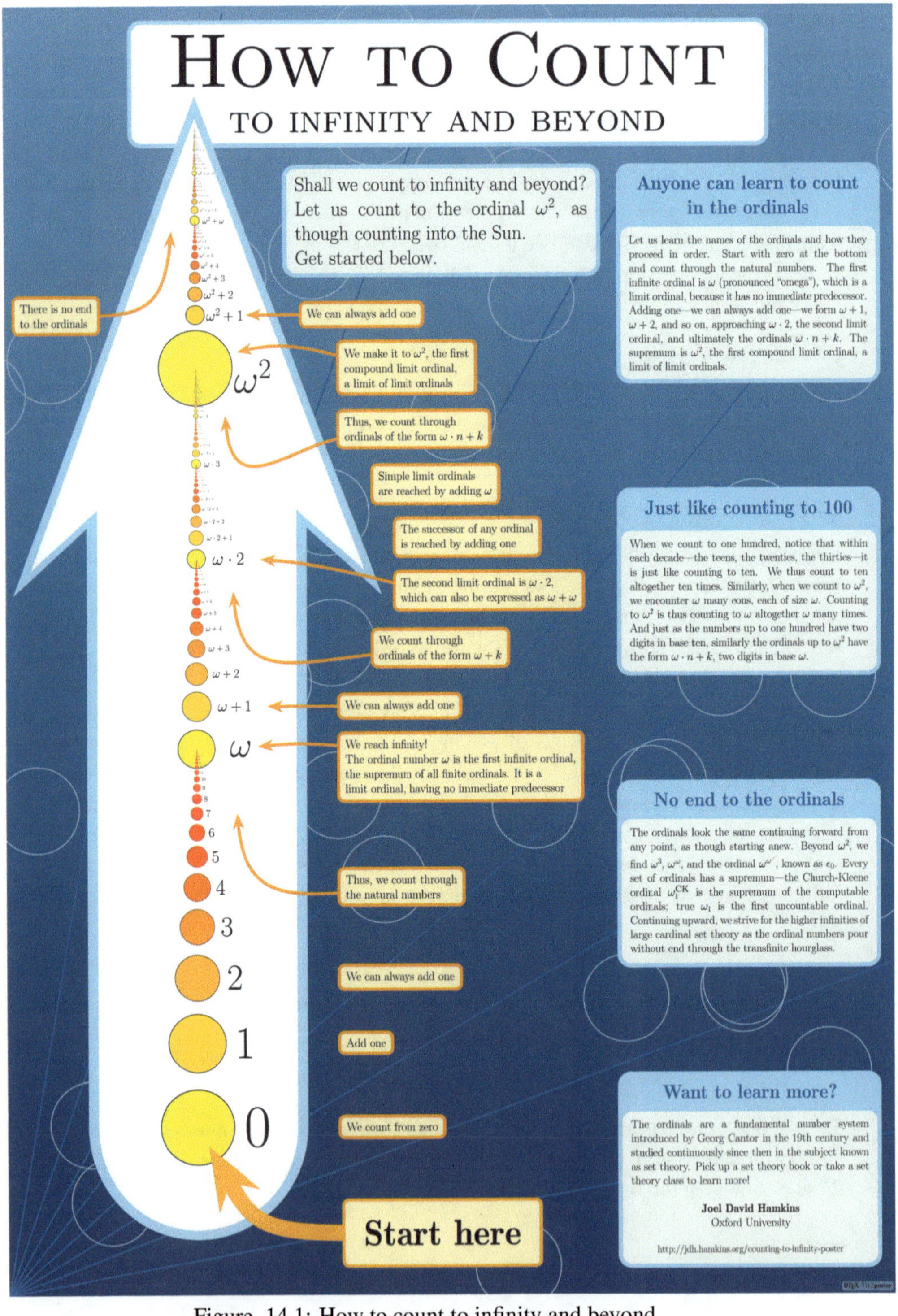

Figure 14.1: How to count to infinity and beyond

In the transfinite realm, therefore, the distinction between ordinal and cardinal numbers is far more salient. All the various ordinals above ω that we have seen so far are countable. They are bigger than ω as ordinals, but their cardinality is the same countable infinity—they would fit into Hilbert's Grand Hotel.

The ordinals represent the *well ordered* order types, the orders in which every nonempty set has a least element. It is precisely this feature of the ordinals that means that there will always be a least next ordinal after any set of ordinals. The well ordered nature of the ordinals is precisely what makes it possible to count coherently in the ordinals. However high you have counted—for any set of ordinals already counted—there will always be a next ordinal afterward in the era just beyond.

14.3 Ordinal Arithmetic

The ordinals have a natural arithmetic, which we have already been using informally.

Ordinal Addition

The ordinal $\alpha + \beta$ is the unique ordinal that is order-isomorphic to a copy of α followed by a copy of β:

$$\alpha + \beta \quad =$$

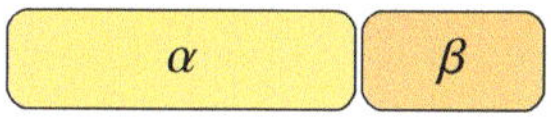

What I mean by this identity is that $\alpha + \beta$ is the unique ordinal whose predecessors are order isomorphic to the order consisting of a copy of α followed by a copy of β.

For example, the ordinals below $\omega + 3$ consists of a copy of ω followed by three more ordinals:

$$0 \quad 1 \quad 2 \quad 3 \quad 4 \quad 5 \quad \cdots \quad \omega \quad \omega+1 \quad \omega+2.$$

Further, counting to $\omega + \omega$ consists of counting through two copies of ω:

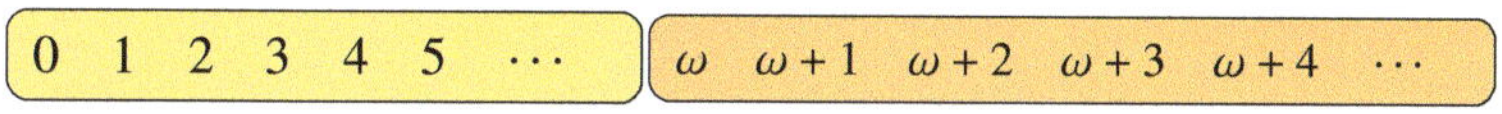

Before continuing, let me point out a certain peculiarity of ordinal addition in the transfinite realm. Namely, we had defined $\alpha + \beta$ as the ordinal corresponding to a copy of α followed by a copy of β. But did it matter which was first? Could we have taken a copy of β followed by a copy of α?

Question. Is ordinal addition commutative? That is, if α and β are ordinals, does it follow that $\alpha + \beta = \beta + \alpha$?

Consider the definition carefully and think about this on your own before reading further.

Think about it...

The fact of the matter is that ordinal addition is not commutative—in some cases, $\alpha + \beta$ can be a different ordinal than $\beta + \alpha$. Perhaps the simplest counterexample occurs with $\omega + 1$ versus $1 + \omega$. The ordinal $\omega + 1$, after all, consists of a copy of ω followed by one more point on top of this. But the ordinal sum $1 + \omega$ consists of one point followed by ω many more points, which altogether has the same order type as a single copy of ω. In short, $1 + \omega = \omega$, but $\omega + 1$ is strictly larger.

Ordinal Multiplication

The ordinals also support a natural notion of multiplication. The product $\alpha \cdot \beta$ is the ordinal whose predecessors have the order of β many copies of α:

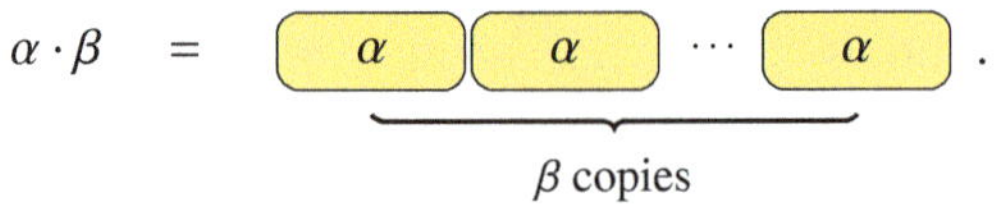

This is precisely why it is correct for us to write the following, where we have two copies of ω:

$$\omega \cdot 2 = \omega + \omega.$$

Similarly, with three copies of ω, we have $\omega \cdot 3 = \omega + \omega + \omega$. The goal of our earlier counting had ω many copies of ω:

$$\omega^2 = \omega \cdot \omega.$$

We can inquire again about commutativity in the case of ordinal multiplication.

Question. Is ordinal multiplication commutative? In other words, if α and β are ordinals, do we know $\alpha \cdot \beta = \beta \cdot \alpha$?

Give it some thought on your own before reading further.

Think about it...

Ordinal multiplication is not commutative. A counterexample is provided by the case $\omega \cdot 2$ versus $2 \cdot \omega$. The ordinal $\omega \cdot 2$ is equal to $\omega + \omega$, with two copies of ω—two complete eras of infinity. But $2 \cdot \omega$ consists of ω many copies of 2, like this:

And this is the same order type as ω itself, so $2 \cdot \omega = \omega$. When counting to ω^2, I was careful to describe the smaller ordinals that we encountered along the way as taking the form $\omega \cdot n + k$. It would have been incorrect for me to have written $n \cdot \omega + k$ since this is the same as $\omega + k$ for finite nonzero n.

Ordinal Exponentiation

The ordinals also support a natural exponentiation α^β. The basic rule is that exponentiation is iterated multiplication, defined by

$$\alpha^{\beta+1} \quad = \quad \alpha^\beta \cdot \alpha,$$

starting of course with the initial value of $\alpha^0 = 1$. This definition implies that $\omega^2 = \omega \cdot \omega$, which consists of ω many copies of ω, just as we used this notation before. But we may now also consider ω^3, which can be calculated as $\omega^3 = \omega^{2+1} = \omega^2 \cdot \omega$, consisting of ω many copies of ω^2. And the same goes for ω^4, ω^5, and so on. We carry ordinal exponentiation to infinite exponents simply by taking the supremum of earlier values:

$$\alpha^\omega \quad = \quad \sup_n \alpha^n.$$

Thus it is sensible to speak of ω^ω. More generally, we define exponentiation for any limit ordinal λ by

$$\alpha^\lambda \quad = \quad \sup_{\beta<\lambda} \alpha^\beta.$$

Thus we may speak of ω^{ω^ω} and far more.

14.4 Is Infinity Even or Odd?

I find it natural to interpret the question of whether infinity is even or odd in terms of the ordinal numbers, asking whether the first infinite ordinal ω, for example, is even or odd. Which is it? To answer this, of course, we shall need to know exactly what it means for a given ordinal number to be even or odd. What does it mean?

Question. What does it mean to say that an ordinal number is even or odd?

In the context of transfinite ordinals, the usual definition is that an ordinal number α is *even* if it can be realized as a succession of pairs—that is, if it is $2 \cdot \beta$ for some ordinal β. The ordinals of the form $2 \cdot \beta$, after all, are precisely those that are β many copies of 2, or in other words, β many pairs in sequence. Let us refer to this as the *sequence-of-pairs* conception of even ordinals. If an ordinal is not even, then we say that it is *odd*.

One attractive feature of this definition of even/odd is that the ordinals will proceed in an alternating even/odd pattern that continues the familiar parity pattern on the finite numbers. Namely, if α is even, it follows that $\alpha + 1$ will be odd, and $\alpha + 2$ will be even again.

The first infinite ordinal ω is even according to this definition since we have already observed that having ω many pairs in sequence is order-isomorphic to ω itself:

$$2 \cdot \omega = \omega.$$

The next ordinal $\omega + 1$ is odd, then $\omega + 2$ is even again, and so on in this even/odd pattern. One can see that indeed every limit ordinal is even according to this definition. So $\omega + \omega$ will be even, as will be ω^2, ω^3, and so on.

A Confounding Alternative Conception of Even

That analysis seems perfectly satisfactory, but let me mention a confounding competitor concept of what it means for an ordinal to be even—namely, the *cut-in-half* concept. An ordinal α is even according to the *cut-in-half* concept, if it takes the form $\beta + \beta$ for some ordinal β. The ordinal $\beta + \beta$, of course, is the same as $\beta \cdot 2$, so the two definitions are simply multiplying by 2 on different sides, which seems natural enough, but since ordinal multiplication is not commutative, these might not be the same.

Indeed, we may notice that ω is not even on the cut-in-half concept, since if $\omega = \beta + \beta$, then β would have to be strictly less than ω and hence finite, but in this case, $\beta + \beta$ would also be finite, so it couldn't be equal to ω. So there is no ordinal β with $\omega = \beta + \beta$, and the two definitions of even consequently have opposite opinions on the question of whether ω is even or odd.

Similar reasoning shows that $\omega + 1$ also is not cut-in-half even, since if we could realize $\omega + 1$ as $\beta + \beta$, then β would have to be either finite or ω itself, but neither possibility works out since if β is finite, then $\beta + \beta$ also is finite and thus not equal to $\omega + 1$; and if β is ω, then $\beta + \beta = \omega + \omega$, which is much bigger than $\omega + 1$.

Notice that this observation means that both ω and $\omega + 1$ are odd according to the cut-in-half concept, so this concept does not obey the alternating even/odd pattern. In fact, none of the ordinals of the form $\omega + k$ for finite k are cut-in-half even, and the first infinite cut-in-half even ordinal is $\omega + \omega$. The next cut-in-half even ordinal after that is $\omega + \omega + 1$, in light of the following calculation:

$$(\omega + 1) + (\omega + 1) = \omega + (1 + \omega) + 1 = \omega + \omega + 1.$$

So the cut-in-half concept of even allows for two successive ordinals to be cut-in-half even. Is this disqualifying for a concept of even?

14.5 Indecomposable Ordinals

Consider the ordinal ω^2, the first compound limit ordinal. This ordinal has the interesting property that if ordinals α, β are both below ω^2, then they are both below some $\omega \cdot n$ for some large enough finite n. Consequently, we may bound the sum $\alpha + \beta$ as follows:

$$\alpha + \beta \quad < \quad \omega \cdot n + \omega \cdot n \quad = \quad \omega \cdot (n + n) \quad < \quad \omega^2.$$

In other words, if α and β are smaller than ω^2, then $\alpha + \beta$ also is smaller than ω^2—the ordinal ω^2 is closed under addition—and this is precisely what it means to say that ω^2 is *additively indecomposable*. Another equivalent way to say this is that it is not possible to express ω^2 as a sum of smaller ordinals $\alpha + \beta$; we cannot decompose the ordinal ω^2 as a finite sum of smaller ordinals.

Using the method of transfinite induction, one can prove that the additively indecomposable ordinals are exactly those ordinals of the form ω^γ for an ordinal γ. This includes the trivial case of the ordinal $1 = \omega^0$ as 1 is technically additively indecomposable since you cannot express 1 using sums of zeros $(0 + 0)$. But further, if ω^γ is additively indecomposable, then the next one after this will be $\omega^{\gamma+1}$ since we should be able to add ω^γ to itself as many finite times as we want below it. And the limit of indecomposable ordinals is indecomposable, since it will still be closed under the sums of smaller ordinals.

The ordinal ω^2 is not multiplicatively indecomposable, however, since it is precisely $\omega \cdot \omega$. The first multiplicatively indecomposable ordinal above ω will be ω^ω. Indeed, the multiplicatively indecomposable infinite ordinals will be exactly the ordinals of the form ω^{ω^γ}, essentially since multiplying ordinals corresponds to adding the exponents, so we want the exponent to be additively indecomposable.

Counting to Epsilon Naught

Let us try to count a little higher in the ordinals. We have already reached ω^2, and by repeating that process ω many times, we reach ω^3. By repeating *that* process ω many times, we reach ω^4, and so on. In this way, we are reaching toward ω^ω, which is the supremum of ω^n for all finite numbers n.

If we count to ω^ω altogether ω many times, we shall reach $\omega^\omega \cdot \omega$, which is the same as $\omega^{\omega+1}$. So we still have a lot of work to do to get to $\omega^{\omega+2}$, and then to $\omega^{\omega+3}$, and so on. Eventually, we shall reach $\omega^{\omega \cdot 2}$.

By then perhaps we can see our way to $\omega^{\omega \cdot 3}$ and $\omega^{\omega \cdot 4}$, on the way to ω^{ω^2}. Eventually, of course, we shall arrive at ω^{ω^3} and ω^{ω^4} and so on, up to ω^{ω^ω}.

In time, with patience, we shall reach further milestones:

$$\omega \quad \omega^\omega \quad \omega^{\omega^\omega} \quad \omega^{\omega^{\omega^\omega}} \quad \omega^{\omega^{\omega^{\omega^\omega}}} \quad \cdots$$

The supremum of these finite-stack tetrations is known as the ordinal ε_0. In a sense, ε_0 is the ω stack tetration of ω, and we might represent it as

$$\varepsilon_0 \quad = \quad \omega^{\omega^{\omega^{\omega^{\cdot^{\cdot^{\cdot}}}}}} \Big\} \omega.$$

Every ordinal that is less than ε_0 is representable as a finite sum of ordinals involving only finite stacks of ω at most.

The typical ordinal less than ε_0 has finitely many terms, each consisting of a finite exponential tower of ωs, modified by various coefficients, like this:

$$\omega^{\omega^{\omega^2 \cdot 5 + \omega} \cdot 2} + \omega^{\omega^3 \cdot 9} + \omega^{\omega^3 + \omega^2 \cdot 7} + 1.$$

This class of ordinals is closed under ordinal sum, ordinal product, and ordinal exponentiation since the exponential of two finite stacks of ω is still merely a finite stack of ωs, which will be less than ε_0. So ε_0 is indecomposable with respect to addition, multiplication, and exponentiation.

In particular, ε_0 is a fixed point of the exponentiation function:

$$\varepsilon_0 = \omega^{\varepsilon_0}.$$

One can begin to see why this must be true in light of the calculation

$$\omega^{\varepsilon_0} = \omega^{\left(\omega^{\omega^{\omega^{\cdot^{\cdot^{\cdot}}}}}\right)} = \omega^{\omega^{\omega^{\omega^{\omega^{\cdot^{\cdot^{\cdot}}}}}}} = \varepsilon_0.$$

Thus we have reached a mind-bending milestone in the ordinals. But of course, we can always add 1, forming $\varepsilon_0 + 1$, $\varepsilon_0 + 2$, and so on. There is no end to the ordinals.

We can make the next exponentiation fixed-point by considering the sequence of ordinals:

$$\varepsilon_0 + 1 \quad \omega^{\varepsilon_0+1} \quad \omega^{\omega^{\varepsilon_0+1}} \quad \omega^{\omega^{\omega^{\varepsilon_0+1}}} \quad \cdots$$

The supremum of these is ε_1, which is bigger than ε_0 and is a fixed point:

$$\varepsilon_1 = \omega^{\varepsilon_1}.$$

By iterating this construction, we find all the epsilon ordinals ε_α for every α. These are precisely the fixed points of exponentiation with base ω, and also the same as the exponentially indecomposable numbers (along with 2 and ω).

The progression of ideas continues, for the fixed points of the ε_α operation itself are known as the *zeta* numbers ζ_α, and these are the tetration indecomposable ordinals. The *Feferman-Schutte ordinal* Γ_0 is the first ordinal above ω that is $\uparrow^n$ indecomposable for every finite n, using the ordinal analogue of the Knuth up-arrow notation. These are some truly very large ordinal numbers.

Cantor Normal Form

Cantor proved that every ordinal α can be uniquely expressed as a finite sum:

$$\alpha = \omega^{\beta_n} + \cdots + \omega^{\beta_0},$$

where $\beta_n \geq \cdots \geq \beta_0$. This is the *Cantor normal form* representation of the ordinal.

To prove this, suppose that the normal-form property holds for ordinals below α. If α is additively indecomposable, then it takes the form ω^γ, which is already in Cantor normal form. And if it is not indecomposable, then it can be written as the sum of two smaller ordinals. But those ordinals have normal forms by the induction hypothesis, so α itself arises as the sum of their forms. Thus, whenever the Cantor normal form holds below an ordinal, it holds at that ordinal, so by transfinite induction, it must hold for all ordinals.

Note that when adding these normal forms together, smaller terms on the left are simply absorbed by larger terms on the right. For example,

$$\omega + \omega^2 = \omega^2$$

and

$$\omega^2 + \omega^3 = \omega^3.$$

The uniqueness of the Cantor normal form also can be proved by transfinite induction—one shows that there can be no smallest ordinal with two representations. Let us leave the details of this aside; interested readers can learn more about it by studying more deeply in the subject of set theory.

14.6 Counting to Uncountability

All the particular ordinals that we have encountered so far, including the ordinals ω^2, ω^ω, the epsilon numbers ε_0, ε_1, the zeta numbers ζ_0, ζ_1, the Feferman-Schutte ordinal Γ_0, and so on, are nevertheless countable—they would fit into Hilbert's Grand Hotel. They are big as countable ordinals, with increasingly large order types, but in terms of equinumerosity, they all have the same countably infinite size as the natural numbers.

To be sure, the countable ordinals are an endless cosmos of complexity, with natural order types arising not just from elementary arithmetic and addition, multiplication, exponentation, tetration, and so on, but from other parts of mathematics.

The ordinal $\omega_1^{\mathfrak{Ch}}$, for example, is the *omega one of chess*, the supremum of the game values that arise for positions in infinite chess, introduced in Evans and Hamkins (2014). The ordinal ω_1^{CK} is the *Church-Kleene ordinal*, which is the supremum of the ordinals that arise as the order type of a computable relation on the natural numbers. The ordinal ω_1^L is the omega one of Kurt Gödel's constructible universe, which under certain strong axioms is a very large countable ordinal. The countable ordinals grow ever taller, realizing new conceptions of the infinite.

Uncountable Ordinals

Can we count up to an *uncountable* ordinal—one that is no longer equinumerous with the natural numbers? Indeed we can. The set of countable ordinals is ordered naturally by the ordinal order, and it must be uncountable since otherwise the order type of this order would be represented by one of the ordinals within it, which is impossible because its successor would be a still larger countable ordinal. So the order type of the set of countable ordinals is exactly the first uncountable ordinal, known as ω_1.

Like ω, the ordinal ω_1 is an *initial ordinal*, which means that it is an ordinal that is not equinumerous with any smaller ordinal. The ordinal ω_1 is therefore the first uncountable infinity, the first uncountable ordinal.

Of course, we can play the same trick again, considering all the ordinals that have a size that is at most ω_1. This will include $\omega_1+\omega_1$ and $\omega_1\cdot\omega_1$ and ordinal exponentiation $\omega_1^{\omega_1}$ and so on, all much bigger as ordinals but still equinumerous with ω_1. But the collection of all such ordinals of cardinality at most ω_1 will be naturally ordered by the ordinal order, and this ordinal cannot itself be represented in that class. So it will be the next initial ordinal, denoted ω_2, the first ordinal above ω_1 of strictly larger size with respect to equinumerosity. We can continue with ω_3 and indeed ω_α for every ordinal α.

There is no end to the ordinals, which continue on and on and on without end, falling like grains of sand through the transfinite hourglass of ordinal time.

Questions for Further Thought

14.1 Devise a supertask manner of counting to ω^2. At which moment will you pronounce the ordinal numbers $\omega \cdot n + k$ that you encounter along the way, so as to be able to pronounce ω^2 at the 1-hour mark? Can you generalize your scheme to ω^3?

14.2 The ordinals up to ω^2 take the form $\omega \cdot n + k$. Can you find an analogous representation of the ordinals up to ω^3? Classify them into zero, successor ordinal, simple limit, and compound limits. How about ω^4? How far can you go?

14.3 Draw a picture of the ordinal $\omega^3 + 5$. Draw a picture of ω^ω. How about $\omega^{\omega+2}$?

14.4 Which ordinal is the first limit ordinal? The first limit of limit ordinals? The first limit of limits of limit ordinals? The first limit of limits of limits of limit ordinals?

14.5 Explain the difference in the two conceptions of even ordinals, the succession-of-pairs conception and the cut-in-half conception. Exactly which ordinals up to ω^2 are even or odd with respect to the two conceptions? Which conception is more successful or robust as a generalization of the familiar even/odd conception for finite numbers to the transfinite realm of ordinals?

14.6 Is every compound limit ordinal additively indecomposable? How about the converse—is every additively indecomposable ordinal a compound limit ordinal?

14.7 Is ω_1 additively indecomposable?

14.8 Can you compare ω_1 and the set of real numbers $\mathbb{R}$ in size with respect to equinumerosity? Which is bigger, or are they the same size?

Credits

The poster in figure 14.1 appeared first in my blog post http://jdh.hamkins.org/counting-to-infinity-poster/, where a free, large-scale, poster-size PDF is available. I also posted a walk-through explanation on my YouTube channel at https://youtu.be/CTyYppKlkjg. Other parts of this chapter are adapted from my book-in-progress, *A Panorama of Logic*, Hamkins (2024a). The material on infinity being odd or even was adapted in part from my answer on Math.stackexchange at Hamkins (2011).

15 Transfinite Recursive Constructions

Abstract. Reaching beyond the finite, we shall undertake various transfinite recursions—constructing ordinal arithmetic, the alephs, the beths, the cumulative hierarchy, and more. We shall construct by transfinite recursion several surprising geometric decompositions of space as an interlocking union of disjoint unit circles or as an assemblage of skew lines.

15.1 Finite Recursion

Can you see the underlying rule of this familiar sequence?

$$1 \quad 1 \quad 2 \quad 3 \quad 5 \quad 8 \quad 13 \quad 21 \quad 34 \quad 55 \quad \cdots$$

This is the *Fibonacci* sequence—each number is the sum of the previous two, starting with two 1s. This sequence is thus an example of a *recursive* definition, where the values of a function are defined in terms of earlier values of the same function. The Fibonacci sequence has two *anchor* values:

$$f_1 = 1 \qquad f_2 = 1,$$

and then there is the *recursion* rule, expressing that the sum of two successive values is the subsequent value:

$$f_n + f_{n+1} \quad = \quad f_{n+2}.$$

This information is sufficient to calculate any given value f_n of the Fibonacci function since we can simply iterate the recursion rule as far out as necessary. If we want to know what f_{73} is, we can simply generate the sequence out that far to find out what the value is. In this way, the recursive definition encapsulates a complete logical account of the function.

Recursions Express Fundamental Relations

Recursion rules often arise when one is analyzing the fundamental nature of a process that is determined by a sequence of successive atomic interactions. Fibonacci, for example, was considering a population of reproducing rabbits. According to his simple model, we start with one pair of rabbits, one male and one female, who are initially juveniles, but they mature and mate after one month, producing one new pair of rabbits each month thereafter. And those rabbits reproduce according to the same model.

So we start with 1 pair of juvenile rabbits at the beginning of the first month. At the beginning of the second month, the female has become pregnant, but we still have only 1 pair of rabbits. At the start of the third month, however, we find the original pair plus the newly born pair, making 2 pairs altogether.

More generally, if f_n is the number of pairs of rabbits at the start of the nth month, then the model predicts the recursion

$$f_n + f_{n+1} = f_{n+2}$$

precisely because the number of rabbit pairs at month $n + 2$ is the number at the previous month f_{n+1} plus the number of newborn rabbit pairs, which arose from the rabbits at the month before that, of which there are f_n many. So $f_{n+2} = f_n + f_{n+1}$.

In this way, the recursive rule expresses this fundamental atomic interaction in the model, and we can subsequently use this recursion rule to mount a mathematical analysis. (Let us set aside the fact that this particular model is biologically unrealistic in several respects—it predicts that rabbits never die, for example, and that they always produce one male and one female every month without fail; and the model does not account for resource limitations or predation.)

Many Familiar Functions Are Realized by Recursion

Many of our most familiar arithmetic functions, which one perhaps might not ordinarily think of as defined by recursion, nevertheless can be seen as the result of a recursive definition. These recursive definitions often express something fundamental about these functions, which can furthermore often be used fruitfully to prove things about them by induction.

Let us define the addition operation $x + y$ as recursively adding 1, taking the successor operation $\mathrm{S}x = x + 1$ as primitive, as in Dedekind arithmetic. The recursion is specified by an anchor case and a recursion rule:

$$\begin{array}{rrcl} \text{Anchor:} & x + 0 & = & x \\ \text{Recursion rule:} & x + (y + 1) & = & (x + y) + 1. \end{array}$$

This is a recursive definition of $x+y$ in terms of successor because it defines the next value $x+(y+1)$ in terms of the previous value $(x+y)$ using only the primitive successor operation of adding 1. In short, to add y is repeatedly to take the successor y times.

Similarly, we can define multiplication as repeated addition by recursion:

$$\begin{array}{rrcl} \text{Anchor:} & x \cdot 0 & = & 0 \\ \text{Recursion rule:} & x \cdot (y + 1) & = & (x \cdot y) + x. \end{array}$$

Thus we have defined multiplication $x \cdot y$ recursively in terms of addition.

Exponentiation is similarly defined by recursion as follows:

$$\begin{aligned} \text{Anchor:} \quad x^0 &= 1 \\ \text{Recursion rule:} \quad x^{y+1} &= x^y \cdot x. \end{aligned}$$

Can you define tetration by recursion? And pentration? Why not also define the Knuth up-arrows $x \uparrow^n y$, as well as $x \Uparrow y$, $x ⤊ y$, $x ⟰ y$, and so on?

Why Are Definitions by Recursion Legitimate?

Why does recursion work? Perhaps it seems circular since we are defining the function in terms of itself. The essence of any recursion, after all, is to specify the value of a function in terms of earlier values of the same function. One can define a function f on the natural numbers $\mathbb{N}$ by defining every value $f(n)$ in terms of the previous values $f(i)$ for $i < n$. Equivalently, we specify the anchor value $f(0)$ and how the successor values $f(n+1)$ are determined from the sequence of prior values $\langle f(i) \mid i \leq n \rangle$.

Nevertheless, definition by recursion on the natural numbers is not circular. The *recursion theorem*, proved by Richard Dedekind, asserts indeed that every such definition by recursion on the natural numbers is legitimate—the recursion has a unique solution. Dedekind's recursion theorem can be proved by using the method of mathematical induction to show that for every n, there is a unique solution to the recursion that works up to n. If this is true below n, then we simply apply the recursive rule to that partial solution to discover the value at n, which shows existence at n, and since all solutions agree below n, they will also agree at n, which shows uniqueness. So the overall solution of the recursion is the function consisting of these coherent partial solutions of it. So every recursive definition on the natural numbers defines a unique function.

A Failure of Recursion

Does recursion work on other domains, such as the real numbers? Suppose that I would like to define a function f recursively on the nonnegative real numbers $\mathbb{R}^{\geq 0} = [0, \infty)$ by specifying its value at every point in terms of the values taken at earlier points. For example, let us take $f(0) = 0$ as the anchor and then specify the recursion rule

$$f(x) = \sup_{y<x} f(y)$$

for every $x > 0$. Does this recursion define a function on $\mathbb{R}^{\geq 0}$?

Well, we might notice that the constant 0 function $f(x) = 0$ obeys the recursive definition that we have provided, since it has the correct value of 0 at the anchor, and it also obeys the recursion since the supremum of a bunch of 0s is 0. So the recursion has a solution.

The problem, however, is that there are also many other solutions of this recursion. For example, the identity function $f(x) = x$ obeys $f(0) = 0$ and $f(x) = sup_{y<x} f(y)$ since every real number is the supremum of the real numbers below it. And the function $f(x) = x^2$

also obeys the recursion. In fact, any increasing continuous function through the origin will solve this recursion. So it doesn't actually succeed as a definition of a function.

Meanwhile, other recursive rules on the real numbers have no solution. Consider the anchor $f(0) = 0$ with the recursion rule $f(x) = \sup_{y<x} f(y) + 1$ for $x > 0$. That is, every value $f(x)$ of the function for $x > 0$ is supposed to be the supremum of the prior values plus 1—the supremum of $f(y) + 1$ for $y < x$. But there is no function solving this rule, I claim, since whenever $y < x$, it must be that $f(y) + 1 \leq f(x)$, which would require a jump discontinuity at every point. But this is impossible since for any $x > 0$, we can fit a tiny chain below it $0 < y_1 < y_2 < \cdots < y_n < x$, and the necessary jumps of 1 at each point y_i would push the value of $f(x)$ bigger than n for every n. So there is no solution to this recursion.

These examples show that a naive approach to recursion on the real numbers simply doesn't work. The recursion rule may have multiple solutions or no solution. The main issue is that the order on the real numbers is not a well-order.

Recursion on the Real Numbers

Nevertheless, let me briefly explain that there is a way to make recursion work on the real numbers using the ordinary real order, even though this is not a well-order. The key to making it work is to require that the recursion rule should specify a continuation of the function at every point, not merely a single value—even a very tiny nonzero continuation will be fine, so long as these various continuations do not contradict each other.

Suppose that we have specified a recursion rule, which tells us at any point x and any function $f \upharpoonright [0, x)$ defined below x how to continue it to x and a little beyond, to a function $f \upharpoonright [0, x+\delta)$ for some $\delta > 0$. Suppose furthermore that these continuations cohere with one another, in that if the rule were applied to some x' between x and $x + \delta$, that the resulting continuation at x' would agree with the continuation defined at x up to δ, at least as far as both are defined. In other words, the continuation that we have specified should be coherent with the continuations that would be constructed at every point on that continuation.

The conclusion, I claim, is that every such recursive coherent system of continuations has a unique solution fulfilling it. To prove this, let me argue that for each real $x \geq 0$, there is a unique function defined on $[0, x]$ that obeys the continuation system for all values in that interval. This is certainly true for $x = 0$. And if it is true below x, then it is true at x and up to some $x + \delta$, simply by applying the continuation that is available at x. Thus there can be no least upper bound to the values of x where the uniqueness property holds, so there can be no such upper bounds at all. So there is a unique global solution.

15.2 Transfinite Recursion

Shall we climb into the higher-infinite heavens? Let us generalize recursion beyond the natural numbers by continuing our recursions into the transfinite realm of infinite ordinals. Because the ordinals are well ordered, every recursive definition on the ordinals defines a unique function—the recursion theorem of Dedekind generalizes from natural numbers to the ordinals. Every recursion rule specifying the value of a function $f(\alpha)$ at an ordinal α by referring to the earlier values $f(\beta)$ for $\beta < \alpha$ will have a unique solution. This method of definition—abstract, powerful, elegant—is called *transfinite recursion.*

Transfinite recursion is often conveniently undertaken by separating the recursive rule into three cases—namely, the anchor value of the function at 0, the value of the function at successor ordinals $\alpha + 1$, and the value of the function at limit ordinals λ.

Ordinal Arithmetic

Ordinal arithmetic can be easily defined by recursion. For addition, we specify the following:

$$\begin{array}{rrcll} \text{Anchor:} & \alpha + 0 & = & \alpha & \\ \text{Successors:} & \alpha + (\beta + 1) & = & (\alpha + \beta) + 1 & \\ \text{Limits:} & \alpha + \lambda & = & \sup\limits_{\beta<\lambda} \alpha + \beta, & \text{for limit ordinals } \lambda. \end{array}$$

One can then prove by induction on β that the ordinal $\alpha + \beta$ has the order type of a copy of α followed by a copy of β, just as we had defined it in chapter 14:

$\alpha + \beta \;=$ 

Multiplication can similarly be defined by recursion:

$$\begin{array}{rrcll} \text{Anchor:} & \alpha \cdot 0 & = & 0 & \\ \text{Successors:} & \alpha \cdot (\beta + 1) & = & (\alpha \cdot \beta) + \alpha & \\ \text{Limits:} & \alpha \cdot \lambda & = & \sup\limits_{\beta<\lambda} \alpha \cdot \beta, & \text{for limit ordinals } \lambda. \end{array}$$

One can then prove by induction that this agrees with the "β many copies of α" definition made earlier:

$\alpha \cdot \beta \;=$ 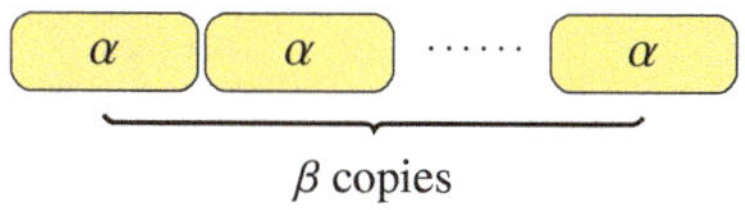

Ordinal exponentiation is similarly defined by recursion:

$$\begin{array}{rrcll} \text{Anchor:} & \alpha^0 & = & 1 & \\ \text{Successors:} & \alpha^{\beta+1} & = & (\alpha^\beta) \cdot \alpha & \\ \text{Limits:} & \alpha^\lambda & = & \sup\limits_{\beta<\lambda} \alpha^\beta, & \text{for limit ordinals } \lambda. \end{array}$$

The Alephs and the Beths

Let us now develop a little more of Georg Cantor's theory of the transfinite. We have already seen that his argument shows that there are infinitely many different sizes of infinity, and indeed there are more different sizes of infinity than any given infinity. Let us give some names and notation to these various infinities.

Cantor defined the sequence of *aleph* numbers, representing all possible infinite well ordered cardinalities:

$$\aleph_0 < \aleph_1 < \aleph_2 < \cdots < \aleph_\omega < \aleph_{\omega+1} < \quad \cdots \quad < \aleph_\alpha < \quad \cdots$$

The definition proceeds by transfinite recursion, beginning with the smallest infinity, $\aleph_0 = \omega$, the countable infinity, which is the smallest ordinal equinumerous with an infinite set. At successor stages, the next aleph number is the *successor cardinal* of the previous one:

$$\aleph_{\alpha+1} = (\aleph_\alpha)^+,$$

meaning the smallest ordinal above $\aleph_\alpha$ that is not equinumerous with it. At limit stages, we simply take the supremum:

$$\aleph_\lambda = \sup_{\alpha<\lambda} \aleph_\alpha \qquad \text{for limit ordinals } \lambda.$$

The cardinal $\aleph_\omega$, for example, is the supremum of the cardinals $\aleph_n$ for finite numbers n. This cardinal may seem quite large, as it is uncountable and larger than infinitely many other uncountable cardinals; indeed, it is a *limit cardinal*, which means that there is no immediately preceding infinity.

Meanwhile, the cardinal $\aleph_\omega$ has another curious property, which may make it seem comparatively accessible. Namely, because it is the supremum of the cardinals $\aleph_n$ for $n < \omega$, it is therefore a *singular* cardinal, the supremum of a comparatively small set of comparatively small cardinals. That is, $\aleph_\omega$ is the supremum of fewer than $\aleph_\omega$ cardinals each of size less than $\aleph_\omega$. With this ladder, we thus climb easily up to $\aleph_\omega$. Notice that, in contrast, the cardinal $\aleph_0$, the countably infinite cardinality, is a *regular* cardinal—one that admits no such comparatively short ladder. That is, every finite sequence of finite numbers has a finite supremum. In this sense, the cardinal $\aleph_0$ is inaccessible from below.

The aleph number $\aleph_\alpha$ is simply the αth infinite cardinal. If the axiom of choice holds, then the aleph numbers $\aleph_\alpha$ represent all possible infinite cardinalities since every infinite set X is well orderable and hence equinumerous with some minimal ordinal, which will be the αth such ordinal for some α, so the cardinality of X will be this $\aleph_\alpha$.

Cantor also considered the alternative related sequence of cardinals, the *beth hierarchy*, which arises by iterating the power set operation at each step:

$$\beth_0 < \beth_1 < \beth_2 < \cdots < \beth_\omega < \beth_{\omega+1} < \quad \cdots \quad < \beth_\alpha < \quad \cdots$$

We start as before with the smallest infinite cardinal $\beth_0 = \aleph_0 = \omega$, and at each stage, we apply the power set operation:

$$\beth_{\alpha+1} = 2^{\beth_\alpha} = |P(\beth_\alpha)|,$$

meaning that we take the cardinality of the power set of the previous cardinality. (Warning: The $2^{\beth_\alpha}$ here is *cardinal* exponentiation, defined as the cardinality of the power set, which makes a clash of notation, since this is much larger than ordinal exponentiation.)

At limits, we take the supremum as follows:

$$\beth_\lambda = \sup_{\alpha<\lambda} \beth_\alpha \qquad \text{for limit ordinals } \lambda.$$

The continuum hypothesis, a principle that we shall explore more fully in chapter 17, is the assertion that $\aleph_1$ and $\beth_1$ are the same infinite cardinal, expressing that our principal method of producing uncountable cardinalities—taking the power set of a countably infinite set—also happens to produce the very next uncountable cardinality. Continuing the pattern further, the *generalized continuum hypothesis* is the assertion that the two hierarchies are identical all the way up, so $\aleph_\alpha = \beth_\alpha$ for every ordinal α.

The cardinal $\beth_\omega$ is quite interesting, for it is a *strong limit* cardinal, meaning that it is closed under the power set operation, in that every set of size less than $\beth_\omega$ also has its power set of size less than $\beth_\omega$; this is because if a set has size less than $\beth_\omega$, then it has size at most $\beth_n$ for some natural number n, in which case its power set has size at most $\beth_{n+1}$, which remains smaller than $\beth_\omega$. In this sense, the cardinal $\beth_\omega$ is inaccessible by means of power sets. Yet, like $\aleph_\omega$, it is singular; it admits a short ladder, a short cofinal sequence, for it is the supremum of the $\beth_n$ for $n < \omega$, making it the supremum of a small set of small cardinals.

The Cumulative Hierarchy

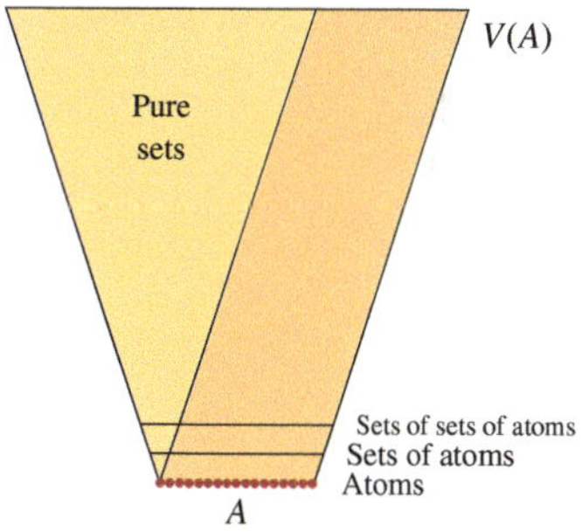

For another set-theoretic recursion, let us construct the *cumulative hierarchy*, providing a vision of how the set-theoretic universe is formed in a stratified hierarchy of levels. We begin at the bottom layer with the collection A of *urelements*, the atomic objects out of which we will form the set:

$$V_0(A) = A.$$

Each subsequent layer is then simply the set of all subsets of the previous layer. In other words, we take the power set of the previous layer:

$$V_{\alpha+1}(A) = P(V_\alpha(A)) = \{ X \mid X \subseteq V_\alpha \}.$$

At limit ordinals λ, we accumulate everything we have seen so far:

$$V_\lambda(A) = \bigcup_{\alpha<\alpha} V_\alpha(A), \qquad \text{for limit ordinals } \lambda.$$

Thus the entire set-theoretic universe $V(A)$ emerges as a hierarchy of sets stratified by ordinal levels, known as the *cumulative hierarchy*.

Inaccessible Cardinals and the Large Cardinal Hierarchy

Could there be a truly *inaccessible* cardinal? This would be an uncountable cardinal κ that was (1) inaccessible by power sets, in that every set of size less than κ also had its power set of size less than κ; but also (2) a regular cardinal, one admitting no short cofinal sequence, no sequence of fewer than κ many cardinals, each of size less than κ, whose supremum was κ. The smallest infinity $\aleph_0$, although countable, has both of those latter properties since the power set of a finite set is finite and the supremum of a finite sequence of finite cardinals is also finite. Thus, to ask for an inaccessible cardinal is to ask for a higher infinity, an uncountable cardinal that looks to the smaller cardinals, including $\aleph_0$, just as $\aleph_0$ looks to the finite cardinals. Can it happen?

The answer is complex and fascinating. The hypothesis that there is an inaccessible cardinal is one of the early *large cardinal* axioms, a strong axiom of infinity, hypothesizing the existence of an infinite cardinality with certain precise largeness features. Here is a list of some of the large cardinal notions studied by experts, listed in order of strength from bottom to top:

Superhuge cardinal
Huge cardinal
Almost huge cardinal
Supercompact cardinal
Strongly compact cardinal
Superstrong cardinal
Strong cardinal
Tall cardinal
Measurable cardinal
Ramsey cardinal
Ineffable cardinal
Ethereal cardinal
Unfoldable cardinal
Totally indescribable cardinal
Indescribable cardinal
Weakly compact cardinal
Mahlo cardinal
Hyperinaccessible cardinal
Inaccessible cardinal
Worldly cardinal

Each large cardinal notion is defined by a specific combinatorial property in infinite combinatorics, which is sometimes quite technical. For example, a cardinal κ is *measurable* if and only if there is a κ-complete nonprincipal ultrafilter on κ, and this turns out to be equivalent to the existence of a certain truth-preserving mapping $j : V \to M$ of the set-theoretic universe V into a transitive class M, which fixes every set below κ but has $\kappa < j(\kappa)$.

The large cardinal axioms, asserting the existence of these various kinds of infinity, are the strongest known axioms in mathematics. Just as the inaccessible cardinals are defined by generalizing properties that hold of the countable infinity $\aleph_0$ to higher infinities, many of the other large cardinal notions arise by positing an uncountable infinity with other features exhibited by $\aleph_0$.

One of the central lessons is that these large cardinals are so enormous that the realm of objects that exist below them is closed under all the ordinary mathematical structure-building operations. It follows from this that it is consistent with the usual axioms of set theory that the large cardinals do not exist. Therefore we cannot prove on the basis of those axioms that the large cardinals exist.

But is that disqualifying for a mathematical theory? Would you find it reasonable to study a mathematical object or structure that we cannot in principle prove to exist?

That is a profound question, but an important part of the answer is that we know from Gödel's incompleteness theorem that mathematics is necessarily replete with true but unprovable statements, reaching ever higher in the hierarchy of consistency strength. This was proved by Gödel first for strange self-referential statements of logic and metamathematics. How wonderful and surprising to find instantiations of the strong statements amongst natural statements of infinite combinatorics, the large cardinal hypotheses, with their own intrinsic interest and fascination. For this reason, set theorists have found them worthy of much further investigation.

15.3 Transfinite Recursive Constructions

Climbing back down to Earth, I should like to illustrate the power and conceptual simplicity of transfinite recursion by using it to realize various surprising geometric arrangements with remarkable properties.

Space Is the Union of Disjoint Circles

I claim that there is a decomposition of three-dimensional space $\mathbb{R}^3$ into circles—this would be a tightly packed tangle of circles oriented in space that completely fills the space, without missing any points and without any two circles ever touching. Every point will find itself on exactly one of the circles, which are each perfectly round and planar, although they may be situated at some odd orientation in space. The circles will be intermingling with each other in a complex, densely packed assemblage that is perhaps difficult to imagine in its entirety.

We shall build the circle arrangement via a transfinite recursive process. First, we enumerate the points of space in a transfinite list of length continuum:

$$p_0 \quad p_1 \quad p_2 \quad p_3 \quad \cdots \quad p_\alpha \quad \cdots \qquad (\alpha < \mathfrak{c}).$$

So we have the αth point p_α for every ordinal α below the continuum $\mathfrak{c}$. The existence of such an enumeration is a consequence of the axiom of the choice and the well-order principle, which we shall consider further in chapter 21. Any minimal-length well-ordering of the real numbers provides such an enumeration. Our construction will proceed for $\mathfrak{c}$ many steps. At each stage $\alpha < \mathfrak{c}$, we shall ensure that the αth point p_α lies on some circle in the assemblage, adding a new circle c_α specifically for this purpose if p_α has not yet been handled. In short, we shall handle the points one at a time in this transfinite progression.

Suppose that the previous circles c_β have already been chosen for $\beta < \alpha$. If p_α already occurs on one of those circles, then there is nothing to do, and we leave c_α undefined. If p_α does not yet occur on a circle, however, then we shall find a new circle c_α that contains it and is disjoint from the earlier circles. Since $\alpha < \mathfrak{c}$, there are fewer than continuum many circles c_β yet specified for $\beta < \alpha$. But there are continuum many different planes through p_α, since we can take any such plane and give it a kick to wobble it through continuum many angles. So there must be such a plane P through p_α that is not coplanar with any previously defined circle c_β for any $\beta < \alpha$. Each such prior circle c_β, therefore, punctures P at at most two points, so there are fewer than continuum many forbidden points in P.

But there is a nested arrangements of continuum many circles in P all containing p_α but otherwise disjoint, as illustrated in the figure here. Since there are fewer than continuum many forbidden points in this plane P (the red dots in the figure), there must be some circles in the nested arrangement that do not contain any forbidden point, like the blue circle shown here, which slips just between all the forbidden points. So we may choose a circle c_α in plane P containing p_α while omitting all the forbidden points in P. Circle c_α will therefore be disjoint from all the prior defined circles.

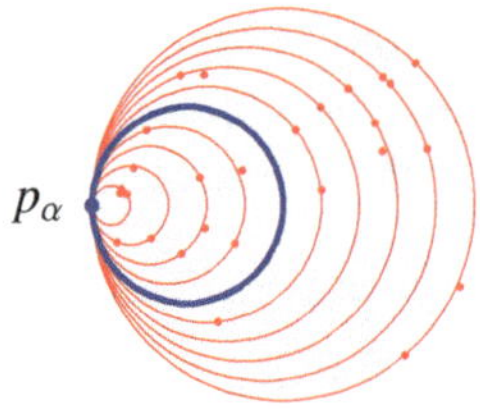

By carrying out this construction for all stages α up to the continuum, we have therefore found an arrangement of circles in space such that every point in space lies on exactly one of them. They partition the space into disjoint nonoverlapping circles.

Space Is the Union of Disjoint Circles with Different Sizes

A simple modification of the construction allows us also to ensure that no two circles in the arrangement have the same size. The reason is that at stage α, there are fewer than continuum many previous circles c_β for $\beta < \alpha$ yet chosen, so there are fewer than continuum many forbidden sizes. So we can find the circle c_α at stage α just as before by avoiding all the forbidden points, while also avoiding the forbidden sizes, since there are continuum many circles containing p_α to choose from, all with different sizes. Thus, space is the union of disjoint circles, all with different sizes.

Space Is the Union of Disjoint Circles, All the Same Size

Let us next see how a modified argument can find an arrangement using circles, all of which are exactly the same size. Let us use unit circles, with radius 1. Again we enumerate the points p_α of space in order type continuum. At stage α, circles c_β for many $\beta < \alpha$ have already been defined, but there are fewer than continuum many of them. So we can find a plane P through p_α that is not coplanar with any of them, just as before. Each prior circle therefore intersects this plane P in at most two points, so P has fewer than continuum many forbidden points.

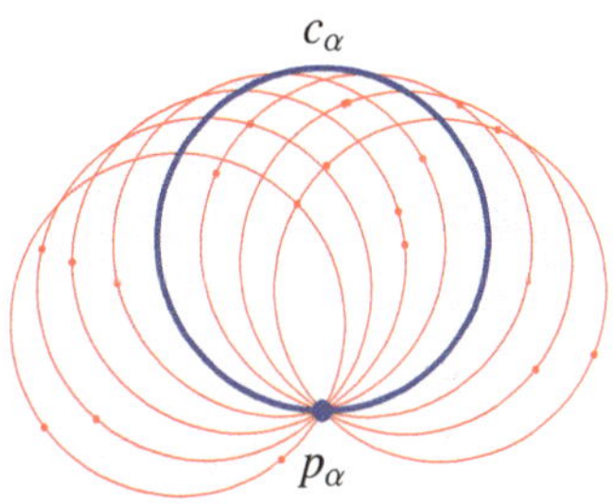

Consider all possible unit circles in plane P that contain the point p_α. Each forbidden point rules out at most two circles. Since we have continuum many circles, some of them must not contain any forbidden point. So there is a unit circle in plane P containing p_α and omitting all the forbidden points, slipping just past them. Thus we have found a unit circle c_α, disjoint from all previous circles c_β and containing the point p_α. By running the construction through all continuum many stages, we thereby construct the desired arrangement of unit circles. So space is the disjoint union of unit circles.

Space Is the Disjoint Union of Triangles, Squares, Parabolas, and Other Curves

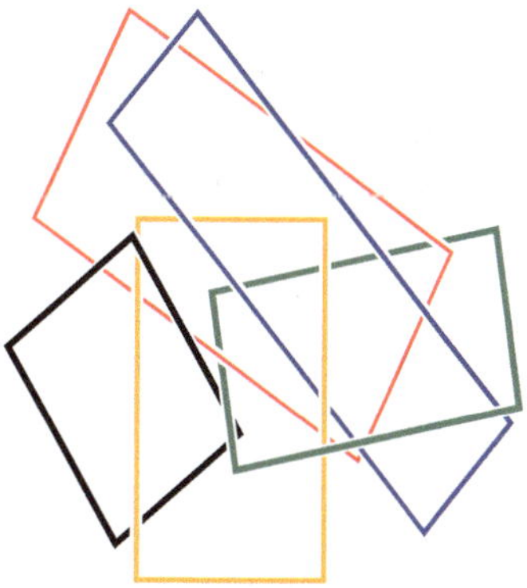

The previous constructions work well with many other planar figures and algebraic curves, including ellipses, triangles, squares, rectangles, parabolas, hyperbolas, and so on. All that is needed is that one can find a plane through the target point p_α that is not coplanar with any of the previous curves, that each previous curve intersects that plane in finitely many points (countably many would also be fine), and that there is a foliation of the curves in that plane with continuum many candidates, as we had with the circles, such that each forbidden point rules out at most countably many of the candidates. In this way, at stage α, we can find a curve in the plane that contains p_α and avoids all the forbidden points, so it will be disjoint from the previously-chosen curves. For many kinds of curves, one can even omit the planarity requirement.

Thus, space admits decompositions into disjoint ellipses, all with the same fixed eccentricity and size, or all with different sizes and eccentricities, or parabolas, triangles, squares, or what have you. The construction is truly flexible.

Space Is the Union of Skew Lines

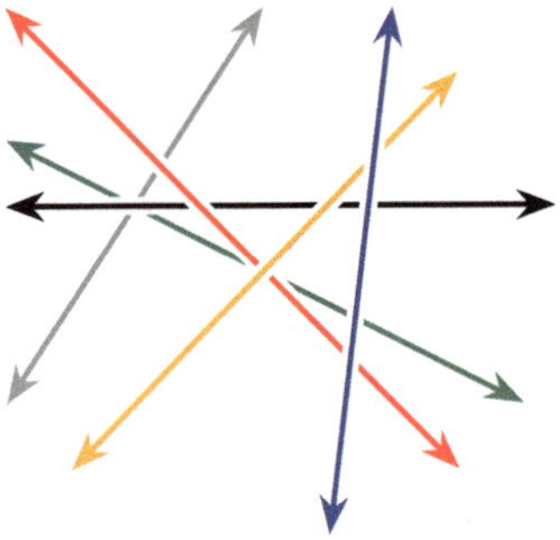

A similar construction method can realize space as an infinite tangle of pick-up sticks, a union of disjoint skew lines, meaning that no two of them are coplanar. (In particular, no two of the lines intersect.) Enumerate the points of space again in order type continuum, with a point p_α for every ordinal up to the continuum $\alpha < \mathfrak{c}$, such that every point in space is some p_α. At each stage α, we have specified fewer than continuum many lines ℓ_β at the previous stages $\beta < \alpha$.

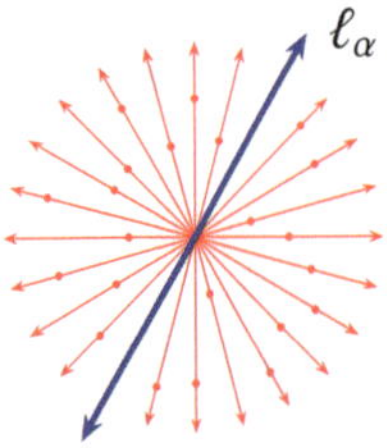

If the point p_α already appears on one of the prior chosen lines ℓ_β, then we have already handled that point, and there is nothing to do. If instead p_α has not yet appeared on any of the previous lines, then observe that there are fewer than continuum many planes determined by a previous line ℓ_β and the point p_α. So we may find a plane P containing p_α and not containing any prior defined line ℓ_β. Each previous line punctures this plane P at at most one point, so there are fewer than continuum many forbidden puncture points in P. But there are continuum many angles proceeding outward from p_α, so there must be a line in P containing p_α and missing all the forbidden points in P. Choose one as ℓ_α, which is therefore a line through p_α that is skew to all previous lines. By continuing the construction through all continuum many stages, we thereby have our arrangement of lines, providing a partition of space into skew lines.

A Collection of Points, with Exactly Two on Every Line

Let us next construct a collection of points in space $\mathbb{R}^3$ such that every line ℓ contains exactly two of them. In particular, no three of the points will be collinear. To construct this arrangement, enumerate the collection of all lines ℓ_α in space $\mathbb{R}^3$ using ordinals α up to the continuum $\mathfrak{c}$, and we shall proceed to specify the points by transfinite recursion.

At stage α, we consider line ℓ_α. Since fewer than continuum many points have been specified thus far, we may consider all possible lines through any two previously specified points. There are fewer than continuum many such pairs of previous points, and thus fewer than continuum many such lines are specified in this way. And thus, since these lines intersect the line ℓ_α in fewer than continuum many points, there are fewer than continuum many points on ℓ_α that would be forbidden by being part of a collinear triple. Since the line otherwise has continuum many points, we may make sure to include two points from ℓ_α in our collection without violating the collinear-triple requirement.

Thus, by carrying out this construction for continuum many steps, we have succeeded in finding a collection of continuum many points in space, such that every line contains exactly two of them. Variations of the construction produce collections of points in space such that every line contains exactly three of them, or exactly four, or exactly countably infinitely many, and so on.

A Collection of Points, Such That Every Circle Has Exactly Three of Them

A variation of the previous construction shows that there is a collection of points in space $\mathbb{R}^3$ such that every circle in space contains exactly three of them. We enumerate the circles c_α in space, for every ordinal α up to the continuum $\mathfrak{c}$. At stage α, we have chosen fewer than continuum many points. Let us assume inductively that no four of the previous points lie on any one circle. Any three noncollinear points determine a unique circle, so there are fewer than continuum many circles determined by the prior points. If the current circle c_α contains three points that are already chosen, then there is nothing more to do. But if it doesn't, then since any two distinct circles intersect in at most two points, there are fewer than continuum many points on the circle c_α that would cause a problem with the at-most-three-on-any-circle inductive requirement. So we may choose points from c_α so as to have three altogether, in such a way that still no four points from our collection lie on a circle. By continuing this construction through all stages up to the continuum, we will have ensured that every circle contains exactly three points from our collection.

15.4 Explicit Constructions Without Choice

For a few of the geometric arrangements we considered previously, but not all of them, it turns out that there are alternative explicit constructions, which achieve the desired arrangement while making no use of transfinite recursion or the axiom of choice.

For example, one can realize space $\mathbb{R}^3$ as a collection of skew lines by taking a single vertical line, surrounded by an infinite nested foliation of hyperboloids, whose angle is slightly changing as the hyperboloids get farther from the origin. Every hyperboloid is a union of straight lines, no two of which are coplanar, and lines from different hyperboloids are also skew, because of the changing angle. So space is a union of skew lines in this entirely constructive manner.

It is also possible to realize three-dimensional space $\mathbb{R}^3$ as a disjoint union of circles by an explicit construction appearing in Jonsson and Wästlund (1998). The construction is concrete, but it is not entirely simple, so let me omit the details.

How shall we compare the value of the two construction methods? To my way of thinking, the transfinite constructions are conceptually clear and simple. Because they rely on the axiom of choice to make a transfinite sequence of arbitrary choices, however, one is left ultimately unsure of exactly what the partition is. Are we to view the axiom of choice constructions as proofs of existence? They seem to provide many different instances of the desired partition, depending on how the choices are made as the construction proceeds, since they show how to undertake the construction for any succession of choices. And yet a more radical axiom-of-choice skeptic might deny that there is after all a coherent manner of making the required choices, and this perspective would prevent the construction from being seen even as a proof of existence.

The concrete constructions, in contrast, although often finicky and detailed, nevertheless provide completely explicit constructions. One seems to know in the end exactly which circles (or lines or other kinds of objects) are part of it.

For most of the cases, meanwhile, the fact of the matter is that we simply have no explicit constructions, and it is an open question whether there are any. So how are we to think about whether indeed there are decompositions of the desired kind?

Questions for Further Thought

15.1 Formulate a version of recursive definition that works on the integers, including both positive and negative integers. That is, we want to define a function f on the integers by specifying the value $f(0)$, and then ... what exactly? Can you prove that such definitions succeed in defining a unique function?

15.2 Show that there is a collection of points in the plane $\mathbb{R}^2$ such that every line contains exactly two of them.

15.3 Show that there is a partition of space $\mathbb{R}^3$ into disjoint unit closed line segments, each with a different direction.

15.4 Discuss the comparative value of the recursive constructions using the axiom of choice and the explicit constructions that achieve the same result. Do the explicit constructions carry extra value purely in virtue of being explicit? Does the difficulty of the constructions affect how we view their comparative value?

15.5 Does transfinite recursion count as a concrete construction method? Suppose that we undertake a recursion for which each step is definite and definable, without any use of the axiom of choice, say, but it proceeds transfinitely. Is the result of the recursion concrete, or is it inherently abstract? Do you give the same analysis for recursion on the natural numbers?

Further Reading

Jonsson and Wästlund (1998). A thorough but technical account of the method of finding partitions of $\mathbb{R}^3$ into curves.

Credits

The material on the alephs and the beths was adapted from Hamkins (2021b, p. 109). I asked the question of whether we can exhibit or refute explicit constructions for the various geometric arrangements on MathOverflow at Hamkins (2012).

16 Slaying the Hydra

Abstract. We shall fight the fearsome Hydra now, and in defeating it, we shall see how the existence of transfinite numbers can have surprising consequences within the finite realm.

16.1 Goodstein Sequences

Let us undertake a certain fantastical number process. Begin with any positive number, such as the number 41, and write it in *hereditary* base 2, which means to write it in base 2 as a sum of powers of 2, but also to write the exponents and iterated exponents in this way. We denote this initial number by a_2, in part to indicate this initial choice of base:

$$a_2 \quad = \quad 41 \quad = \quad 32+8+1 \quad = \quad 2^5+2^3+1 \quad = \quad 2^{2^2+1}+2^{2+1}+1.$$

We form the next number a_3 simply by replacing all the 2s in this representation with 3s and subtracting 1, writing the result in hereditary base 3:

$$a_3 \quad = \quad 3^{3^3+1}+3^{3+1}+1-1 \quad = \quad 3^{3^3+1}+3^{3+1}.$$

This number a_3 is 22876792455042 in decimal, or about $2 \cdot 10^{13}$, which is much larger than our original number 41. The succeeding number a_4 is formed similarly by replacing all 3s with 4s and subtracting 1:

$$\begin{aligned} a_4 &= 4^{4^4+1}+4^{4+1}-1 \\ &= 4^{4^4+1}+3\cdot 4^4+3\cdot 4^3+3\cdot 4^2+3\cdot 4+3. \end{aligned}$$

In decimal notation, a_4 is the following number:

5363123171977038839829609999282338450991746328236957351089424577488705612029418790720749719266761371076012743274594420341501553124778627978573459602433740 7

which is about $5 \cdot 10^{155}$, a big step up from a_3. We form a_5 similarly by replacing 4s with 5s in the hereditary base 4 representation and subtracting 1, and so on. Starting from any given initial value a_2, we generate in this way a sequence:

$$a_2 \quad a_3 \quad a_4 \quad a_5 \quad \cdots$$

This is known as a *Goodstein sequence* in light of the remarkable theorem about them proved in 1944 by Reuben Goodstein (1912–1985). The amazing fact is that for any natural number starting value a_2, if we form the corresponding Goodstein sequence according to the process that we have described, then at some point n, the value will become zero: $a_n = 0$.

What?! It had seemed that the sequence would always grow larger, with each next value a_{n+1} much bigger than the previous value a_n. How can we reach $a_n = 0$? Won't a_{n+1} always be larger than a_n? Actually, no. According to Goodstein's theorem, the initial impression that the numbers grow ever larger is simply mistaken. Eventually, the numbers start getting smaller, and they will eventually reach 0.

The only way that the numbers could get smaller, of course, is because of the subtracting 1 part, so what must eventually happen is that the replacing n with $n+1$ part of the process somehow doesn't occur. This would happen if the process somehow catches up with the base since then the number would be smaller than the base, and there were no n to replace with $n + 1$. If that were to happen, then from that point on, we would just keep subtracting 1 until we hit zero. But does the process catch up with the base? It is difficult to imagine how this would happen.

But it does happen. To prove this, Goodstein's argument uses a remarkable application of the infinite within the finite realm. Let us see how his argument works. What we shall do is use the structure and nature of the transfinite ordinals up to ε_0 to prove that the Goodstein sequence must eventually hit zero.

The main idea of the proof is to use the ordinals to measure the complexity of the numbers that arise and to prove that this complexity strictly descends with each step. At each stage, we shall associate an ordinal with the number a_n using the hereditary base n representation. We simply replace every occurrence of n in the expression with the ordinal ω and interpret the result using ordinal arithmetic. For example, the ordinals associated with the previous Goodstein sequence begin like this:

$$\begin{aligned} a_2 &= 2^{2^2+1} + 2^{2+1} + 1 &&\mapsto \omega^{\omega^\omega+1} + \omega^{\omega+1} + 1 \\ a_3 &= 3^{3^3+1} + 3^{3+1} &&\mapsto \omega^{\omega^\omega+1} + \omega^{\omega+1}. \end{aligned}$$

At the next step, comparing with the hereditary base 4 representation of a_4 given previously, we accommodate the noncommutative nature of ordinal multiplication by placing the coefficients on the right, leading to

$$a_4 \mapsto \omega^{\omega^\omega+1} + \omega^\omega \cdot 3 + \omega^3 \cdot 3 + \omega^2 \cdot 3 + \omega \cdot 3 + 3.$$

The key observation about these associated ordinals is that they are getting smaller. The reason why they get smaller is that the associated ordinals are not affected by the operation of replacing n with $n + 1$, since both of these numbers will turn into ω, but they do get smaller during the subtracting 1 part of the process.

It is a fundamental feature of the ordinals, however, that there is no infinite descending sequence. You cannot go down forever in the ordinals because they are *well ordered*, which means that every nonempty set of ordinals has a least element. As a consequence, since the ordinals arising in the Goodstein sequence cannot go down forever, they must eventually hit zero, and the only way that this happens is if the number a_n itself is zero.

One can see how the phenomenon occurs by considering in detail the move from a_3 to a_4. Although the number did get much larger in that case, nevertheless, the new number was simpler by the ordinal measure. Specifically, to accommodate the minus 1, we had to break the term 4^{4+1} into a sequence of multiples of 4^4 and lower, rather than 4^{4+1}. Thus, while a_3 had an ordinal term $\omega^{\omega+1}$, the associated ordinal for a_4 had merely $\omega^\omega \cdot 3 + \omega^3 \cdot 3 + \omega^2 \cdot 3 + \omega \cdot 3 + 3$, which is a much smaller ordinal. It is precisely when the subtracting 1 operation requires one to break up complex terms and use numbers less than the base (such as 3s instead of 4s in the representation of a_4) that the associated ordinal makes a big drop. Eventually, the proof shows that the complexity of the expression is broken up again and again until it becomes low enough that the base exceeds the number. From this point on, one is just subtracting 1 until 0 is reached.

Thus, we have used our understanding of the infinite to prove a consequence about Goodstein sequences, a purely finite process. It is truly remarkable.

Representation of Ordinals up to Epsilon Naught

The hereditary base n notation shows us in fact how to describe these various infinities in a purely finitary syntactic manner, a concrete notation system for all the ordinals up to ε_0. Namely, every ordinal up to that limit is represented by some finite number m written in some hereditary base n, but where we think of the instances of n in that representation of m as a stand-in for ω, in just the same way that we used in the proof of Goodstein's theorem.

Take a typical ordinal below ε_0, such as the ordinal

$$\omega^{\omega^{\omega^3+2}\cdot 5} + \omega^{17}\cdot 7 + \omega + 3.$$

This ordinal has the same coeffient and base structure as the number

$$25^{25^{25^3+2}\cdot 5} + 25^{17}\cdot 7 + 25 + 3.$$

Because of this, we can think of this number, together with the specification of base 25, as providing a representation of that infinite ordinal—it is a finitary, essentially syntactic way of speaking about and dealing with the infinite ordinal. And indeed, every ordinal below ε_0 can be represented in this manner—the ordinals below ε_0 consist of various finite height towers of iterated exponential expressions of ω, augmented with various finite coefficients as further exponents or multiples.

Furthermore, all the basic operations of ordinal arithmetic—that is, ordinal addition, multiplication, and exponentiation—can be performed easily on these representations of the ordinal with natural numbers. In this sense, the finite numbers can serve as stand-ins for the infinite numbers that they are representing for all the purposes that we might want of those infinite numbers. We have a finitary computable denotation system for this realm of infinite ordinals.

16.2 The Hydra Game

The fearsome multiheaded Hydra stands before us—can we defeat it? If we should strike at it, perhaps cutting off a head, more heads will grow according to the rules of the *Hydra game*, introduced by Laurie Kirby and Jeff Paris (1982). The Hydra will not merely grow new heads, but it will duplicate entire parts of itself according to the rule.

Specifically, in the Hydra game, the Hydra has a branching tree structure, as pictured next. You may cut off any head where it is attached to the node below, but upon doing so, the Hydra will duplicate itself with n additional copies (on turn n) of the entire remaining part of the Hydra that contains the neck section proceeding from the node one level lower up to the cutting node. That is, if you cut a head, you look at the section of neck one level below it, and whatever remains of the Hydra growing up from this neck section (and perhaps beyond) is duplicated n times. If the head was already attached directly at the bottom, then there is no such lower neck section, and no duplication occurs.

To illustrate, here are a few moves in a sample instance:

We start with the three-headed Hydra at left. On the first turn, the head at the upper left is cut, so the duplication occurs one level down, creating a new copy of that remaining length-two neck and head combination, shown in orange in the second Hydra. On the second turn, a head at the top is cut, but what remains one level down is just a single neck, so this is duplicated with two additional copies, shown in orange in the resulting third Hydra. On the third turn, a head is cut, but what remains of the Hydra through the neck one level down has a big assemblage above it, so three additional copies of that entire assemblage are created. At any point, the head connected directly to the bottom could be cut without duplicating anything. Can you win? Is there a strategy enabling us to cut off every head?

Think about it...

The amazing fact is that *every* strategy is a winning strategy! No matter how you proceed to cut off the heads, eventually, all of them will be cut off.

What? It seems incredible that this could be true. After all, the Hydra seems to be growing increasingly out of control as one proceeds. How shall we be able to cut off all the heads at all? But indeed, not only can we cut off all the heads, but it doesn't even matter how we proceed to cut—the theorem is that it is inevitable that we shall eventually win.

One can begin to see how this might be true by noticing that although the Hydra does initially get larger, nevertheless the complexity of branching in the Hydra is slightly simplifying in various ways. For example, the first Hydra here has a branching node at level 3, but after the first cut, this will never arise again; similarly, the third Hydra has a level-four head growing out of a degree four node, but no subsequent Hydra will have such a node.

We shall prove the theorem by associating each Hydra with a countable ordinal below ε_0, as a measure of its complexity, much as we did with the Goodstein sequences. For every Hydra that arises, we assign an ordinal value to each of its nodes according to the following procedure. The heads, which are terminal nodes in the tree, are assigned a value of 0. The main iterative idea is that if a node has value α, then this will contribute value ω^α to the node just below it. Specifically, if the successors of a node have value $\alpha_n \geq \cdots \geq \alpha_0$, placed in decreasing order like this, rather than the order in which they may appear in the Hydra presentation, then the node itself is assigned value $\omega^{\alpha_n} + \cdots + \omega^{\alpha_0}$, as in the Cantor normal form.

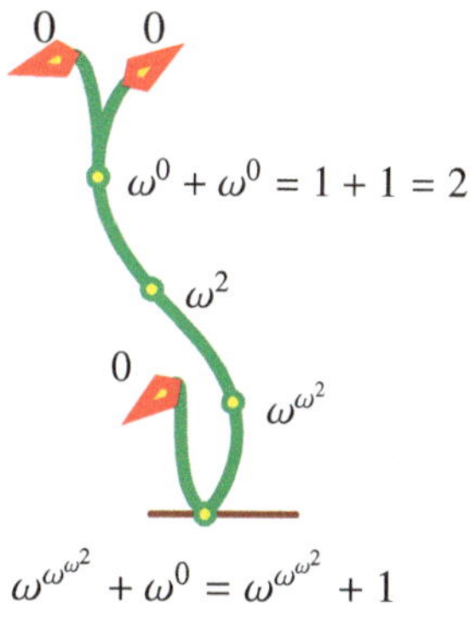

Let us calculate for illustration the ordinal value of the initial Hydra that we considered earlier. The assignments are shown worked out here. The heads all have value 0, which means that the branching node just below the two top nodes has value $\omega^0 + \omega^0$, which is the same as $1 + 1$, or value 2. Below that, therefore, is ω^2, and below that ω^{ω^2}. The extra head near the bottom has value 0, which contributes an extra $\omega^0 = 1$ to the value of the root node, but added on the right, for an overall value of $\omega^{(\omega^{\omega^2})} + 1$ for this Hydra.

Similarly, we can see how the ordinals are calculated after the various cutting moves that we made earlier. The ordinals in the first tree are just as we calculated here. In the second tree that follows, notice that the two heads at the top have value 0, so the nodes underneath them each have value ω^0, which is 1, so the node under that has value $\omega^1 + \omega^1$, which is the same as $\omega + \omega = \omega \cdot 2$. This node therefore changed its value from ω^2 in the original Hydra to $\omega \cdot 2$ after the cut and regeneration process. The node underneath thus has value $\omega^{\omega \cdot 2}$, and ultimately $\omega^{\omega^{\omega \cdot 2}} + 1$ for the whole second Hydra. In the third Hydra, the heads all get value 0, and then the lone node below the very top head gets value $\omega^0 = 1$. The branching node below that sits below nodes with values 0, 1, 0, 0, so after placing these in order, it gets value $\omega^1 + \omega^0 + \omega^0 + \omega^0$, which is $\omega + 1 + 1 + 1 = \omega + 3$. Similarly, the nodes in the fourth Hydra get value $\omega + 2$ as indicated because the nodes above them have value 1, 0, 0, so one calculates $\omega^1 + \omega^0 + \omega^0 = \omega + 1 + 1 = \omega + 2$. The node below therefore gets value $\omega^{\omega+2} \cdot 4$, since there are four nodes with value $\omega + 2$ above, leading to the final value as indicated:

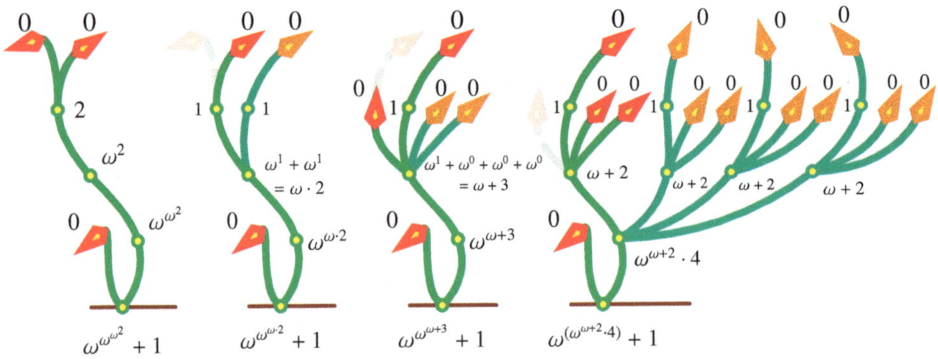

The key observation is that as we cut heads, the associated ordinal values descend. In the example here, the Hydras had values descending like this:

$$\omega^{\omega^{\omega^2}} + 1 \quad > \quad \omega^{\omega^{\omega \cdot 2}} + 1 \quad > \quad \omega^{\omega^{\omega+3}} + 1 \quad > \quad \omega^{\omega^{\omega+2} \cdot 4} + 1.$$

In fact, the ordinals will always descend after any cutting of a Hydra head. To see this, consider first that because the head itself has value 0, it contributes $\omega^0 = 1$ to the value of the node from which it grows, which must therefore have value $\alpha + 1$ for some ordinal α. When the head is cut, this node will afterward have value α, but there will now be n copies of this node (and the accompanying structure above) at the same level, joined at the node one level down. Before the cut, the contributed value to the lower node was $\omega^{\alpha+1}$, but after the cut, the combined contributions of the duplicated parts will be $\omega^\alpha \cdot n$ because there are n copies. The main point is that regardless of the value of n, this is less than $\omega^{\alpha+1} = \omega^\alpha \cdot \omega$ because $n < \omega$, and this is why the ordinals go down with each step. In short, whenever we cut a head, we replace a node contributing value $\omega^{\alpha+1}$ with n nodes each contributing ω^α, and because $\omega^\alpha \cdot n < \omega^{\alpha+1}$, the total value has gone down after the cut and regeneration. Consequently, the root node of the Hydra will have progressively lower ordinal value as the cutting proceeds, regardless of how the cuts are made. And since there is no infinite descending sequence of ordinals, it follows that eventually the ordinal value of the Hydra must become 0, which means that the Hydra was defeated. In other words, every strategy of fighting the Hydra by cutting heads will eventually win.

Thus again we used the nature and existence of infinite ordinals to prove something about a purely finite process.

16.3 Logical Independence

The basic fact of Goodstein's theorem and its Hydra variation is certainly surprising enough. But this theorem truly packs a one-two punch because not only is the theorem itself surprising, but there is a deep follow-up theorem.

Namely, Kirby and Paris proved that Goodstein's theorem is not provable in the standard first-order theory of arithmetic, the theory known as *Peano arithmetic*. The statement of Goodstein's theorem is *independent* of the theory of Peano arithmetic, which means that it is neither provable nor refutable in that theory. We had proved Goodstein's theorem in set theory, using the infinite ordinals, but it is not possible, Kirby and Paris showed, to prove Goodstein's theorem in the more elementary theory of arithmetic. Similarly, they proved that the related claim about the Hydra game also is not provable in Peano arithmetic.

What Is Peano Arithmetic?

Peano arithmetic is the standard first-order axiomatization of arithmetic, a formal theory expressible in the language of arithmetic, which allows for addition, multiplication, the order $<$, specific reference to 0 and 1 (and therefore also to any particular number as $1 + 1 + \cdots + 1$), and quantification over the numbers.

The axioms of Peano arithmetic include all the familiar fundamental arithmetic principles, such as associativity of addition and multiplication:

$$x + (y + z) = (x + y) + z \qquad x \cdot (y \cdot z) = (x \cdot y) \cdot z;$$

the commutativity of addition and multiplication:

$$x + y = y + x \qquad x \cdot y = y \cdot x;$$

the distributivity law:

$$x \cdot (y + z) = x \cdot y + x \cdot z;$$

the identity and absorption properties of 0 and 1:

$$x + 0 = x \qquad x \cdot 0 = 0 \qquad x \cdot 1 = x;$$

the transitivity and irreflexivity order axioms:

$$\neg\, x < x \qquad\qquad (x < y < z \implies x < z)$$

the trichotomy principle:

$$x < y \text{ or } x = y \text{ or } y < x;$$

and the interaction of the order with arithmetic:

$$x < y \implies x + z < y + z$$

$$(x < y \text{ and } 0 < z) \implies x \cdot z < y \cdot z$$

$$x < y \implies \exists z\, x + z = y$$

$$0 < 1 \qquad 0 \leq x \qquad \neg\exists z\, 0 < z < 1.$$

In addition, the theory of Peano arithmetic includes the first-order principle of induction. That is, for any expressible property φ, we include the axiom

$$[\varphi(0) \text{ and } \forall x\, (\varphi(x) \implies \varphi(x+1))] \implies \forall x\, \varphi(x).$$

This principle says that if φ is true at 0 and is propagated from every number x to $x + 1$, then indeed it holds for all x.

Roman Kossak describes the theory as follows: All the arithmetic facts you can think of in 10 minutes, plus the induction scheme. The point is that the theory is quite robust, and equivalently axiomatized with many different axiom sets, provided that one has remembered to include induction. In particular, the axiom list given here is highly redundant, and we can prove many of these principles by induction from very weak instances of them.

The main point to make about the first-order theory of Peano arithmetic is that it suffices for nearly the entirety of the usual development of elementary number theory. One can prove the fundamental theorem of arithmetic, the infinitude of the primes, the theory of irrational numbers such as $\sqrt{2}$, and much, much more. Essentially every purely finitary combinatorial argument in mathematics can be carried out in Peano arithmetic, including results about finite graphs, finite groups, finite geometries, and more. Peano arithmetic encapsulates in a highly robust manner the content of finitary mathematics.

In light of this, the fact that Goodstein's theorem is not provable in Peano arithmetic substantiates the view of this theorem as an instance where infinity has had an effect on the finite. We proved Goodstein's theorem not by using purely finitary reasoning, but rather infinitary reasoning with the transfinite ordinals up to ε_0, and the point is that it is not possible to prove the result in the purely finitary theory of Peano arithmetic. It is truly remarkable, even profound.

First-Order Peano Arithmetic Versus Second-Order Dedekind Arithmetic

Let us compare the theory of Peano arithmetic with the theory of Dedekind arithmetic that we discussed in chapter 11. Dedekind arithmetic was stated in the language with just successor, and it may have seemed much more primitive for that reason, except for the fact that it used the *second-order* induction axiom, stating the induction principle not merely for finitely expressible properties φ as we did for Peano arithmetic, but stating induction for arbitrary sets of numbers. This difference is enormous, and one could reasonably view Dedekind arithmetic as an infinitary theory on account of it, since it requires one to quantify over all sets of natural numbers, a truly uncountable realm.

We had mentioned in that earlier chapter that in fact, one can define addition and multiplication, and indeed all the usual arithmetic functions within Dedekind arithmetic and prove all the usual properties about them. In particular, Peano arithmetic should be seen as part of Dedekind arithmetic, rather than the reverse. Further, Dedekind arithmetic is strictly stronger—the two are definitely not equivalent—since for deep reasons connected with the model theory of first-order logic, no first-order theory with an infinite model can be categorical. In particular, the theory of Peano arithmetic must admit *nonstandard* models, other than the standard model $\langle \mathbb{N}, +, \cdot, 0, 1, < \rangle$. Indeed, it is precisely because there is such a nonstandard model in which Goodstein's theorem is false that we know the theorem is not provable in Peano arithmetic. In this way, logical independence is wrapped up with the existence of nonstandard realms in which our favored theories are otherwise true.

Further Independence in Higher Realms

In chapter 25, we shall encounter the famous incompleteness theorem of Kurt Gödel (1906–1978), which shows the necessary ubiquity of independence in arithmetic—every theory we can write down will admit true but unprovable assertions. The independence of Goodstein's theorem over Peano arithmetic is exactly an instance of this phenomenon predicted by Gödel's theorem.

Meanwhile, it is in the higher realm of set theory where the independence phenomenon has taken stronger hold. Set theory overflows with hundreds (perhaps thousands) of natural examples of fundamental assertions that are now known to be independent of the fundamental axioms of set theory. This includes prominent questions, such as the continuum hypothesis and essentially every nontrivial assertion in infinite combinatorics, as well as the axiom of choice, assertions about cardinal characteristics of the continuum, and, if consistent, the existence of any of the various large cardinals that we shall later encounter. The central discovery of twentieth-century research in set theory is a pervasive independence phenomenon in set theory.

Meanwhile, another profound set-theoretic discovery is that of a continuing logical flow of consequences from the higher infinite down to the lower realms of the mathematical universe, including the purely finitary arithmetic realm. The strongest known axioms in mathematics are the large cardinal axioms, asserting the existence of enormous infinities realizing a vast uncountable structure theory. In every case, these axioms have direct finitary consequences, even in arithmetic. For example, each of the large cardinal axioms implies that certain polynomial equations have no integer solutions and certain sets of polygonal tiles can tile the plane, but these assertions are not provable without those strong assumptions. The effects of the infinite world on the finite are continuing and robust. It is amazing.

Questions for Further Thought

16.1 Try the Goodstein sequence with very small starting values, such as 1, 2, 3, 4, or 5. Exactly how long does it take to reach zero in these cases?

16.2 Compute the Goodstein sequence with a larger initial number, for a dozen or more steps, and provide the associated ordinals, observing that indeed they descend during the process.

16.3 Draw some very small Hydra instances and determine how long it takes to cut off all the heads.

16.4 Draw a more substantial Hydra instance and carry out several steps of the Hydra game. Calculate the associated ordinals and verify that indeed the ordinals are descending.

16.5 The Kirby-Paris theorem shows that for any Hydra, all strategies for cutting off heads will eventually win. But do all strategies take the same amount of time?

16.6 Can you prove that the axiomatization of Peano arithmetic I gave is redundant? Prove one of the axioms from another on the list.

Further Reading

Houston-Edwards (2017a), (2017b). Excellent videos providing an accessible account of Goodstein's theorem with ordinal values, as well as the Kirby-Paris independence result on Goodstein's theorem.

Credits

Goodstein's theorem was proved by Reuben Goodstein (1944), and the Kirby-Paris theorem and the Hydra result appears in Kirby and Paris (1982). Some of the material in this chapter was adapted from Hamkins (2021b, p. 249).

17 The Continuum Hypothesis

Abstract. A confounding question lies at the center of set theory: Is there any infinity strictly between the infinity of the natural numbers $\mathbb{N}$ and the infinity of the real numbers $\mathbb{R}$?

The infinity of the set of real numbers $\mathbb{R}$ is uncountable—it is strictly larger than the infinity of the set of natural numbers $\mathbb{N}$:

$$\mathbb{N} \quad < \quad \mathbb{R}$$

Question. Is there any infinity strictly between them?

This innocent and very natural question, asked by Georg Cantor in the early days of set theory shortly after proving the uncountability of $\mathbb{R}$, was the origin of a sweeping mathematical research effort spanning the twentieth century. The *continuum problem*, as the question was called, became one of the most prominent open questions in mathematics, one leading furthermore to vibrant philosophical disputes concerning fundamental issues in the foundations of mathematics and the nature of mathematical truth. Let me relate the story here in this chapter.

Cantor himself had become truly obsessed with the question, although it staunchly resisted all attempts at resolution. He formulated his ideas and expectations for what the answer might be with the *continuum hypothesis*, the principle asserting that indeed there are no such intermediate infinities.

Continuum hypothesis. *There is no infinity strictly between* $\mathbb{N}$ *and* $\mathbb{R}$.

But Cantor had no proof. He struggled with the question his entire life, at times searching for counterexamples but mostly striving to establish the truth of the continuum hypothesis. He observed in many instances that particular sets conform with the continuum hypothesis—indeed, in every case in which he was able to determine the outcome, the result conformed with the continuum hypothesis—but he lacked a general proof that this would always be the case.

Cantor laid out a strategic program for proving the continuum hypothesis—he proposed to establish the hypothesis first for simple sets and then proceed to sets of higher logical complexity. He successfully carried out the beginning steps of this program, but alas, he was unable to carry it further. Regrettably, he died without learning an answer. And it was the same for his mathematical contemporaries, which included many prominent, distinguished mathematicians—none at the time could find a resolution of the continuum problem.

The eminent mathematician David Hilbert had placed the continuum problem at the very top of a list of twenty-three open mathematical problems that he had announced at the 1900 International Congress of Mathematicians, a list that went on to become extremely influential, inspiring numerous huge mathematical research programs. These were considered the most important mathematical questions of that time—several of the problems remain unsolved to this day—and Cantor's continuum problem was at the very top of the list. Thus the continuum problem was arguably the most prominent open question in mathematics.

What is your view? Can you produce a set that realizes an intermediate size of infinity between $\mathbb{N}$ and $\mathbb{R}$? Can you prove that there is no such intermediate infinity?

Think about it...

17.1 Ubiquitous Equinumerosity

Let us gain some familiarity and context for the continuum problem by exploring the cardinalities of various naturally arising sets.

Intervals in the Real Numbers

We have already seen with Galileo in chapter 10 that any two line segments can be placed in one-to-one correspondence:

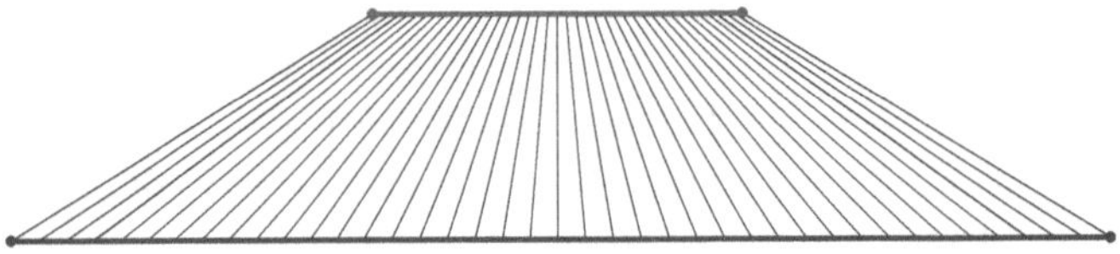

This correspondence shows that any two nontrivial intervals in the real numbers are equinumerous. Since the correspondence associates endpoints of the first interval with endpoints of the second interval, what it shows more precisely is that any two nontrivial closed intervals $[a, b]$ in the real numbers are equinumerous, as are any two open intervals (a, b) or half-open $[a, b)$, $(a, b]$.

We also saw with Galileo that any nontrivial open interval (a, b) is equinumerous with the whole real line:

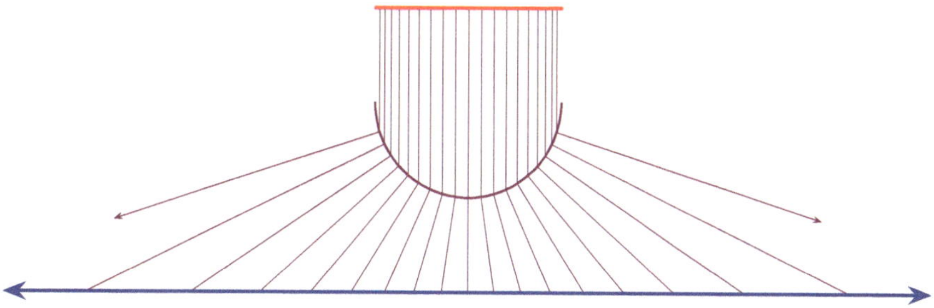

For an alternative proof of this, consider the arctangent function, climbing gracefully from the lower horizontal asymptote to the upper:

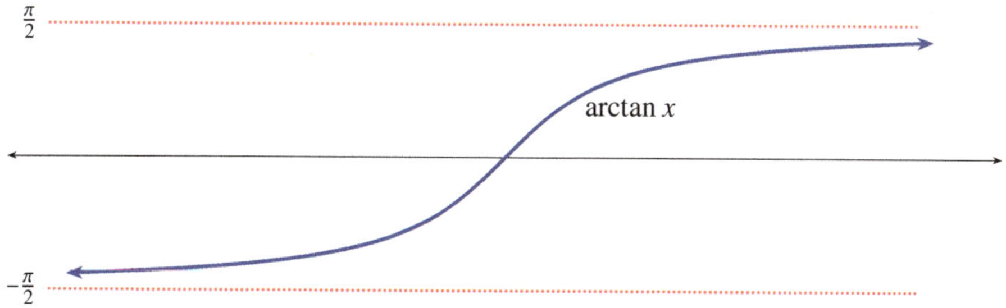

This function thus provides a one-to-one correspondence between the real line $\mathbb{R}$ and the open interval $(-\frac{\pi}{2}, \frac{\pi}{2})$, thereby showing them to be equinumerous. By scaling and shifting, $\mathbb{R}$ is thus equinumerous with any open interval (a, b).

We can similarly provide an equinumerosity between the real line $\mathbb{R}$ and the ray of positive real numbers with a graph of the function

$$y = e^x.$$

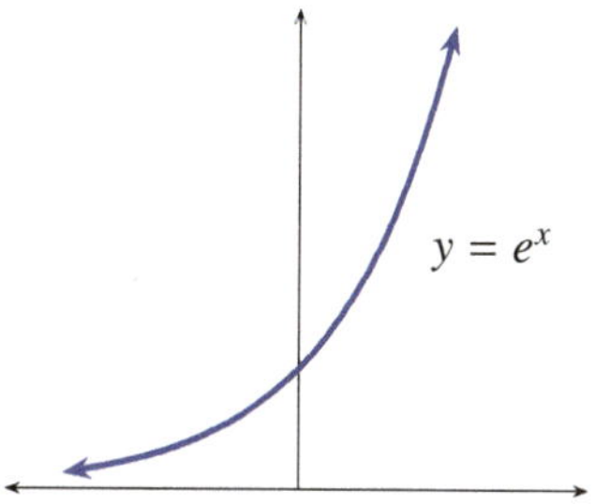

It follows from all these results and the Cantor-Schröder-Bernstein theorem (see page 137) that all nontrivial intervals in the real numbers—whether open or closed or half-open, whether bounded or unbounded—can be placed in one-to-one correspondence:

$$[a, b] \quad (a, b) \quad [a, b) \quad (a, b] \quad (a, \infty) \quad [a, \infty) \quad (-\infty, b) \quad (-\infty, b]$$

They are all equinumerous.

Cantor Space and Power Sets

Let us show next that the following two sets are equinumerous:

$$\begin{aligned} \text{Power set } P(\mathbb{N}) &= \text{set of all subsets of } \mathbb{N} = \{A \mid A \subseteq \mathbb{N}\} \\ \text{Cantor space } 2^{\mathbb{N}} &= \text{set of all infinite binary sequences.} \end{aligned}$$

We can exhibit this equinumerosity by associating every subset $A \subseteq \mathbb{N}$ with its *characteristic sequence*, also known as the *indicator sequence*:

$$s_A = s_0 s_1 s_2 s_3 \ldots$$

where s_n is 1 if $n \in A$ and otherwise 0. The pattern of 0 and 1 in s_A is exactly the pattern of membership of elements $n \in A$. For example, here are the characteristic sequences for the even numbers, the odds, the primes, the whole set $\mathbb{N}$, and the empty set $\varnothing$:

$$\begin{aligned} s_{\text{evens}} &= 10101010101010101010101010101\ldots \\ s_{\text{odds}} &= 01010101010101010101010101010\ldots \\ s_{\text{primes}} &= 00110101000101000101000101000\ldots \\ s_{\mathbb{N}} &= 11111111111111111111111111111\ldots \\ s_{\varnothing} &= 00000000000000000000000000000\ldots \end{aligned}$$

Altogether this association is a one-to-one correspondence between sets $A \subseteq \mathbb{N}$ and binary sequences s_A because every set determines a unique sequence and vice versa. Thus the power set $P(\mathbb{N})$ is equinumerous with the Cantor space $2^{\mathbb{N}}$ of all infinite binary sequences.

Meanwhile, it is clear that $2^{\mathbb{N}} \leq \mathbb{R}$ in size because any given binary sequence can be interpreted as a real number in decimal notation. It is also easy to see that $\mathbb{R} \leq P(\mathbb{Q})$ since every real number is determined by the set of rational numbers below it:

$$r \mapsto A_r = \{q \in \mathbb{Q} \mid q < r\}.$$

This is the *Dedekind cut* of that real, the cut that it makes in the rational numbers:

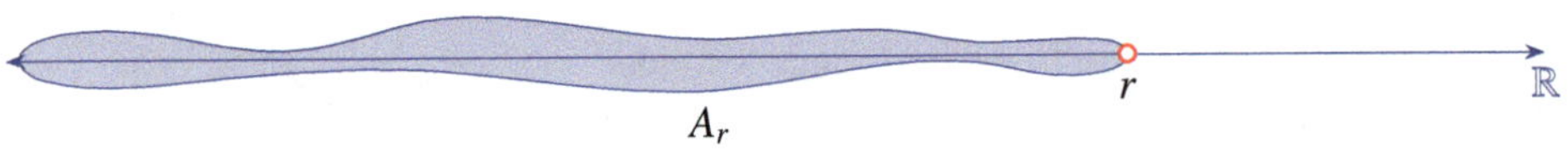

Since $\mathbb{Q}$ is countably infinite, however, there is a one-to-one correspondence of $\mathbb{Q}$ with $\mathbb{N}$, and this correspondence induces a correspondence of the power sets $P(\mathbb{Q})$ and $P(\mathbb{N})$, which are therefore equinumerous. Putting all this together, we have

$$P(\mathbb{N}) \leq 2^{\mathbb{N}} \leq \mathbb{R} \leq P(\mathbb{Q}) \leq P(\mathbb{N}),$$

and so by the Cantor-Schröder-Bernstein theorem, they are all equinumerous.

The Real Plane and the Real Line Are Equinumerous

Using his ideas about infinity and equinumerosity, Cantor made a shocking discovery—the infinity of the real line $\mathbb{R}$ is the same as the infinity of the plane $\mathbb{R}^2$, and also of space $\mathbb{R}^3$ and indeed hyperspace $\mathbb{R}^n$ in any finite dimension. What? Crazytown! The line and the plane have the same number of points? Yes, indeed. This shocking result led Cantor to write to Dedekind in 1877 (see Dauben 1979, p. 55):

> I see it, but I don't believe it!

Let us give the argument. Consider any point (x, y) in the first quadrant of the plane; that is, with $0 \leq x, y$. Let z be the real number that interleaves the digits of x and y in their decimal expansions:

$$(3.14159\ldots,\ 2.71828\ldots) \mapsto 32.1741185298\ldots$$

There is a subtle issue caused by the fact that some real numbers have two decimal representations, such as with $0.999\ldots = 1.000\ldots$, so for the interleaving process, let us decide always to use the representations of x and y that do not end in all 9s. The resulting map $(x, y) \mapsto z$ is one-to-one since from z, we can recover the decimal digits of x and y (note that z will never end in all 9s). This association will not be onto the set of real numbers, however, since, for example, $z = 0.69696969\ldots$ would have to have come from $x = 0.66666\ldots$ and $y = 0.99999\ldots$, but we have said to use the eventual 0 representation of y, so this number z is not actually associated with any pair by our correspondence.

What the argument shows that is that first quadrant of $\mathbb{R} \times \mathbb{R}$ is equinumerous with a set of real numbers. Since the real numbers are equinumerous with the positive real numbers, it follows that the whole plane $\mathbb{R} \times \mathbb{R}$ is at most $\mathbb{R}$ in size. And since the converse relation is obvious, it follows by the Cantor-Schröder-Bernstein theorem that the real plane $\mathbb{R} \times \mathbb{R}$ is equinumerous with the real line $\mathbb{R}$. An essentially similar argument generalizes this to show that the real line $\mathbb{R}$ is also equinumerous with the real number space $\mathbb{R}^n$ in any finite, or indeed countable dimension. The line, the plane, space, higher-dimensional hyperspace—they are all equinumerous with each other. It is truly remarkable.

Comparing Two Function Spaces

If you wanted the bigger set, would you rather have the set of all functions from the real numbers to the integers:

$$\mathbb{Z}^{\mathbb{R}} = \{f \mid f : \mathbb{R} \to \mathbb{Z}\}$$

or the set of all functions from the integers to the real numbers:

$$\mathbb{R}^{\mathbb{Z}} = \{g \mid g : \mathbb{Z} \to \mathbb{R}\}?$$

Which of these function spaces is larger? Are they the same size?

Think about it...

Consider the second function space first. Every function $g : \mathbb{Z} \to \mathbb{R}$ is determined by its course of values, which amounts to a countable list of real numbers since the domain $\mathbb{Z}$ is countable. But every such list can be encoded up in a single real number by the interleaving-digits trick that we just used previously. Namely, given any list of real numbers r_1, r_2, r_3, and so on, simply place the digits of the first real r_1 on the even places, then the digits of the next real number r_2 on every other odd place, the next real r_3 on every other place of what remains, and so on. Alternatively, we may place the kth digit of r_n at position $3^n \cdot 5^k$ of the coding real, and use 0s everywhere else. In this way, every function from a countable set to the real numbers is coded up with a single real, so the number of such functions is no more than the real numbers themselves. In light of the constant functions, there are at least that many functions, so the function space of all functions from the integers to the real numbers is the same as the number of real numbers.

We can also argue more compactly like this. We know that $\mathbb{R}$ has the same size as $P(\mathbb{N})$, which has the same size as Cantor space $2^{\mathbb{N}}$, the functions from $\mathbb{R}$ to the two-element set $2 = \{0, 1\}$. We now simply calculate

$$\mathbb{R}^{\mathbb{Z}} \quad \simeq \quad (2^{\mathbb{N}})^{\mathbb{Z}} \quad \simeq \quad 2^{\mathbb{N}\times\mathbb{Z}} \quad \simeq \quad 2^{\mathbb{N}} \quad \simeq \quad \mathbb{R}.$$

One needs to verify, of course, that the familiar exponentiation rules work for this kind of equinumerosity calculation, but this is a simple matter of carrying out the equinumerosity associations. So the number of functions from the integers to the real numbers is the same as the number of real numbers itself.

The functions from the real numbers to the integers, in contrast, is the space $\mathbb{Z}^{\mathbb{R}}$, which is at least as large as $2^{\mathbb{R}}$, the functions from $\mathbb{R}$ to the two-element set $\{0, 1\}$, which is equinumerous with the power set $P(\mathbb{R})$, which is strictly larger than the set of real numbers.

In conclusion, there are far more functions from the real numbers to the integers than from the integers to the real numbers:

$$\mathbb{R}^{\mathbb{Z}} \quad < \quad \mathbb{Z}^{\mathbb{R}}.$$

So the first space that we mentioned in this discussion is the bigger space.

How Many Functions from the Real Numbers to the Real Numbers?

Consider the full function space of all real-valued functions on the real numbers:

$$\mathbb{R}^{\mathbb{R}} \quad = \quad \{ f \mid f : \mathbb{R} \to \mathbb{R} \}.$$

Every such function $f : \mathbb{R} \to \mathbb{R}$ is determined by its *graph*, which is a subset of the plane:

$$\text{graph}(f) \quad = \quad \{ (x, f(x)) \mid x \in \mathbb{R} \}.$$

So the number of functions f is at most the number of subsets of $\mathbb{R} \times \mathbb{R}$. That is, the space of all functions is at most $P(\mathbb{R} \times \mathbb{R})$. But since $\mathbb{R} \times \mathbb{R}$ is equinumerous with $\mathbb{R}$, this is the same as $P(\mathbb{R})$ in size.

Conversely, for every subset $A \subseteq \mathbb{R}$, we may define the characteristic function

$$s_A(x) = \begin{cases} 1 & x \in A \\ 0 & x \notin A. \end{cases}$$

This function completely determines A, so $P(\mathbb{R})$ has the same size as $2^{\mathbb{R}}$. Putting these observations together, we see that

$$\mathbb{R}^{\mathbb{R}} \quad \leq \quad P(\mathbb{R} \times \mathbb{R}) \quad \simeq \quad P(\mathbb{R}) \quad \simeq \quad 2^{\mathbb{R}} \quad \leq \quad \mathbb{R}^{\mathbb{R}},$$

so they all have the same size by the Cantor-Schröder-Bernstein theorem.

How Many Continuous Functions Are There on the Real Numbers?

Consider the set of all continuous functions $f : \mathbb{R} \to \mathbb{R}$. How many are there? How does this infinity compare with the set of real numbers $\mathbb{R}$ or with the set of all functions from $\mathbb{R}$ to $\mathbb{R}$?

Think about it...

What I claim is that the number of continuous functions on the real numbers is the same as the number of real numbers themselves. In short, not very many. Of course, in light of the constant functions, there are at least as many continuous functions as there are real numbers since we may associate every real number with the constant function having that constant value. But the claim is that the number of continuous functions altogether isn't more than this.

To see this, let us argue for the converse relation. Consider any continuous function $f : \mathbb{R} \to \mathbb{R}$. Let F be the set of rational points below the graph of the function—that is, the set of points (p, q) for which both p and q are rational and $q < f(p)$:

$$F = \{(p, q) \mid p, q \in \mathbb{Q} \text{ and } q < f(p))\}.$$

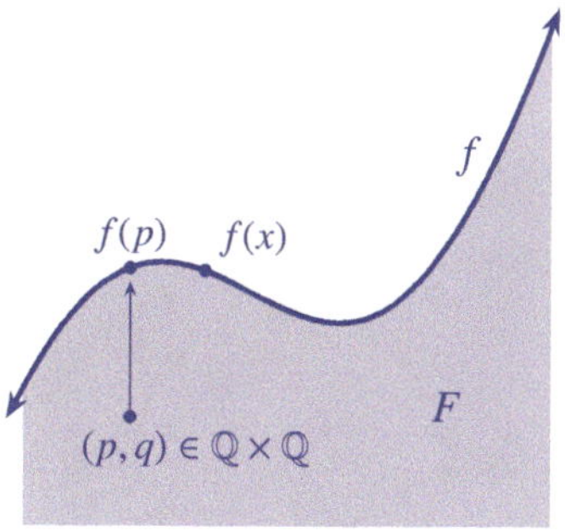

In particular, $F \subseteq \mathbb{Q} \times \mathbb{Q}$. Observe that from F, we can recover the function f as the envelope of F because for any rational number p, the value of $f(p)$ will be the supremum of the section of F at coordinate p, and then by continuity, any $f(x)$ is the limit of the values $f(p)$ as $p \to x$. Thus, the set F determines the continuous function f. So the number of continuous functions is at most the number of such sets F. But F is a subset of the countable set $\mathbb{Q} \times \mathbb{Q}$, so there are at most $P(\mathbb{Q} \times \mathbb{Q})$ many such sets F. But since $\mathbb{Q} \times \mathbb{Q}$ is countable, the set $P(\mathbb{Q} \times \mathbb{Q})$ is equinumerous with $P(\mathbb{N})$, which is equinumerous with ℝ. So the number of the continuous functions on the real numbers is the same as the number of real numbers themselves.

17.2 The Continuum Hypothesis Holds for Closed Sets

Let us consider the closed sets of real numbers in light of the Cantor-Bendixson process, investigated by Cantor from 1872 and independently by Ivar Otto Bendixson (1861–1935). Cantor realized that his analysis shows in fact that closed sets always conform with the continuum hypothesis—every closed set of real numbers is either countable or is equinumerous with the whole set of real numbers.

The Cantor-Bendixson Process

Several key set-theoretic ideas germinate in the Cantor-Bendixson theorem, for whose proof and analysis Cantor develops his set-theoretic ideas on arbitrary sets of real numbers, the ordinals, well-orders, and transfinite recursion. The theorem grows out of Cantor's 1872 investigation of the process of iteratively casting out isolated points from a closed set of real numbers. A set is *closed* if it contains every point x that is a limit of points from the set. A point is *isolated* in a set if it stands alone in some small neighborhood as the only element of the set; or equivalently, if it is not a limit of other points from the set. We form a new set, the *derived* set, by casting out all isolated points from the original set. The derived set of a closed set is also closed because every limit point of the derived set must be in the original set, but it cannot have been isolated there. The curious phenomenon to observe with derived sets is that, although we cast out all isolated points from the original set, the derived set can exhibit newly isolated points—points that were not isolated in the original set but become isolated in the derived set. And if we cast these out in a next step, the resulting set could again exhibit freshly isolated points; again and again we might cast out isolated points, iterating the process many times.

Consider the set C_0 shown here on the left:

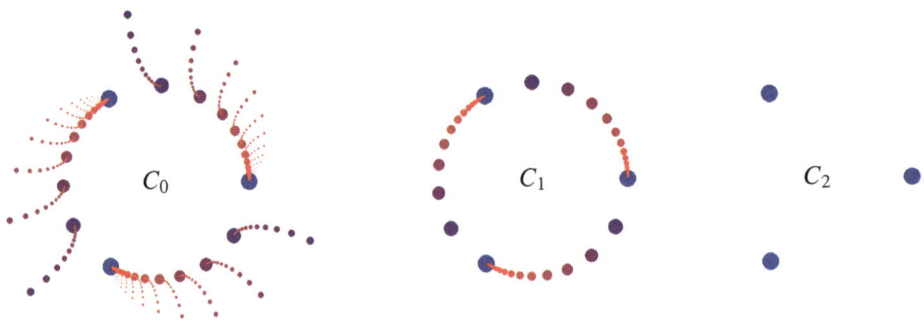

This set has many isolated points on the spiral fringes waving out from the central circle, but the points on the main circle of C_0, being limits of those fringe points, are not isolated. The derived set C_1, therefore, has cast out all the fringe points but keeps the central circle points. Most of these points, however, have now become isolated in C_1, so in the next step of the Cantor-Bendixson process, they are cast out, leaving us with just three points in the derived set C_2. In the general case, with another more complex set, we might continue similarly to cast out isolated points again and again, forming the further derivatives C_3, C_4, and so on. The process stops only when we reach a *perfect* set, which is a closed set with no isolated points (perhaps empty).

The *rank* of the original closed set is the number of steps that it takes for the process to terminate. The set shown previously has rank 3, because after three steps, there are no remaining isolated points. In general, however, sets can have much higher ranks. Perhaps the reader can construct a set of rank 7, or rank n for any finite number n. Cantor realized that a set can have infinite rank. Imagine an initial set C_0 with a sequence of points x_0, x_1, x_2, and so on, converging to a point x, but x_n has rank n, and all the points converging to x disappear by some finite stage; this would leave x isolated at the limit stage $C_\omega = \bigcap_n C_n$, the set of points that have survived through all the finite stages.

How confusing and strange this might seem. We cast out isolated points again and again, but even after doing so infinitely many times, we still have isolated points with what is left in the limit set C_ω. Since this limit set is closed, however, it makes sense to try yet again, to form the next Cantor-Bendixson derivative $C_{\omega+1}$, then perhaps $C_{\omega+2}$, and so on. The Cantor-Bendixson process appears to be inherently transfinite. Will we ever complete the process? What is the nature of this iteration? What are the stages of this construction? It proceeds by a recursive process, defining the later derived sets from the earlier ones, but the recursion is transfinite, proceeding beyond the finite stages.

Cantor realizes that to describe the construction, he needs a way to talk independently about the stages of the construction. The stages themselves form a certain kind of order structure. He needs a number system transcending the finite numbers to serve as the names or indices for the stages in this transfinite recursive process. This is exactly what the ordinals provide. Cantor invents the ordinals specifically to make sense of the iterative Cantor-Bendixson process. He comes to the concepts of well-orders, ordinals, and transfinite recursion—concepts at the very center of set theory as it is known today. Ultimately, for any closed set C_0, one defines all the Cantor-Bendixson derivatives C_α, for every ordinal α, in a transfinite recursive process: Given C_α, the next derivative $C_{\alpha+1}$ is obtained by casting out isolated points, and at limit ordinals, C_λ is the intersection of all earlier C_α.

The Cantor-Bendixson theorem is the claim that for any closed set of real numbers (and in any finite or indeed countable dimension), there is a countable ordinal stage α at which the derived set C_α has no isolated points; that is, the Cantor-Bendixson process terminates, after countably many steps, in a perfect set. To prove this, simply imagine that every cast-out isolated point x is associated with a rational isolating neighborhood, a reason why it was isolated—that is, a rational interval (p, q) containing x and no other points of the derived set at that stage; we shall never use the same isolating neighborhood again because it can no longer isolate another point at some later stage, as there are none left in it. Since there are only countably many rational intervals, it follows that there will also be only countably many points cast out altogether. Therefore, there must be some countable ordinal stage of the construction by which we will have achieved a perfect set.

The Continuum Hypothesis Holds for Closed Sets

Cantor makes a remarkable conclusion from this observation. Namely, the theorem shows that every closed set of real numbers is the union of a countable set (the isolated points cast out during the construction) and a perfect set (the points remaining at the end, if any). And since Cantor had already proved that every nonempty perfect set of real numbers is equinumerous with the whole real number line, he concludes that every closed set of real numbers is either countable or equinumerous with the whole real number line. Thus, he proves that the continuum hypothesis holds for closed sets of real numbers.

Cantor's Strategy for Settling the Continuum Hypothesis

Cantor takes his analysis of closed sets as the first step of a program aimed to establish the truth of the continuum hypothesis by establishing it for more and more complicated sets, beginning with the closed sets. Of course, the continuum hypothesis also holds for open sets since every nonempty open set of real numbers contains an interval and thus has size continuum.

The open and closed sets lead to the family of all *Borel sets*, which are obtained from the closed sets by iteratively closing under the process of complements and countable unions. Thus, the countable unions of closed sets are Borel (but not always closed) and their complements are Borel, and the countable unions of *those* kinds of sets are Borel, and their complements, and so on transfinitely. Cantor thus aims to establish the continuum hypothesis for all Borel sets, and then ultimately for sets beyond the Borel hierarchy. Indeed, this is proved by Mikhail Yakovlevich Suslin (1894–1919), who establishes that the continuum hypothesis holds for all *analytic* sets, which are the continuous images of closed sets of real numbers, and this includes the Borel sets.

One naturally moves on to the higher levels of complexity into the *projective hierarchy*, which closes the Borel sets under both continuous images and complements. The projective sets form the expansive hierarchy of sets that you can create by defining sets of real numbers in a language that allows basic arithmetic concepts and quantification over the integers and over the real numbers. These sets are the core focus of the subject known as *descriptive set theory*, which analyzes sets of real numbers with respect to the logical complexity of their descriptions. The Cantor program aimed to establish the continuum hypothesis by handling more and more complicated sets of real numbers, climbing into the projective hierarchy and ultimately beyond it.

Unfortunately, the program stalled after Suslin found his result on analytic sets. No additional positive instances were proved at that time, but also no counterexamples were found. This must have seemed mysterious and confusing at the time—how odd to have neither a proof nor a refutation. Matters didn't become well understood until much later.

Developments in the late twentieth century, however, brought Cantor's program to life again, for set theorists proved that if there are certain sufficient kinds of large cardinals that fulfill very strong axioms of infinity, then indeed the continuum hypothesis holds not only for the Borel and analytic sets, but for all sets in the projective hierarchy. In this way, the large cardinal assumptions offered a partial fulfillment of Cantor's program.

17.3 What Is Cantor's Continuum Hypothesis?

Where do our various equinumerosity observations leave us? On the one hand, we seem to have many natural instances of countably infinite sets:

> The set of natural numbers $\mathbb{N}$; the set of integers $\mathbb{Z}$; the set of rational numbers $\mathbb{Q}$; the set of finite binary sequences $2^{<\mathbb{N}}$; the set of integer polynomials $\mathbb{Z}[x]$; the set of algebraic real numbers; the set of computable real numbers.

And we also seem to have many natural instances of sets of size continuum:

> The set of real numbers $\mathbb{R}$; the set of transcendental real numbers; nontrivial intervals in the real numbers (a,b), $[a,b]$, $[a,b)$, (a,∞), $[a,\infty)$, $(-\infty,b)$, $(-\infty,b]$; nonempty open sets of real numbers; uncountable closed sets of real numbers; the Cantor set obtained by omitting middle thirds; the set of complex numbers $\mathbb{C}$; Cantor space $2^{\mathbb{N}}$ (the set of infinite binary sequences); the power set of the natural numbers $P(\mathbb{N})$; the space of continuous functions $f:\mathbb{R}\to\mathbb{R}$.

But we seem to have no sets provably of intermediate cardinality between $\mathbb{N}$ and $\mathbb{R}$. Despite vigorous investigation, no natural candidates have arisen for a strictly intermediate cardinality set of real numbers. Moving to higher dimensions doesn't change matters since $\mathbb{R}$ is equinumerous with $\mathbb{R}^n$ in any finite or even countable dimension.

As we have mentioned, Cantor formulated the continuum hypothesis to express the answer that he expected.

Continuum hypothesis. *There is no infinity strictly between* $\mathbb{N}$ *and* $\mathbb{R}$.

What is the status of this hypothesis? We can't seem to prove it beyond the closed sets and the analytic sets. And yet we also can't seem to produce any counterexamples. The explanation for this combination of facts was gradually revealed in a series of profound mathematical developments occuring over several decades in the middle of the twentieth century.

The Continuum Hypothesis Is Consistent

Namely, first, Kurt Gödel proves in 1938 that if the axioms of Zermelo-Fraenkel set theory are consistent—these are the standard axioms formalizing the most fundamental principles of set theory—then this theory is also consistent with both the axiom of choice and the continuum hypothesis being true.

It follows that assuming that the continuum hypothesis is true cannot lead one into any contradictions that one could not arrive at without that assumption. In this sense, Gödel had established that it was safe to assume both that the axiom of choice and the continuum hypothesis are true.

This result explains why Cantor was never able to find a set with intermediate cardinality between $\mathbb{N}$ and $\mathbb{R}$, and indeed why nobody was able to find such a set. It is consistent with set theory to suppose that there is no such set.

Gödel proves his theorem with the method of *model construction*—he builds an alternative mathematical reality in which the continuum hypothesis is true. Specifically, he shows that if there is any model in which the axioms of set theory are true, then there is another such model constructed from it, known as the *constructible universe*, in which the continuum hypothesis also is true. Thus, if set theory is consistent, then it is consistent with the continuum hypothesis being true.

This model-construction method is at bottom the same method used to establish the consistency of non-Euclidean geometry. Namely, we know that we cannot prove the parallel postulate from the other axioms of Euclidean geometry because there are models of non-Euclidean geometry, such as the Poincaré disc, in which all the other axioms of geometry are true but the parallel postulate is false.

The Gödel constructible universe plays a role in set theory analogous to the role played by the Poincaré disc in geometry. In geometry, all the axioms of geometry of true in the Poincaré disc, except that the parallel postulate fails. In set theory, all the axioms of set theory are true in the constructible universe, along with the continuum hypothesis. In each case, we know that a particular combination of axioms is relatively consistent because they are true in the alternative model that was constructed. Because of Gödel's constructible universe, we cannot expect ever to refute the continuum hypothesis.

The Failure of the Continuum Hypothesis Is Also Consistent

There matters stood for twenty-five years, until another truly remarkable development in 1963. Namely, Paul Cohen (1934–2007) proves that if the axioms of Zermelo-Fraenkel set theory are consistent, then they are also consistent with the axiom of choice and the *negation* of the continuum hypothesis. In this sense, it is also safe to assume that the continuum hypothesis is false. This result explains why Cantor was never able to find a proof that it was true since it is in fact consistent with the axioms of set theory that it is not.

To prove his result, Cohen introduces the method known as *forcing*, a powerful tool for set-theoretic model building. Given any set-theoretic universe V, he describes how to construct an alternative universe—an alternative mathematical reality, the *forcing extension* $V[G]$—in which set-theoretic truth is different. Using this forcing method, he shows that forcing extensions can be found in which the continuum hypothesis is false.

The forcing method was subsequently developed far beyond the applications that Cohen had in mind, and we are now able to construct diverse forcing extensions realizing all different kinds of set theoretic truths. The situation is as though we have found sci-fi portals from our mathematical universe to alternative mathematical realities realizing alternative mathematical truths. Furthermore, these deviations in truth can be precisely predicted and controlled to achieve particular configurations. It is a remarkably powerful mathematical method, revealing a kind of fundamental indeterminism in mathematical truth.

The Independence of the Continuum Hypothesis

The two developments of Gödel and Cohen, taken together, show that the continuum hypothesis is *independent* of the axioms of set theory—it is neither provable nor refutable on the basis of those axioms. If the axioms are consistent, then it is impossible either to prove that the continuum hypothesis is true and also impossible to prove that it is false. Let me emphasize that it isn't just a matter that *we don't yet know*. Rather, the situation is that *we know that we cannot know* whether the continuum hypothesis is true on the basis of the standard axioms.

But is it true? Perhaps the fact that the continuum hypothesis is independent of the axioms simply shows the weakness of that theory—the axioms are too weak. Perhaps we need a better axiomatic framework. With better axioms, perhaps we shall be able to settle the continuum problem.

Gödel's Hope to Settle the Continuum Hypothesis

Gödel hoped that new natural axioms would settle the continuum problem. He had specifically hoped that the new strong axioms of infinity—the large cardinal axioms—would settle the continuum hypothesis one way or the other. In fact, he expected the continuum hypothesis to fail, with the continuum having value $\aleph_2$, the next cardinal after the first uncountable cardinal $\aleph_1$, leaving exactly one intermediate cardinality. Since these new strong axioms of infinity were also attractive in many other respects, he hoped that the large cardinal axioms would thus finally settle the continuum problem.

Unfortunately, however, it was later proved by Lévy and Solovay that in principle, none of the large cardinal axioms would be able to settle the continuum problem. The reason for this was that Cohen's method of forcing could be used to show that each large cardinal axiom was preserved to forcing extensions in which the continuum hypothesis was true and others in which the continuum hypothesis was false. Thus the large cardinal axiom itself cannot determine it either way.

17.4 Debate on Pluralism in the Foundation of Mathematics

The unfinished nature of the answer to the continuum problem has led to a robust philosophical debate in the philosophy of set theory, one engaging with fundamental questions on the nature of mathematical truth. Is there a fact of the matter about whether the continuum hypothesis is true or not? Is every mathematical assertion either definitively true or definitively false? If so, by what grounds would that be the case, and how would we know it?

The Universe View

Some philosophers of set theory defend the *universe view*, taking the axioms of set theory to be about a singular conception of set, one giving rise to a unique intended interpretation of the concept of set, a unique, all-encompassing, intended set-theoretic universe, containing all the possible sets that there could be. On the universe view, every mathematical and set-theoretic assertion has a definitive final truth value in this set-theoretic universe. Our task is to determine for each whether it is true or false.

The fact that the continuum hypothesis is independent of the usual axioms of set theory is taken on the universe view not as a sign of indeterminacy, but rather merely to indicate the weakness and inadequacy of the underlying theory—with stronger axioms we would be able to settle the matter. Indeed, on the universe view, it is our core task to search for those deeper set-theoretic principles that will explain and ultimately settle all the open questions that currently vex us, including the continuum problem and many others.

The Multiverse View

Critics of the universe view point out that the central discovery of set theory over the past half-century is the rich cosmos of set-theoretic possibility. We have constructed diverse set-theoretic universes exhibiting an abundant range for set-theoretic truth. In some universes, the continuum hypothesis is true and in others it is not, and these various worlds seem perfectly set-theoretic to us. Such a perspective opens the door to pluralism in the foundations of mathematics, showing that the truth of a given mathematical question may depend on which mathematical world one is in.

A Dream Solution to the Continuum Problem

Gödel's hope of settling the continuum problem with large cardinal axioms was a cancelled instance of what I call the *dream solution* template for the continuum hypothesis. Namely, set theorists have dreamed that the continuum problem could someday be settled by finding a "missing" axiom, a perfectly natural and acceptable set-theoretic principle on its face, but which also happens to settle the continuum problem. The idea is that if we were to discover a new fundamental principle, one that was clearly valid for our intended interpretation of sets, but which also proved the continuum hypothesis or proved its negation, then this would be a dream solution. We would adopt the new axiom and take with it the accompanying resolution of the continuum problem.

Gödel's hope that large cardinals would serve as such a dream solution were dashed by the Lévy-Solovay theorem, which showed that none of the large cardinal axioms could have that property since if they were consistent, then they were also consistent with both the continuum hypothesis and its negation. So set theorists still dream, yearning for the solution to this fundamental problem that has defied ultimate resolution for a century.

And yet I argue in Hamkins (2015) that in light of the multiverse view we shall never succeed with the dream solution template. The reason is that if a new axiom candidate is ever shown to settle the continuum hypothesis, then this by itself would immediately serve to undermine its acceptability as a natural axiom since it would conflict with the set-theoretic universe that we have already come to know by forcing, in which the continuum problem had the opposite answer. Those worlds seem perfectly acceptable set-theoretically. In my paper, I write:

> Our situation with CH is not merely that CH is formally independent and we have no additional knowledge about whether it is true or not. Rather, we have an informed, deep understanding of how it could be that CH is true and how it could be that CH fails. We know how to build the CH and ¬CH worlds from one another. Set theorists today grew up in these worlds, comparing them and moving from one to another while controlling other subtle features about them. Consequently, if someone were to present a new set-theoretic principle Φ and prove that it implies ¬CH, say, then we could no longer look upon Φ as manifestly true for sets. To do so would negate our experience in the CH worlds, which we found to be perfectly set-theoretic. It would be like someone proposing a principle implying that only Brooklyn really exists, whereas we already know about Manhattan and the other boroughs. Hamkins (2015)

The Analogy with Geometry

The study of classical geometry had proceeded for thousands of years, from the time of Euclid until the late nineteenth century, as a singular effort concerned with a single domain of knowledge—what we might describe as the one true geometry, which many (including Immanuel Kant) took to be the geometry of physical space. But with the shocking discovery of non-Euclidean geometry in the early nineteenth century, the singular concept of geometry splintered into distinct, separate concepts. From that time, we now have spherical geometry, hyperbolic geometry, and elliptical geometry. It became no longer sensible to ask whether the parallel postulate is simply true or not, full stop. Rather, one asks whether it is true in this geometry or that geometry.

Similarly, on the pluralist view in set theory, it is as absurd to ask whether the continuum hypothesis is really true or not, full stop, as to ask in geometry about the parallel postulate. The parallel postulate is true in some geometries and false in others. The continuum hypothesis is true for some concepts of set and false for others. On the pluralist view, there simply is no fact of the matter.

Questions for Further Thought

17.1 Make a list of half a dozen of your favorite sets of real numbers, and classify them (with argument) as to whether they are countable or size continuum—or do they constitute an intermediate cardinality, thus refuting the continuum hypothesis?

17.2 Argue that the continuum hypothesis holds for open sets of real numbers.

17.3 Argue that if the continuum hypothesis holds for two sets A and B, so they are both either countable or size continuum, then it also holds for the union set $A \cup B$. Indeed, if the continuum hypothesis holds for countably many sets A_0, A_1, A_2, and so on, then it holds for the union set $\bigcup_n A_n$.

17.4 Draw sets of points in the plane with Cantor-Bendixson rank 4, rank 5, and rank 6. Can you describe a set of rank ω? How about $\omega + 1$ or $\omega + 2$?

17.5 Is there a fact of the matter about whether the continuum hypothesis holds?

Further Reading

Hamkins (2021b). Chapter 8 contains a survey of several matters in the philosophy of set theory, including the continuum hypothesis and large cardinals.

Credits

Parts of this chapter were adapted from Hamkins (2021b, chapter 3, 8).

18 Throwing Darts at the Real Line

Abstract. In an attempt to resolve the continuum problem, mathematical philosophers have proposed thought experiments suggesting various resolutions of the issue—we are to throw darts randomly at the real line or to imagine a somewhat different mathematical history. Can we hope legitimately to answer mathematical questions with thought experiments in this way?

Although we cannot expect to settle the continuum hypothesis on the basis of proof from the standard Zermelo-Fraenkel (ZFC) axioms of set theory—it is independent of those axioms, after all, neither provable nor refutable—nevertheless we might hope to settle it on other grounds. In this chapter, we shall consider two mathematical thought experiments proposed by mathematical philosophers as a way to gain insight into what our attitude toward the continuum hypothesis should be or could have been.

18.1 Throwing Darts at the Real Line

Chris Freiling (1986), for example, proposes a thought experiment harnessing our intuitive ideas about probability and randomness to arrive at a conclusion about the continuum hypothesis. To begin, he supposes that we have associated every real number x with a certain countable set of real numbers A_x. In this situation, Freiling invites us to imagine throwing two darts at the real line. The first one lands at a real number x, and since the set A_x associated with that number is countable, almost surely the second dart will land at some place y that is not in A_x. The second point y is associated with its own set A_y:

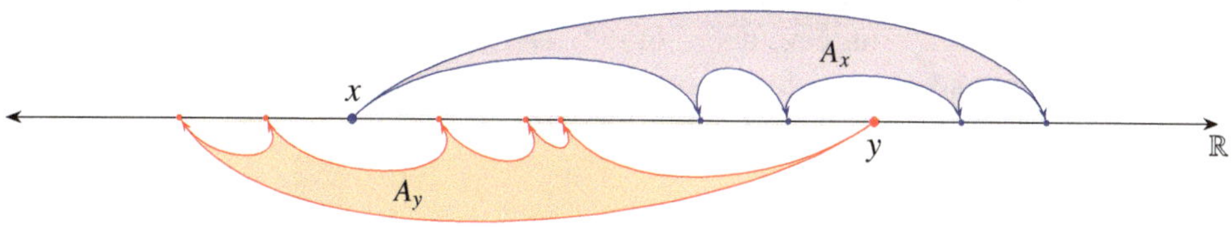

By the symmetry of the situation, Freiling argues, we should similarly expect that x is not in A_y—it simply shouldn't matter which dart we think of as being first. This line of thinking leads to what Freiling calls the axiom of symmetry.

Axiom of symmetry. *For every association $x \mapsto A_x$ mapping real numbers to countable sets of real numbers, there are distinct real numbers x, y with*

$$y \notin A_x \quad \textit{and} \quad x \notin A_y.$$

By throwing darts at the real line and appealing to primitive notions of probability, Freiling thus offers us a philosophical justification of his axiom.

The Axiom of Symmetry Is Equivalent to the Negation of the Continuum Hypothesis

The principal mathematical observation to be made about the axiom of symmetry is that it is equivalent to the negation of the continuum hypothesis.

Let me elaborate. First, I shall argue that the continuum hypothesis implies the failure of the axiom of symmetry. Suppose that the continuum hypothesis holds. So there is a well-ordering $\lhd$ of the set of real numbers $\mathbb{R}$ with order type ω_1, the first uncountable ordinal. The essential feature of such a well-ordering is that it is uncountable, and yet every proper initial segment is countable—just like the ordinal ω_1 itself, which is uncountable, and yet every ordinal below it is countable. At bottom, this is what the continuum hypothesis asserts—that one can place the real numbers into an uncountable well-ordering in which every number has only countably many predecessors below it. We may therefore associate with each real number x the set of real numbers appearing earlier in this order:

$$A_x = \{y \in \mathbb{R} \mid y \lhd x\}.$$

As we said, this is a countable set. But notice that if x, y are distinct real numbers, then either $x \lhd y$ or $y \lhd x$, since $\lhd$ is a linear order, and therefore we must have either $x \in A_y$ or $y \in A_x$, which contradicts the axiom of symmetry. Thus, we have proved that if the continuum hypothesis holds, then the axiom of symmetry cannot be true.

Conversely, suppose that the continuum hypothesis fails. Consider any set $B \subseteq \mathbb{R}$ of size ω_1, which by our assumption is less than the continuum. Let us close the set B under the operation $x \mapsto A_x$. What I mean by this is that we take B as B_0 and then form a larger set B_1 by adding to B_0 all the elements of A_x for any $x \in B_0$; Then we form B_2 by adding all the elements of A_x for any $x \in B_1$; and so on, defining B_{n+1} by adding to B_n all elements of A_x for any $x \in B_n$. In this way, the set $B^* = \bigcup_n B_n$ will contain all the sets B_n, and thus also all the elements of A_x for any $x \in B^*$. In short, it will be *closed* under this operation. Since the original set B had size ω_1 and all the sets A_x are countable, it follows that all the sets B_n and also B^* still have a size of only ω_1. Because we have assumed that the continuum hypothesis is false, this is not all the real numbers. So we may find some real number $y \notin B^*$. Since A_y is a countable set but B^* is uncountable, having size ω_1, there must be some $x \in B^*$ with $x \notin A_y$. But A_x will be contained in B^* because B^* was closed under the operation, so in particular, $y \notin A_x$. Thus we have found distinct points x, y with $x \notin A_y$ and $y \notin A_x$, which verifies this instance of the axiom of symmetry. Therefore, the axiom of symmetry is equivalent to the failure of the continuum hypothesis—that is, the size of the continuum must be at least $\aleph_2$.

Has Freiling Refuted the Continuum Hypothesis?

How successful is the argument? Has Freiling given us reason to deny the continuum hypothesis? Has he provided a dream solution to the continuum problem?

Freiling (1986) describes his argument as containing "a simple philosophical 'proof' of the negation of Cantor's continuum hypothesis." He begins with some philosophical intuition-building, "subject[ing] the continuum to certain thought experiments involving random darts," before ultimately landing at his axiom of symmetry, presented as a true and natural axiom, an "intuitively clear axiom," whose truth, Freiling argues, follows from our strongest prereflective ideas about symmetry and likelihood. And then, ultimately, he proves as we did in this discussion that his axiom implies the failure of the continuum hypothesis.

This pattern of argument follows exactly the dream solution template we discussed earlier (page 226), in which one presents a new natural axiom, argues that it is manifestly true, and then proves that it settles the continuum problem.

In my personal experience, mathematicians often readily find the dart-throwing thought experiment appealing upon the first presentation, and often thereby become favorably inclined toward the axiom of symmetry, although this is generally more true for younger mathematicians and much less so with more experienced mathematicians. Mathematicians favorably inclined to the axiom of symmetry in this way, however, are often surprised to learn subsequently that it is equivalent to the failure of the continuum hypothesis, and this often leads to a revision in their initial judgement of the axiom.

Reception of Feiling's Argument

Was Freiling's argument received as the longed-for dream solution to the continuum problem? No. Many mathematicians object that Freiling's argument implicitly assumes for a given assignment $x \mapsto A_x$ that various sets are measurable. That is, Freiling assumes that the concepts of probability and likelihood are applicable to these sets, whereas we have extensive experience in mathematics with sets that are badly behaved with respect to measure and probability. Although he does not consider it directly, the relevant set for Freiling is the two-dimensional set $\{(x, y) \mid y \in A_x\}$ in the plane, which has every vertical section being countable, since the xth slice consists of the points (x, y) with $y \in A_x$, which has only countably many instances. In that sense, when considered from the vertical perspective, the set seems rather thin. And the symmetry argument is supposed to lead us to think that it should also be thin from the horizontal perspective.

But the troubling aspect is that we already know how it can happen that a badly behaved set can appear very thin from one perspective and not from others. Every well-ordering of $\mathbb{R}$ in the order type of the continuum, for example, gives rise to such a set in plane by taking A_x as the set of predecessors of x with respect to the well-order. Such a set appears very thin when considered via vertical stripes since every real number would have fewer than continuum many real numbers below it, but it seems very thick when considered via horizontal stripes for the dual reason since every real number is below all the rest of the numbers in the well-order. Do such examples undermine the naive intuitions in which Freiling's thought experiment is grounded? To be sure, he uses this very example as a philosophical argument against the idea that $\mathbb{R}$ admits a well-ordering in the first place.

The objection that Freiling has presumed his sets to be well behaved with respect to probability and measure strikes to the heart of his argument, which relies on our intuitions concerning two independent events—the two dart throws—and depends fundamentally on the robustness of our probability and likelihood concepts for this two-dimensional set, rather than for the one-dimensional slices of it that are explicitly mentioned in the argument. Do we really have direct intuitions about probability and likelihood for the two-dart set A in the plane determined by an arbitrary, and perhaps extremely chaotic function $x \mapsto A_x$? That is what the argument requires.

Freiling anticipates this objection, answering that the intuitive justification that he offers for the axiom of symmetry can be viewed as prior to the entire mathematical development of measure theory, prior to our knowledge of sophisticated counterexamples to the naive principles. He wants to present his justification as primitive.

Nevertheless, mathematical analysis has revealed the presence of the measure-theoretic monsters—we shall talk much more about nonmeasurable sets in chapter 21 (see page 284). For mathematicians with this knowledge, the prereflective arguments often fall flat. We have become deeply skeptical of any intuitive or naive use of measure concepts precisely because we know the pitfalls—we know now how complicated and badly behaved functions and sets of real numbers can be in their measure-theoretic properties.

For these reasons, Freiling's argument, rather than being accepted as the longed-for solution to the continuum hypothesis, is instead most often presented as a curiosity, an unusual but attractive equivalent formulation of the negation of the continuum hypothesis. Typical presentations of the axiom include a discussion of Freiling's dart-throwing thought experiment, not as evidence that the axiom is true but rather as a warning about the measure-theoretic monsters, a warning that we must take extra care with issues of nonmeasurability, lest we be fooled. Freiling's simple philosophical argument is thus turned on its head, used not as a justification of the continuum hypothesis but rather as a warning about the error that may arise from a naive treatment of measure concepts—a warning that what seems obviously true might still be wrong. In Hamkins (2015), I argue that the entire episode bears out the pattern of response that I predict for any attempted dream solution to the continuum problem, namely, a rejection of the new axiom from a perspective of deep mathematical experience with the contrary.

Higher-Order Axioms of Symmetry

There are natural higher-order versions of the axiom of symmetry. Consider the three-dart axiom, for example, which asserts that for any association of *pairs* of real numbers (x, y) with countable sets of real numbers $(x, y) \mapsto A_{x,y} \subseteq \mathbb{R}$, there are three real numbers x, y, z for which none of them is in the set determined by the other two:

$$x \notin A_{y,z} \quad \text{and} \quad y \notin A_{x,z} \quad \text{and} \quad z \notin A_{x,y}.$$

This axiom can be justified by just the same kind of dart-throwing thought experiment as before—we shall throw three darts instead of two. The last dart is almost surely not in the set determined by the first two, and then by symmetry, it should also be true for the other two combinations, thereby justifying the axiom. It turns out that this three-dart axiom of symmetry is equivalent to the assertion that the continuum has a size of at least $\aleph_3$.

And similarly, the four-dart axiom asserts that for every association $(x, y, z) \mapsto A_{x,y,z} \subseteq \mathbb{R}$ of triples of real numbers with countable sets of real numbers, there are four real numbers w, x, y, z for which none is in the set determined by the other three:

$$w \notin A_{x,y,z} \qquad x \notin A_{w,y,z} \qquad y \notin A_{w,x,z} \qquad z \notin A_{w,x,y}.$$

This axiom is equivalent to the assertion that the continuum has size at least $\aleph_4$. And the pattern continues like this for all finite n. On the basis of these thought experiments, we should expect that the continuum hypothesis fails quite badly, and indeed, Freiling argues, perhaps there should be no well-ordering of the set of real numbers at all.

The Free-Set Lemma

The underlying phenomenon of the axiom of symmetry has been generalized in certain ways to other contexts. In the *free-set lemma*, for example, we have associations $x \mapsto A_x$ of a certain sort with $x \notin A_x$, and we seek not just two or finitely many independent points, but an entire infinite set F of *freely independent* points, meaning one for which $x \notin A_y$ and $y \notin A_x$ for any $x, y \in F$. Diverse forms of this lemma are known, but let me discuss the simple case where we associate with every real number $x \in \mathbb{R}$ a closed set $A_x \subseteq \mathbb{R}$ with $x \notin A_x$. In this case, we may indeed find a free set of size continuum. To prove this, observe that since A_x is a closed set and $x \notin A_x$, we may find a tiny rational neighborhood of x that is disjoint from A_x. That is, we may find rational numbers $p < q$ such that x is in the interval (p, q), but this interval is disjoint from A_x, as shown here:

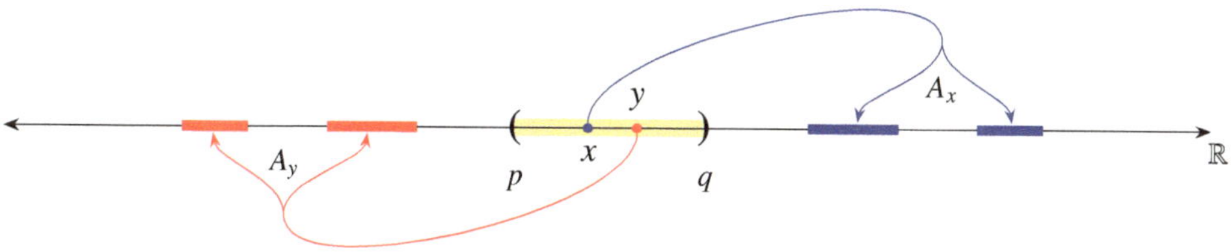

Since there are only countably many pairs of rational numbers (p, q) but there are uncountably many real numbers, it follows that we must very often have chosen the same pair (p, q) for lots of different xs. Indeed, because the countable union of sets of size less than continuum has size less than continuum, there must be some rational pair (p, q) that we had chosen continuum many times. That is, there is a set F of size continuum, such that every $x \in F$ had used the same rational pair (p, q). The point now is that since all elements $x, y \in F$ have chosen the same interval (p, q), it means that x and y are both inside this interval (p, q), which is disjoint from both A_x and A_y, since that is how we chose the neighborhood intervals. In particular, this implies both

$$x \notin A_y \quad \text{and} \quad y \notin A_x.$$

Thus we have found a large, freely independent set F, as desired.

Meanwhile, it turns out that there are many other variations of the free-set lemma proved for other kinds of associations and with other cardinals.

18.2 Another Mathematical Thought Experiment

Let us turn now to a totally different thought experiment, a historical thought experiment of my own—I introduce it in Hamkins (2024b), aiming to show that our attitude toward the continuum hypothesis could easily have been very different than it is. If our mathematical history had been just a little different, I argue, if certain mathematical discoveries had been made in a slightly different order, then we would naturally view the continuum hypothesis as a fundamental axiom of set theory—one furthermore necessary for mathematics and indeed indispensable for making sense of the core ideas underlying calculus.

The thought experiment is this. Imagine that in the early days of calculus Isaac Newton and Gottfried Wilhelm Leibniz had been a little clearer about the nature of infinitesimals. In our actual history, to be sure, the lack of rigor surrounding infinitesimals was famously mocked by Bishop George Berkeley (1685–1753) with withering criticism:

> And what are these same evanescent Increments? They are neither finite Quantities nor Quantities infinitely small, nor yet nothing. May we not call them the ghosts of departed quantities?
> (1734, p. 59)

In my thought experiment, however, I propose that Newton and Leibniz are a little clearer on the nature of infinitesimals, positing that they inhabit a larger realm of numbers, what we might call the *hyperreal* numbers $\mathbb{R}^*$ (to use the contemporary terminology), a larger number realm than the real numbers $\mathbb{R}$. In the thought experiment, Newton and Leibniz posit that every conceivable cut in the real numbers is filled by hyperreal numbers—the infinitesimals occupy the region between 0 and the positive real numbers—and otherwise the hyperreal numbers obey all the same laws as the real numbers.

From a contemporary perspective, these ideas appear as incipient forms of the theory of nonstandard analysis, discovered by Abraham Robinson (1918–1974) in the 1960s, providing a fully robust and rigorous infinitesimal theory of calculus. A glance at the remarkable undergraduate textbook by Jerome Keisler (2000) shows how well one can implement infinitesimal ideas in calculus, replacing the classic $\epsilon\delta$ and limit arguments with fully rigorous infinitesimal analysis, beginning from only basic elementary notions.

In the thought experiment, the hyperreal numbers would become embedded at the center of calculus from an early stage, at first informally but then gradually, with greater rigor. The hyperreal numbers would thus find their place in the mathematical pantheon of number systems—the natural numbers, the integers, the rational numbers, the real numbers, the complex numbers, and in the world of my thought experiment, the hyperreal numbers:

$$\mathbb{N} \qquad \mathbb{Z} \qquad \mathbb{Q} \qquad \mathbb{R} \qquad \mathbb{C} \qquad \mathbb{R}^*.$$

At the end of the nineteenth and early twentieth centuries, mathematicians provided categorical accounts of all the classical number systems in this pantheon. As discussed in chapter 11, Richard Dedekind (1888) proves that the natural number structure $\mathbb{N}$ is characterized up to isomorphism by his theory of the successor operation, leading directly to Giuseppe Peano's elegant development of elementary number theory in that framework. Building on this, mathematicians provide categorical accounts of the integer ring $\mathbb{Z}$ and the rational field $\mathbb{Q}$. Edward Huntington (1874–1952) proves in (1903) that the real field $\mathbb{R}$ is characterized up to isomorphism as the complete ordered field. The complex numbers are characterized as the algebraic closure of $\mathbb{R}$. These categorical characterizations of the core mathematical structures are centrally important for the coherence of the mathematical enterprise, enabling us to refer to the various fundamental mathematical structures by their defining characteristics. Daniel Isaacson (2011), inspired by Georg Kriesel (1923–2015), emphasizes the vital role that categoricity results play in providing for the meaning and basic reality of these fundamental structures.

But what about the hyperreal numbers? The key event in the thought experiment, next, is that a Zermelo-like figure steps up to provide a categorical account of the hyperreal numbers. And indeed, the theory of Zermelo-Fraenkel (ZFC) set theory with the continuum hypothesis is able to prove that there is up to isomorphism a unique smallest countably saturated real-closed field, and this is the field of hyperreal numbers $\mathbb{R}^*$. Thus, in the theory ZFC + CH, we have a foundational account of infinitesimals and the hyperreal numbers, which in the thought experiment is a core structure in the foundations of calculus.

The critical related observation is that one cannot prove this categoricity result without the continuum hypothesis (see further discussion of this point in Hamkins (2024b)). In particular, the phrase "the hyperreal numbers," which one quite commonly hears in mathematical discourse, is not meaningful in ZFC set theory because in ZFC, there is not necessarily just one mathematical structure fitting the description. But with the continuum hypothesis, there is. With the continuum hypothesis, we can specify the hyperreal numbers up to isomorphism as a canonical structure, the unique smallest countably saturated real-closed field. Without the continuum hypothesis, unfortunately, there would be chaos and confusion in the infinitesimal foundations of calculus.

In the thought-experiment world, therefore, the continuum hypothesis would be seen as a necessary part of the fundamental theory providing meaning for the core mathematical structures in the foundations of mathematics, even calculus. We would take the continuum hypothesis as part of the fundamental theory that provides meaning to our mathematical activities. This is how it could have been that we look upon the continuum hypothesis as a fundamental axiom.

18.3 Thought Experiments

So we have two thought experiments—one suggesting that the continuum hypothesis is false and one explaining how we might naturally have come to view the continuum hypothesis as true, and indeed essential for mathematics. Are such thought experiments a reliable guide to truth in mathematics? By what criterion do we accept our most fundamental axioms in mathematics? The reader is invited to contemplate these issues in the questions below.

Questions for Further Thought

18.1 Is there a fact of the matter about whether the continuum hypothesis holds?

18.2 Does Freiling's argument for the axiom of symmetry provide a reason to believe that the continuum hypothesis is false?

18.3 If there is a fact of the matter about the continuum hypothesis, then is any argument that we could have taken CH as a fundamental axiom amount to an argument that CH is in fact already a fundamental axiom? Does the calculus thought experiment provide grounds for that stronger claim?

18.4 Critically assess the question of whether thought experiments of the kind mentioned in this chapter can be a reliable guide to mathematical truth.

Further Reading

Freiling (1986). This is the article in which Freiling mounts his philosophical argument against the continuum hypothesis.

Hamkins (2024b). This is my article in which I introduce the mathematical historical thought experiment regarding Newton and Leibniz and the hyperreal numbers.

Credits

The phrase, "throwing darts at the real line" is due to Freiling (1986). Parts of this chapter were adapted from Hamkins (2015) and Hamkins (2024b). The theorem that the axiom of symmetry is equivalent to the negation of the continuum hypothesis was proved first by Wacław Franciszek Sierpiński, long before Freiling. The free-set lemma in various forms was proved by Andras Hajnal. A shorter version of this chapter will appear as a "Short Stories" column in the AMS Notices as Hamkins (2026b).

19 The Orders of Infinity

Abstract. What is the spectrum of possible behavior at infinity for a real-valued function on the real numbers? What are the possible rates of growth? What is the structure and complexity of these diverse orders of infinity?

Consider the various functions $f : \mathbb{R} \to \mathbb{R}$ on the real numbers and the vastly different behaviors that they might exhibit as they progress to infinity. Some may tend to zero, others settle into a modest linear growth, and still others might grow superexponentially.

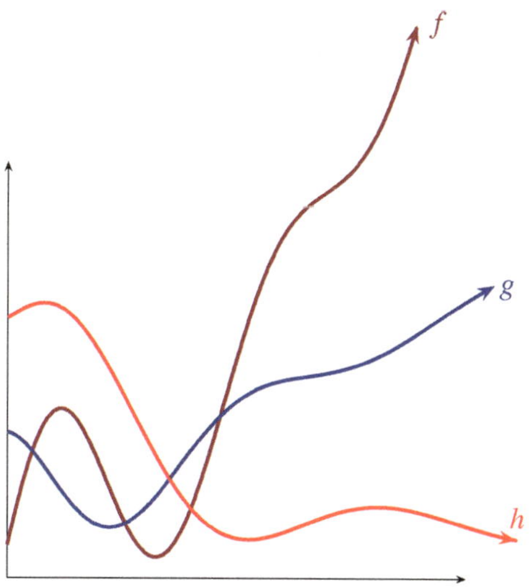

What is the range of possibility? For that matter, what does it mean precisely—and what *should* it mean—to say that two functions exhibit the same behavior at infinity? What are the possible rates of growth? What is the structure and complexity of these "orders of infinity"?

Consider a simple case—a constant function $y = c$. Perhaps a function f approaching it with a horizontal asymptote $\lim_{x\to\infty} f(x) = c$ could be said to exhibit the same behavior "at infinity":

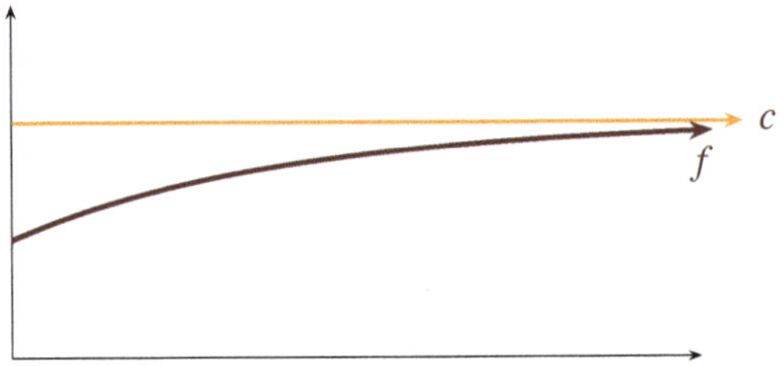

A linear function could similarly be said to have the same behavior at infinity as a function with that line as a slant asymptote:

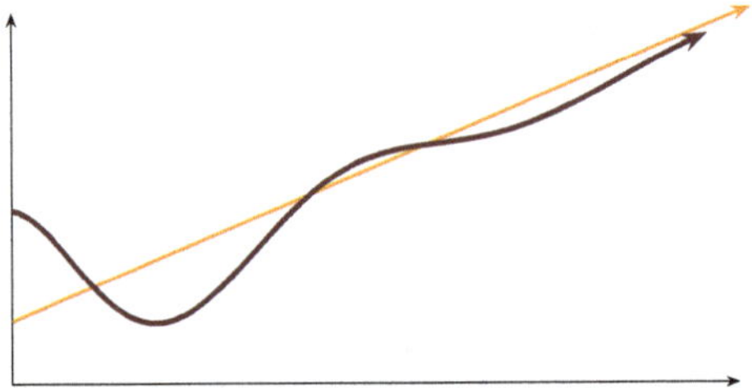

And so on with other functions. To understand the range of possibility, we would like somehow to group together functions that exhibit the same behavior at infinity:

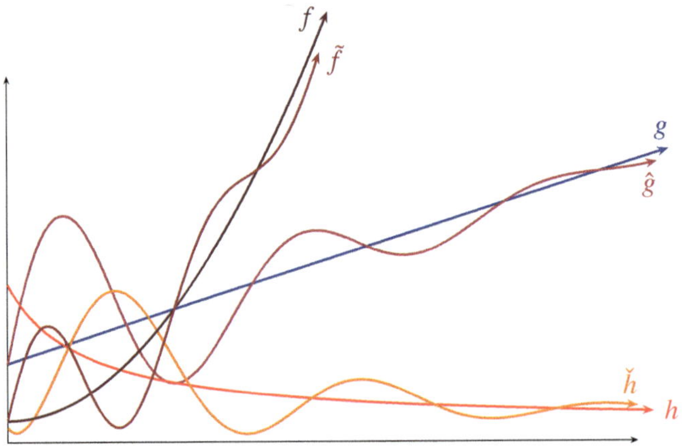

But what does this mean exactly? We would like to be able to express the idea that f and $\tilde{f}$ have the same limiting behavior, or g and $\hat{g}$ are equivalent, or h and $\check{h}$ are. What is the precise criteria that we are seeking?

19.1 The Orders of Infinity

What we are looking for is to define an equivalence relation that captures and expresses the intuitive notion of two functions having the same behavior at infinity. It turns out that there are a variety of natural ways to go about this, depending on whether one is seeking a fine or coarse concept.

Asymptotic Convergence

On the one hand, we might define that functions f and g are *asymptotically convergent* to each other if the limit of their difference goes to zero; that is, if the difference between the functions can be made eventually as small as one likes as x becomes large:

$$\lim_{x\to\infty} |f(x) - g(x)| \quad = \quad 0.$$

This is the notion of limiting behavior that we use when saying that a function f has a horizontal asymptote or a slant asymptote—the difference between the function and the asymptote line itself can be made eventually as small as desired as x becomes large. The concept of asymptotic convergence makes sense even when the asymptote is not linear—indeed, the concept works even when the functions are not continuous, provided that they indeed get close to each other as the input grows to infinity. Asymptotic convergence is a useful, robust notion.

In terms of evaluating rates-of-growth for fast-growing functions, however, this notion is rather strict, and perhaps we may eventually prefer a more relaxed notion. We may ultimately want to say that x^2 and $x^2 + 10$ have essentially the same behavior at infinity, since once x becomes large, the $+10$ part just doesn't matter much. But these two functions are not asymptotically convergent, since they always differ by 10.

Another sense in which asymptotic convergence may not be what we want is that it does not distinguish rates of approach to zero. That is, for functions that become very small at infinity, asymptotic convergence will decree them equivalent even in cases where we might reasonably want to distinguish them.

Consider, for example, the functions $1/x$ and $1/x^2$:

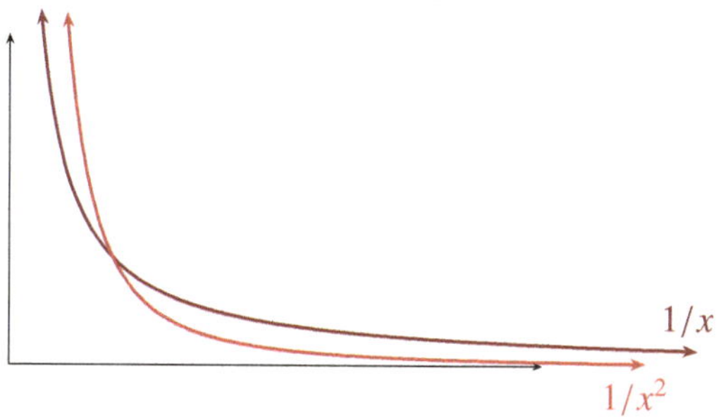

Both functions tend to 0 as $x \to \infty$, so their difference also tends to 0, making them asymptotically convergent. They both converge to the x-axis as a horizontal asymptote.

But the fact is that $1/x^2$ converges to zero much more rapidly—the function $1/x$ is a comparative laggard. When $x = 10$, for example, the value

$$\frac{1}{x} = \frac{1}{10}$$

is 10 times bigger than the value

$$\frac{1}{x^2} = \frac{1}{100}.$$

When x is larger than 1 million, then the function $1/x$ is more than 1 million times larger than $1/x^2$. Their ratio grows indeed without bound:

$$\lim_{x\to\infty} \frac{\frac{1}{x}}{\frac{1}{x^2}} \quad = \quad \lim_{x\to\infty} x \quad = \quad \infty.$$

Is this a reason to say that they have different behaviors at infinity? If so, that may be a reason to soften or perhaps reject the asymptotic convergence concept of equivalence at infinity.

Asymptotic Equivalence

This proportionality way of thinking suggests an alternative concept of equivalence at infinity, where we say that functions f and g are *asymptotically equivalent* if they become proportionally identical in the limit; that is, if the limit of their ratio goes to 1 as x goes to infinity:

$$\lim_{x\to\infty} \frac{f(x)}{g(x)} \quad = \quad 1.$$

By this criterion, we would say that x^2 and $x^2 + 5x$ are equivalent, for example, since the limit of their ratio goes to 1, which one can see by multiplying through by $1/x^2$ like this:

$$\lim_{x\to\infty} \frac{x^2}{x^2 + 5x} = \lim_{x\to\infty} \frac{x^2}{x^2 + 5x} \cdot \frac{\frac{1}{x^2}}{\frac{1}{x^2}} = \lim_{x\to\infty} \frac{1}{1 + \frac{5}{x}} = \frac{1}{1 + 0} = 1.$$

A similar calculation shows that every polynomial like $2x^3 + 200x^2 - 17x + 5$ is asymptotically equivalent to its leading term $2x^3$, thereby providing a welcome simplification of the rate of growth—one that seems to concentrate meaningfully on the most important cause of the growth.

Meanwhile, the concept is not without issues. If g is unboundedly often zero, for example, then the limit is problematic. But also, the concept may still be too strict, still a little too fine-grained. Observe, for example, that this notion of asymptotic equivalence is sensitive to the leading constant—it distinguishes sharply between $2x^3$ and $5x^3$, whose ratio tends to $2/5$ rather than 1.

It may be both useful and natural to prefer a still coarser concept—one that groups all nontrivial linear growth into the same class, all quadratic growth together, and all cubic growth as equivalent. For this, we shall need a different concept of equivalence at infinity.

Big O Notation

The *big O* concept is based on the idea of measuring the rate of growth of a function while ignoring the effects of nonzero constant multiples. Specifically, we say that a function g is in $O(f)$ for $x \to \infty$ if g is eventually at most a constant multiple of f; that is, if there is some constant $c > 0$ such that $|g(x)| \leq |cf(x)|$ for all sufficiently large x. In other words, g is in $O(f)$ if it is eventually bounded by some multiple of f. The big O concept thus aims to provide a flexible and useful notion for an upper bound on the rate of growth of a function rather than directly providing a notion of equivalence.

Dominant Terms

When a function is expressed as a sum of several terms, such as $x^5 + 4x^3 + 3$, then the big O class of the function is determined by its dominating term. For example, $O(x^5 + 4x^3 + 3)$ is the same as $O(x^5)$ since if g is eventually dominated by a multiple of $x^5 + 4x^3 + 3$, then it will also be eventually dominated by a multiple of x^5 since for large values of x, we have $c(x^5 + 4x^3 + 3) \leq c(x^5 + 4x^5 + 3x^5) = 8cx^5$.

In addition, because of the way that it is defined, the big O notation absorbs constant multiple factors. For example, a function g is $O(2x^2)$ if and only if it is also $O(x^2)$ since being eventually bounded by a multiple of $2x^2$ is the same as being eventually bounded by a multiple of x^2—one is just using a bigger multiple.

For this reason, the big O class of a polynomial is the same as for its dominant power alone:

$$\begin{aligned} O\left(5x^3 - 10x^2 + 17x + 1000\right) &= O(x^3) \\ O\left(2x^5 + x^3\sqrt{x} + 10^{100}\right) &= O(x^5) \\ O\left(x^2 + 10^{10^{100}}x\right) &= O(x^2). \end{aligned}$$

Note that $O(f) = O(kf)$ for any positive constant $k > 0$.

Infinitesimal Asymptotics

Although in this chapter we are mainly discussing *infinite asymptotics*, which is the behavior of functions at infinity, let me mention that one also sees the big O notation used for *infinitesimal asymptotics*, which concerns the behavior of functions near zero. As one takes x to 0, the lower-degree terms dominate. Consider, for example, the series expansion of the exponential function:

$$e^x = 1 + x + \frac{x^2}{2} + \frac{x^3}{3!} + \frac{x^4}{4!} + \frac{x^5}{5!} + \cdots$$

The initial terms of this series are a very good approximation near zero, and one can express this precisely by writing

$$e^x = 1 + x + \frac{x^2}{2} + O(x^3) \quad \text{as } x \to 0.$$

What this means is that the difference error between e^x and the approximation $1 + x + x^2/2$ has order $O(x^3)$ as $x \to 0$, which means that the error is bounded by a multiple of x^3 as x tends to 0. Since x^3 goes to zero much more rapidly than the terms of $1 + x + x^2/2$, this is very useful information bounding the error. The big O notation thus provides a compact means of expressing the order of the error precisely:

$$e^x = 1 + x + \frac{x^2}{2} + \frac{x^3}{3!} + \frac{x^4}{4!} + O(x^5) \quad \text{as } x \to 0.$$

Such expressions are very useful in estimation analysis.

An Underlying Equivalence Relation

The big O notation leads directly to a notion of equivalence at infinity. Namely, let us say that two functions f and g are *big O equivalent* at infinity if $O(f) = O(g)$. That is, functions are equivalent when each one stands in the big O relation to the other—each function is eventually bounded by a multiple of the other as x becomes large. This will occur, for example, if the function g eventually remains inside the cone about f, with $bf(x) \leq g(x) \leq cf(x)$ for constants $b, c > 0$, as pictured here:

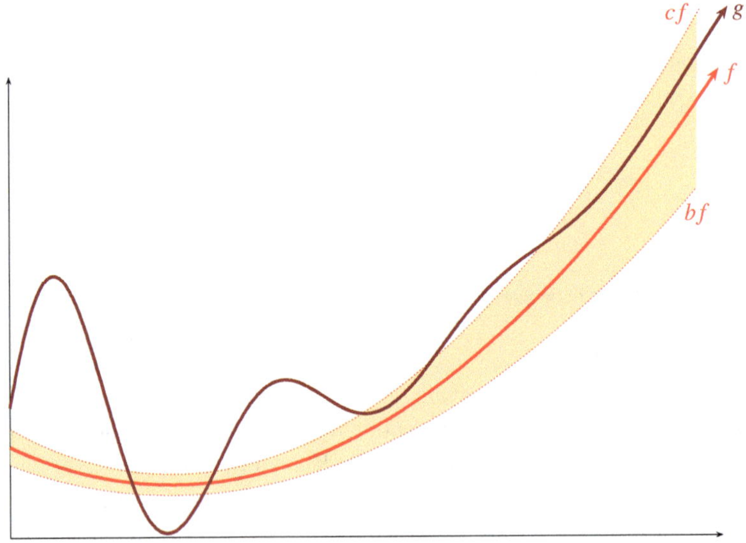

This is a symmetric relation because

$$bf(x) \quad \leq \quad g(x) \quad < \quad cf(x)$$

implies the converse instance

$$\frac{1}{c}g(x) \quad \leq \quad f(x) \quad \leq \quad \frac{1}{b}g(x).$$

In other words, if g eventually enters a cone about f, then also f eventually enters a cone about g.

Little *o* Notation

The little o notation provides a means to express a related strict order relation. Namely, we say that g is $o(f)$, pronounced "little o of f," if for every $\varepsilon > 0$, it is eventually true for large enough x that $|g(x)| < \varepsilon|f(x)|$. In other words, even very tiny, scaled-down versions of f eventually outgrow g. For example, $\sqrt{x}$ is $o(x)$, since for any $\varepsilon > 0$, if we take large enough values of x, we shall see that $\sqrt{x} < \varepsilon x$:

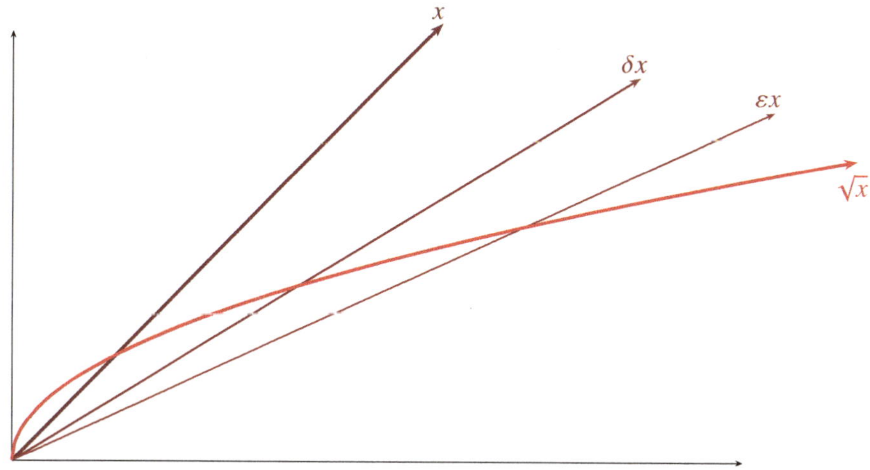

Similarly, x is $o(x^2)$ and every polynomial is $o(e^x)$. If g is $o(f)$, then it follows that g is $O(f)$, but f is not $O(g)$.

19.2 Transcending Rates of Growth

Let us explore the rates of growth, climbing into the hierarchy as high as we might dare, venturing into the upper realm of the orders of infinity. What sights shall we see from the towering transcendental summits that we shall find there?

Perhaps we begin our journey at the bottom of the hierarchy, departing from a base camp settled peacefully in the low flatlands of the constant functions, graphs formed by perfectly horizontal lines. These simple functions are eventually transcended by any line with a positive slope, of course. Even a modest upward slope is sufficient to transcend any constant function:

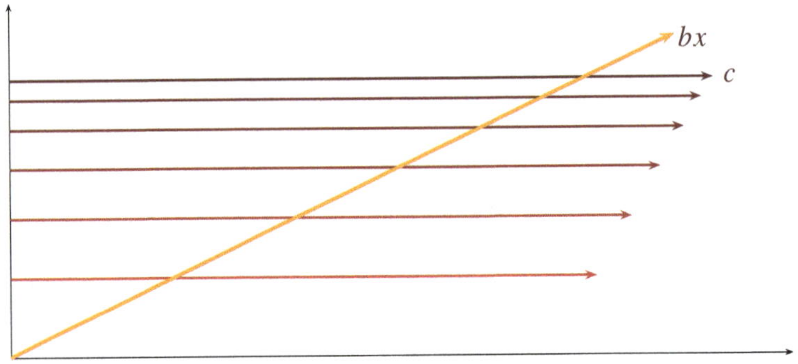

The linear functions $y = bx + c$ have steeper slopes as b becomes larger. Is there a function that climbs faster than all linear functions? Yes. The quadratic function $y = x^2$ eventually beats any given linear function $y = bx + c$. That is, for large enough values of x, the linear function $bx + c$ is eventually transcended by the parabola $y = x^2$:

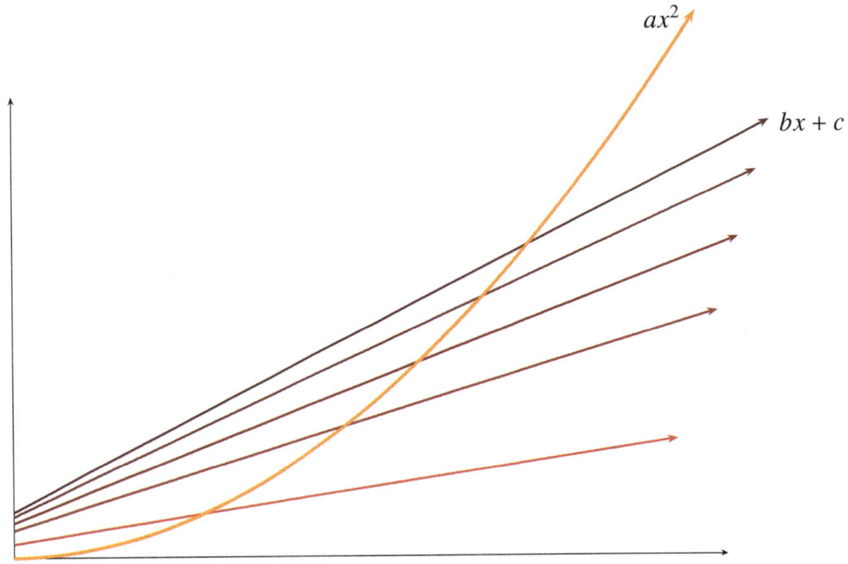

We might consider a family of parabolas, $y = ax^2+bx+c$. Is there a function that transcends all the quadratics?

Yes, the cubic function $y = x^3$ eventually beats any quadratic. Once x is larger than a, for example, then x^3 will be bigger than ax^2:

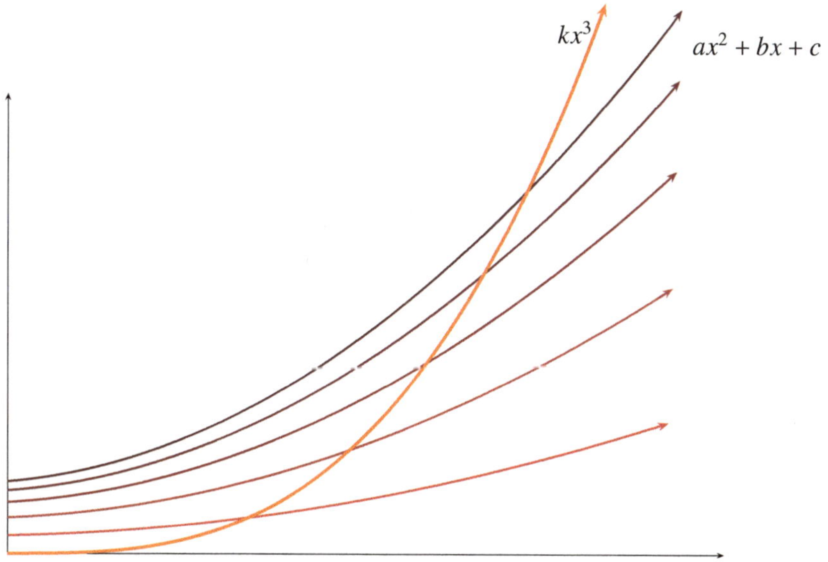

And this phenomenon continues, with the quartic functions $y = x^4$ eventually outgrowing the cubics kx^3, the quintics $y = x^5$ outgrowing the quartics, and so on. In general, the function $y = x^n$ will eventually outgrow any polynomial of lesser degree. If $m < n$, then ax^m will be bested by x^n once $x > a$. Thus we seem to have climbed through the entire hierarchy of polynomial growth in the orders of infinity.

Superpolynomial Growth

Can we climb still higher? Is there a function that outgrows every polynomial function? Yes, indeed, the exponential function $y = e^x$ beats all polynomials. Let us compare e^x with $x^3 + 17x + 2$, for example. We can compute the limit of the ratio by means of L'Hôpital's rule, taking the deriviative in both the numerator and denominator as follows:

$$\lim_{x\to\infty} \frac{e^x}{x^3 + 17x + 2} = \lim_{x\to\infty} \frac{e^x}{3x^2 + 17} = \lim_{x\to\infty} \frac{e^x}{6x} = \lim_{x\to\infty} \frac{e^x}{6} = \infty.$$

Thus, the function e^x can be made eventually many times larger than $x^3 + 17x + 2$—in fact, as many times larger as desired.

That same calculation works with any polynomial since e^x is fixed by the derivative in the numerator, but the derivative of a polynomial in the denominator will drop each time until one gets a constant. In this way, one sees that e^x beats any polynomial. Similar reasoning shows that any exponential function $y = a^x$ for any base $a > 1$ will eventually outgrow any given polynomial.

Superexponential Growth

Can we transcend exponential growth? Are there functions that beat all the exponentials $y = Ka^x$?

Yes, indeed there are. Note that $a^x = (e^{\ln a})^x = e^{x \ln a}$, so changing the base from e to a amounts to considering functions of the form e^{kx}. Can we beat all functions of the form Ke^{kx}? Yes, the function $y = e^{x^2}$ will do it since x^2 beats kx. Similarly, e^{x^3} will beat all functions of the form e^{ax^2+bx+c}. Thus the analysis is merely pushed into the exponent.

We can similarly beat all functions of the form $e^{p(x)}$ for a polynomial $p(x)$ by considering the iterated exponential e^{e^x}. And then comes $e^{e^{e^x}}$, and so on, with stacks of exponentials. We can employ all the tricks of the large tweetable numbers, with stacks of exponentials and tetrations and pentrations and so on—we shall have analogues of the entire Knuth hierarchy in the orders of infinity for functions on the real numbers.

There seems truly to be no end to the rates of growth.

Sublinear Growth

Let us digress a bit by considering a converse question—is there a function that grows more *slowly* than all linear functions? Yes, indeed—we have already seen it with $\sqrt{x}$. And $\sqrt[3]{x}$ and $\sqrt[4]{x}$ grow more slowly still. The function $\log x$ grows more slowly than all the fixed-root functions, for similar reasons why the exponential function outgrows polynomials.

Incomparable Rates of Growth

Are there incomparable rates of growth? What I mean is, are there functions f and g such that neither is bounded in growth by the other—that is, neither is in big O of the other?

$$f \notin O(g) \quad \text{and} \quad g \notin O(f).$$

Can this happen? Yes, indeed it can. Consider the two functions f and g pictured here, which oscillate to increasing heights, but out of phase, so when f is very large, g is very small and vice versa:

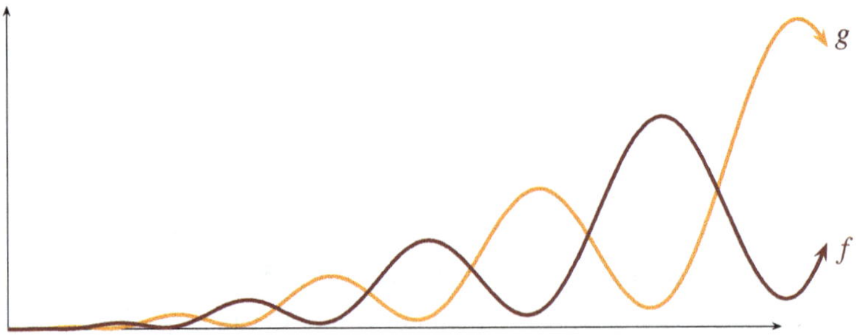

It follows that neither function is bounded by a constant multiple of the other since the oscillations will eventually become large enough to exceed any fixed constant factor.

A more difficult challenge would be to provide incomparability with strictly increasing functions. But this is also possible. One way to achieve it is simply to integrate the previous functions, considering the two definite integral functions:

$$\int_0^x f(x)\,dx \qquad\qquad \int_0^x g(x)\,dx.$$

Since f and g were positive, the integrals add more area as the upper limit x increases, so these definite integral functions are monotone increasing. If the oscillations of f and g grow sufficiently, then the amount of area added with each cycle will also exceed any given constant multiple. For example, the area of each hump in either f or g more than doubles the total previous area, so the peaks of the integrals are growing exponentially, but they are out of phase, so neither will be bounded by a constant multiple of the other.

19.3 The Eventual Domination Order

Let us investigate another approach to the phenomenon of rates of growth, focusing on the *eventual domination* order $f \leq^* g$, which holds of functions $f, g : \mathbb{R} \to \mathbb{R}$ when $f(n) \leq g(n)$ for all *sufficiently large n*, which is a precise, compact way of saying that there is a number N such that $f(x) \leq g(x)$ for all $x \geq N$. In other words, $f \leq^* g$ holds if eventually $g(x)$ is at least as large as $f(x)$. We may similarly define the strict version $f <^* g$, which means that $f(x) < g(x)$ for all sufficiently large x:

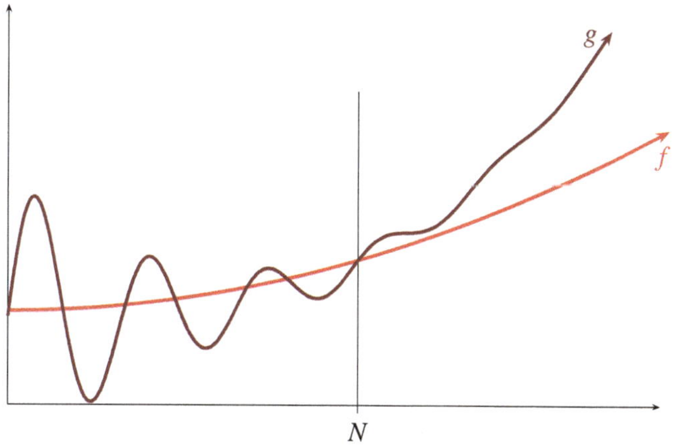

It is also common to consider the eventual domination order for functions on the natural numbers $f : \mathbb{N} \to \mathbb{N}$ rather than on the real numbers, and since every function on the natural numbers extends continuously to the real numbers, the case of $\mathbb{N}$ shares many of the features of the analysis for $\mathbb{R}$.

Every Countable Family Is Dominated

We have seen already that the linear functions are dominated by the quadratics, which are dominated by the cubics and the quartics, and indeed the entire polynomial hierarchy is dominated by the exponentials, which are dominated by the superexponentials, and so on. The hierarchy appears mind-expandingly vast. Can we imagine climbing to the top of it?

Felix Hausdorff (1868–1942) made a remarkable observation about this. He proved that there is no countable sequence of functions that climbs cofinally in the hierarchy. Every countable family of functions is transcended—for every countable list of functions f_0, f_1, $f_2, \ldots$, no matter how fast they grow, there is a function g that eventually dominates each of them: $f_n \leq^* g$ for every n:

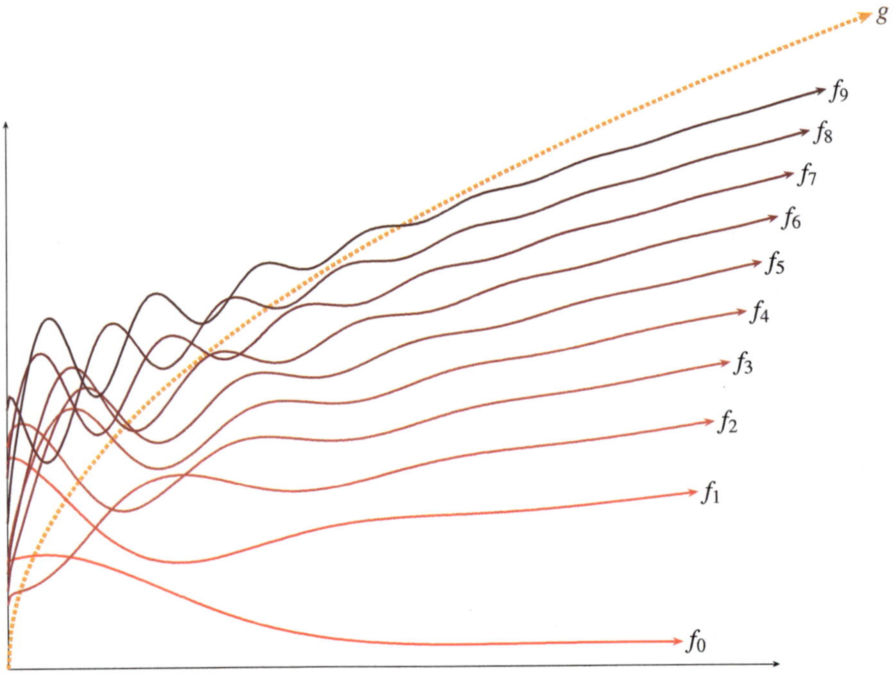

The conclusion is incredible, but the proof is not difficult. For x in the interval $[n, n+1)$, we can simply let $g(x)$ be the supremum of the functions up to f_n:

$$g(x) = \max\{f_0(x), f_1(x), \ldots, f_n(x)\} \qquad \text{for } n \leq x < n+1.$$

In other words, $g(x) = \max_{n \leq x} f_n(x)$. The value of $g(x)$ will therefore be at least as large as $f_n(x)$ whenever $n \leq x$, so this function g eventually dominates every function f_n, as desired.

What about the case of continuous functions? If every function f_n is continuous, the bounding function g that we have just constructed might have discontinuities because at each boundary value n, we suddenly enlarge the set of values that we are maximizing. For example, if f_5 is much larger than the earlier functions, then $g(x)$ will make a big jump up at $x = 5$, when the value $\max\{f_0(x), f_1(x), \ldots, f_5(x)\}$ now includes $f_5(x)$.

Nevertheless, one can modify the construction to find a continuous bound g if all the f_n are continuous. The function g, as defined previously, might have jump discontinuities at $x = n$, but one can simply patch it up by modifying g in the neighborhood below each $x = n$ so as to climb continuously to the next target value. In this way, we see that every countable family of continuous function admits a continuous function that eventually dominates each of them.

I should like to emphasize how strange and mind expanding this bounding property is. We have shown that every countable family in the eventual domination order is bounded. This is a property that is not shared by many of the other natural orders with which you might be familiar. For example, the natural orders on $\mathbb{N}$, $\mathbb{Q}$, and $\mathbb{R}$ do not have this property, since in each case, we can easily find unbounded countable sets of points, which climb cofinally in the order. What Hausdorff's theorem shows, in contrast, is that it is not possible to construct a ladder of functions:

$$f_0 \leq^* f_1 \leq^* f_2 \leq^* f_3 \leq^* \cdots$$

that climbs cofinally or even unboundedly in the eventual domination order. Every such attempt, no matter how high the functions are in the order or how fast they grow, will still admit an upper bound with a function g eventually dominating every function f_n in the ladder. It is truly incredible.

Gaps in the Orders of Infinity

Hausdorff explores a further subtle variation of the phenomenon. Namely, instead of seeking merely to dominate a given countable family of functions by finding a function g on top of them all, let us investigate whether it is possible to find a function h that fills a gap between two countable families of functions. That is, suppose that we have two countable families of functions, a lower family of functions f_n increasing with respect to the eventual domination order $\leq^*$ and a descending upper family of g_n above them all:

$$f_0 \leq^* f_1 \leq^* f_2 \leq^* \cdots \qquad \cdots \leq^* g_2 \leq^* g_1 \leq^* g_0.$$

This situation is called an (ω, ω)-*gap*, since we have an increasing ω sequence from below and a decreasing ω-sequence from above. The question is whether we can find a function h that fills the gap, one for which $f_n \leq^* h \leq^* g_m$ for every n, m.

The answer, proved by Hausdorff, is that indeed every such countable gap can be filled:

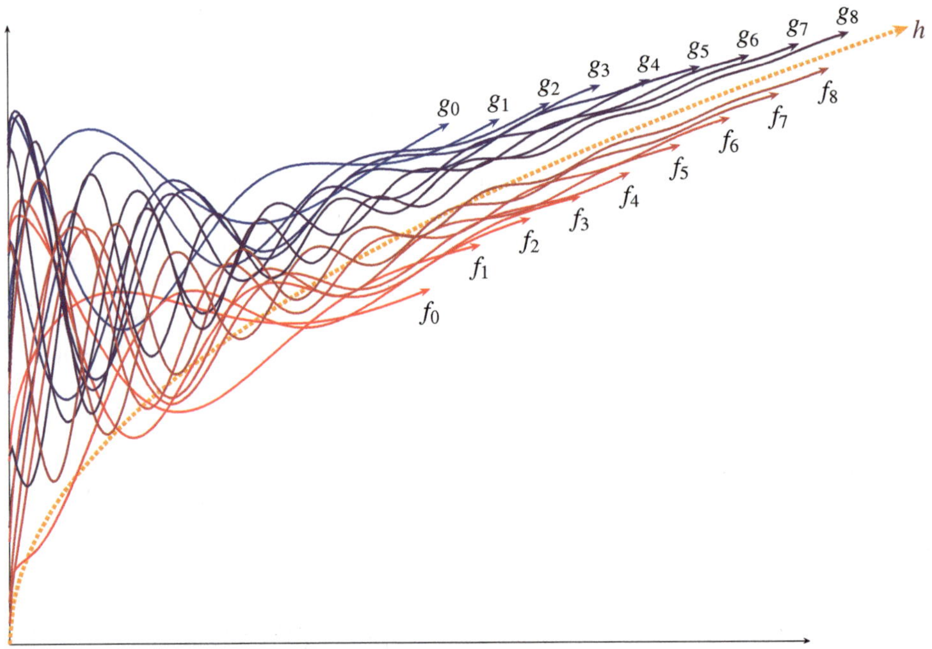

To see this, observe first that there is a number N_0 such that $f_0(x) \leq g_0(x)$ for all $x \geq N_0$. Then there is a larger number $N_1 > N_0$ such that $f_1(x) \leq g_1(x)$ for $x \geq N_1$. Continuing in this way, we find a sequence

$$N_0 \quad < \quad N_1 \quad < \quad N_2 \quad < \quad \cdots$$

such that $f_k(x) \leq g_k(x)$ for all $x \geq N_k$. Now, simply let $h(x) = \max_{k \leq n} f_k(x)$ for any x in the interval $N_n \leq x < N_{n+1}$. This is in each case the maximum of a finite set of values. Notice that $f_n(x) \leq h(x) \leq g_n(x)$ if $x > N_n$, since for these values of x, every $f_k(x)$ is below every $g_k(x)$ for $k = 0, 1, \ldots, n$ and $h(x)$ is the largest of the f-values amongst them. Therefore, $f_n \leq^* h \leq^* g_n$ for every n, so h has filled the gap.[4]

[4] A note on terminology: Many mathematicians use the word "gap" only when the arrangement of functions does not admit any intermediate function h, calling it a *pregap* otherwise. My own preference is for a more relaxed, neutral terminology, speaking of gaps generally, and then of whether they are filled or unfilled. After all, we would still speak of the gap between the bookcase and the wall, even if it should be occupied by a mouse.

19.4 The Bounding and Dominating Numbers

Let me close this chapter by pointing the way toward several deeper ideas. We have mentioned Hausdorff's observation that every countable family of functions $f : \mathbb{R} \to \mathbb{R}$ is dominated by some function. The same analysis applies to functions $f : \mathbb{N} \to \mathbb{N}$, and this observation leads to the question: What is the size of the smallest *unbounded* family of functions, a family of functions $f : \mathbb{N} \to \mathbb{N}$ that is not dominated by any single function?

Hausdorff shows that any such family must be uncountable, and of course, it has size at most continuum, which is the size of the set of all functions $f : \mathbb{N} \to \mathbb{N}$. If the continuum hypothesis holds, therefore, then every unbounded family has size continuum. But if the continuum hypothesis fails, however, then there are plenty of other uncountable cardinals available, which might be the size of an unbounded family of functions.

The *bounding number* $\mathfrak{b}$ is defined as the smallest size of an unbounded family, a family $\mathcal{B}$ of functions $f : \mathbb{N} \to \mathbb{N}$ that is not eventually dominated by any single function $g : \mathbb{N} \to \mathbb{N}$. This is an uncountable cardinal, of size at most continuum. How big is it? A related cardinal number is the *dominating number* $\mathfrak{d}$, the smallest size of a *dominating* family of functions, a family $\mathcal{D}$ of functions $g : \mathbb{N} \to \mathbb{N}$ such that every function $f : \mathbb{N} \to \mathbb{N}$ is eventually dominated by some function $g \in \mathcal{D}$.

Since every dominating family is an unbounded family, it follows that the smallest size of an unbounded family is less than or equal to the smallest size of a dominating family. In other words, the bounding number $\mathfrak{b}$ is less than or equal to the dominating number $\mathfrak{d}$. We may thus place these cardinals into their natural relations as follows:

$$\aleph_1 \quad \leq \quad \mathfrak{b} \quad \leq \quad \mathfrak{d} \quad \leq \quad \mathfrak{c} \quad = \quad |\mathbb{R}|.$$

If the continuum hypothesis holds, then $\aleph_1 = |\mathbb{R}|$, which makes all these cardinals equal. If the continuum hypothesis fails, however, then the lower bound $\aleph_1$ can be less than the continuum $\mathfrak{c}$, and in fact, it is known to be relatively consistent with the Zermelo-Fraenkel axioms of set theory that

$$\mathfrak{b} \quad < \quad \mathfrak{d}.$$

With the method of forcing, one can construct models of set theory in which $\aleph_1 = \mathfrak{b}$ and $\aleph_2 = \mathfrak{d} = \mathfrak{c}$, among many other definite possibilities. In particular, it is relatively consistent with the other axioms of set theory that the size of the smallest unbounded family is strictly less than the size of the smallest dominating family.

These observations are a part of the subject of mathematics known as *cardinal characteristics of the continuum*, in which set theorists study not only the bounding number and the dominating number, but many other such cardinal characteristics, generally defined as the smallest size of a family realizing some kind of infinitary combinatorial characteristic property. The subject is saturated with profound metamathematical results, proving certain definite relations amongst the cardinals, while also observing that a range of other possibilities remain consistent with the Zermelo-Fraenkel axioms of set theory.

This work has revealed a fascinating realm of set-theoretic possibility for the answers to fundamental questions.

Questions for Further Thought

19.1 Provide functions that are asymptotically convergent but not asympototically equivalent, and conversely, functions that are asymptotically equivalent but not asymptotically convergent.

19.2 Provide examples of three functions f, g, and h that have incomparable rates of growth. Can you provide an infinite list f_0, f_1, f_2, and so on, of functions with pairwise incomparable rates of growth? Extra challenge: Is there an uncountable family of functions, with any two incomparable in their rates of growth? Can you do it with monotone increasing functions? (Hint: Consider different periods of oscillation, not rational multiples of each other.)

19.3 Which of the notions—asymptotic convergence, asymptotic equivalence, big O equivalence, or another notion—provides the best account of the behavior of a function at infinity?

19.4 How does behavior at infinity interact with continuity? Using any of the available notions of equivalence at infinity, can a continuous function be equivalent to a discontinuous function? Is every continuous function equivalent to a discontinuous function? Is every discontinuous function equivalent to a continuous function? Consider the question for each of the equivalence conceptions.

19.5 Show that for every countable family of positive-valued functions $f_n : \mathbb{R} \to \mathbb{R}^+$, there is a function $g : \mathbb{R} \to \mathbb{R}^+$ that goes to zero faster than each of them.

Further Reading

Hardy (1910). The classic treatise of Hardy's on the orders of infinity. A fantastic read, highly recommended.

20 The Surreal Numbers

Abstract. All the numbers, great and small. The surreal numbers, systematically generated in a recursive process of completion, unify the real numbers, the transfinite ordinals, and the infinitesimals into a single stupendous number system.

Join me, friend! Let us endeavor together to build the numbers, *all* the numbers, great and small. We shall construct the *surreal* number system, that monumental ediface of numbers, encompassing all the most familiar number systems, to be sure—the natural numbers, the integers, the rational numbers, and the real numbers—but also including the transfinite ordinals, the infinitesimals, and all the strange new numbers that arise in combination from these, like $\sqrt{\omega} - \frac{\pi}{\omega^2}$, mixing the infinitely large with the infinitely small. The surreal numbers unify all these number systems into one colossal yet systematically coherent and graceful new number system.

20.1 Surreal Number Generation

The surreal numbers grow from absolutely nothing into an awesome structure, generated by a single elegant rule—applied recursively, endlessly, and transfinitely—a simple recursive instruction telling us how to create new surreal numbers from the previously created numbers and how to order the surreal numbers amongst each other. The entire surreal number system grows out of this simple process of completion.

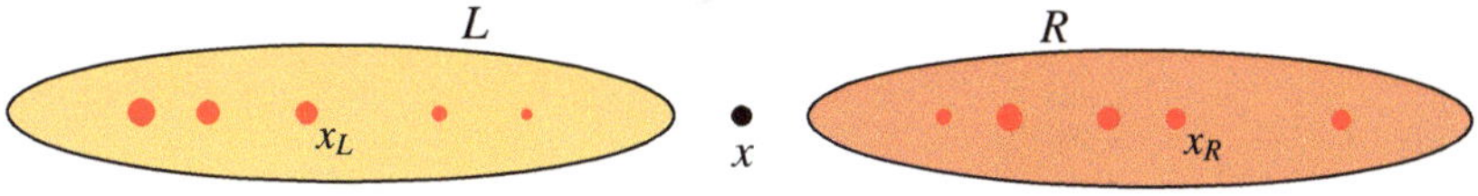

Surreal number generation rule: Iteratively fill all gaps. At every stage, in all possible ways, separate the numbers constructed so far into a lower set and an upper set and create a new surreal number sitting strictly between them, filling this gap.

According to the generation rule, at every stage we divide the current numbers into a lower set L and an upper set R, meaning that every element of L is less than every element of R, and we create a new surreal number x sitting exactly between them, so $x_L < x$ for every x_L in the lower set and $x < x_R$ for every x_R in the upper set. We can denote this newly created number using the *surreal numeral* notation

$$x = \{L \mid R\}.$$

The numeral specifies for this number x exactly the circumstances of its birth.

The Surreal Genesis, the Big Bang of Numbers

But we have gotten a little ahead of ourselves. Let us investigate how the surreal number generation rule works from the very start. At the beginning, of course, we have not created any numbers yet. How are we to divide this nothing into a lower set and an upper set? We simply take both the lower set and the upper set as empty, of course:

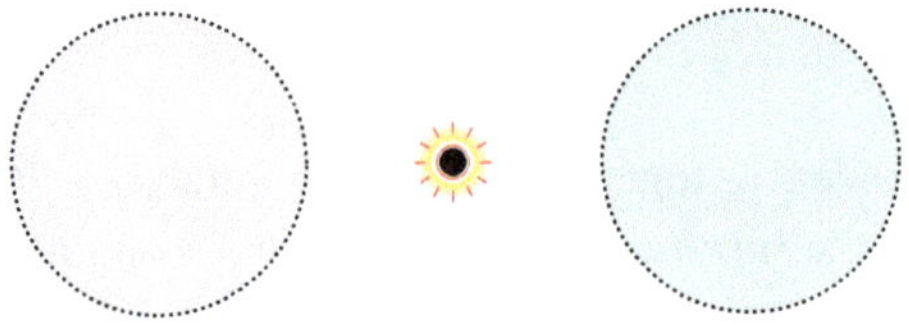

And then a truly remarkable event occurs, for according to the surreal number generation rule, applied even in this trivial instance, a new point is created having so far nothing previous below it and nothing previous above it—a newly created point filling this empty cut. Thus is born the first surreal number, the number *zero*, born on day 0. We have created something from nothing. We are witnessing the surreal genesis, the big bang of numbers:

0

The surreal numeral for zero is written with empty lower and upper sets, like this:

$$0 = \{\ \mid\ \}.$$

Being the very first number, zero has no previously created numbers below it and no previously created numbers above it.

Having created this number 0, we are now able to specify further separations relative to it, either by placing 0 into a new lower set or into a new upper set. Thus on day 1 we create the number 1, the first-born number greater than zero:

$$1 = \{0 \mid \ \}$$

and also the number −1, the first-born number less than zero:

$$-1 = \{\ \mid 0\}.$$

Thus, altogether so far we have created three surreal numbers:

$$-1 \qquad \mathbf{0} \qquad 1$$

We proceed immediately to make four cuts in these numbers, thereby creating four new surreal numbers, born on day 2:

$$-2 \quad -1 \quad -\tfrac{1}{2} \quad \mathbf{0} \quad \tfrac{1}{2} \quad 1 \quad 2$$

The number $\frac{1}{2}$ arises specifically to occupy the gap between 0 and 1 and is therefore denoted by $\frac{1}{2} = \{0 \mid 1\}$, but it is more fully specified as $\{-1 \quad 0 \mid 1\}$. And the number $-2 = \{\ \mid -1\}$ is newly created as the first-born number below −1. Can you write the numerals of the other numbers here?

On day 3, we form further separations of these numbers and create new numbers to fill all the resulting gaps that we are now able to specify relative to them:

$$-3 \quad -2 \quad -\tfrac{3}{2} \quad -1 \quad -\tfrac{3}{4} \quad -\tfrac{1}{2} \quad -\tfrac{1}{4} \quad \mathbf{0} \quad \tfrac{1}{4} \quad \tfrac{1}{2} \quad \tfrac{3}{4} \quad 1 \quad \tfrac{3}{2} \quad 2 \quad 3$$

The elegant surreal number line thus gradually emerges:

$$\cdots -5\ -4\ -\tfrac{7}{2}\ -3\ -\tfrac{11}{4}\ -\tfrac{5}{2}\ -\tfrac{9}{4}\ -2\ -\tfrac{15}{8}\ -\tfrac{7}{4}\ -\tfrac{13}{8}\ -\tfrac{3}{2}\ -\tfrac{11}{8}\ -\tfrac{5}{4}\ -\tfrac{9}{8}\ -1\ -\tfrac{15}{16}\ -\tfrac{7}{8}\ -\tfrac{13}{16}\ -\tfrac{3}{4}\ -\tfrac{11}{16}\ -\tfrac{5}{8}\ -\tfrac{9}{16}\ -\tfrac{1}{2}\ -\tfrac{7}{16}\ -\tfrac{3}{8}\ -\tfrac{5}{16}\ -\tfrac{1}{4}\ -\tfrac{3}{16}\ -\tfrac{1}{8}\ -\tfrac{1}{16}\ \mathbf{0}\ \tfrac{1}{16}\ \tfrac{1}{8}\ \tfrac{3}{16}\ \tfrac{1}{4}\ \tfrac{5}{16}\ \tfrac{3}{8}\ \tfrac{7}{16}\ \tfrac{1}{2}\ \tfrac{9}{16}\ \tfrac{5}{8}\ \tfrac{11}{16}\ \tfrac{3}{4}\ \tfrac{13}{16}\ \tfrac{7}{8}\ \tfrac{15}{16}\ 1\ \tfrac{9}{8}\ \tfrac{5}{4}\ \tfrac{11}{8}\ \tfrac{3}{2}\ \tfrac{13}{8}\ \tfrac{7}{4}\ \tfrac{15}{8}\ 2\ \tfrac{9}{4}\ \tfrac{5}{2}\ \tfrac{11}{4}\ 3\ \tfrac{7}{2}\ 4\ 5 \cdots$$

New numbers are created at each stage, which are then used to form further separations, and therefore still more numbers, a doubling of the points (plus one) at each finite stage. The surreal numbers accumulate around us.

Finite Days

We shall presently go much farther, but at this juncture, let me ask which surreal numbers are created at finite stages. Which surreal numbers have a finite birthday?

It is clear that every natural number n is born on day n, along with the integer $-n$. And we are also systematically creating rational numbers at the various midway points. For these numbers, we are therefore always dividing by 2—and our later analysis of the surreal arithmetic will verify that these numbers indeed deserve those rational names. So the numbers created at finite stages will always have a denominator that is a power of 2, and these are precisely the *dyadic* rational numbers, the rational numbers of the form $\pm m/2^n$. Since every dyadic rational indeed will be created at some finite stage, we see that the surreal numbers having a finite birthday are exactly the dyadic rational numbers.

The First Infinite Day

Day ω is a festive, happy day in the surreal numbers—the first day after all finite stages, a day with many delightful surprises. The first observation is that on this day, all the rest of the real numbers, including $1/3$, $\sqrt{2}$, e and π, and so forth, are all born at once, an explosion of number existence. The reason is that every real number r is determined in the Dedekind manner by the cut that it makes in the dyadic rational numbers, so every real number not yet born comes into existence on day ω as the number filling its cut. In each case, the dyadic rationals on the left and right approach the given real number more closely from each side:

$$\sqrt{2} = \left\{1 \quad 1\tfrac{1}{4} \quad 1\tfrac{3}{8} \cdots \;\middle|\; \cdots 1\tfrac{7}{16} \quad 1\tfrac{1}{2} \quad 2\right\}$$

$$\pi = \left\{3 \quad 3\tfrac{1}{8} \quad 3\tfrac{9}{64} \cdots \;\middle|\; \cdots 3\tfrac{5}{32} \quad 3\tfrac{3}{16} \quad 3\tfrac{1}{4} \quad 3\tfrac{1}{2} \quad 4\right\}.$$

Perhaps it is a little surprising that nondyadic rational numbers like $1/3$ and $5/12$ are born in this way at the same time as irrational algebraic numbers like $\sqrt{2}$ and even transcendental numbers like e and π—all the rest of the nondyadic real numbers are born at once on day ω, determined by their cuts in the dyadic rational numbers.

But there is more. The ordinal number ω, for example, is also born on day ω, filling the cut at the top, above all the natural numbers:

$$\omega = \left\{0 \quad 1 \quad 2 \cdots \;\middle|\; \right\}$$

And similarly, $-\omega$ is born this day as the first-born number below all integers:

$$-\omega = \left\{ \;\middle|\; \cdots -3 \quad -2 \quad -1\right\}.$$

There is still more. The number ε is born on day ω, filling the gap between 0 and the positive dyadic rationals:

$$\varepsilon = \left\{0 \;\middle|\; \cdots \tfrac{1}{8} \quad \tfrac{1}{4} \quad \tfrac{1}{2} \quad 1\right\}.$$

This number ε is positive, being strictly larger than 0, but it is smaller than every positive dyadic rational number—thus it is the first-born *infinitesimal* number (along with its negative $-\varepsilon$, born at the same time). It will turn out when we define the surreal arithmetic later that the numbers ε and ω are multiplicative reciprocals of each other:

$$\varepsilon = \frac{1}{\omega} \qquad \omega = \frac{1}{\varepsilon}.$$

But wait—there are still further surprises occurring on day ω, for we also have the surreal number $\frac{1}{2} + \varepsilon$ born on this day, filling the cut above $\frac{1}{2}$ and below all the other dyadic rationals above $\frac{1}{2}$, like this:

$$\tfrac{1}{2} + \varepsilon = \left\{ \tfrac{1}{2} \;\middle|\; \cdots \; \tfrac{9}{16} \;\; \tfrac{5}{8} \;\; \tfrac{3}{4} \;\; 1 \right\}.$$

Similarly, the numbers $\frac{1}{2} - \varepsilon$, $5 + \varepsilon$, and so forth are born on day ω. The surreal numbers born on day ω, it turns out, are exactly the nondyadic real numbers, all the numbers $r \pm \varepsilon$ for every dyadic rational number r, and both ω and $-\omega$.

Further Stages

On day $\omega + 1$, we start filling the infinitesimal neighborhoods around every real number, creating $\pi + \varepsilon$ and $\sqrt{2} - \varepsilon$, for example, but also adding further infinitesimals around the dyadic rationals r, creating $r + \frac{\varepsilon}{2}$ and $r + 2\varepsilon$, as well as $\pm(\omega + 1)$ and $\pm(\omega - 1)$:

$$\omega - 1 = \left\{ 0 \;\; 1 \;\; 2 \;\; \cdots \;\middle|\; \omega \right\}.$$

Thus the construction of the surreal numbers proceeds, endlessly, transfinitely—it never stops. At every ordinal birthday, new surreal numbers are created on the top, at the bottom, and filling any conceivable cut in the numbers previously created:

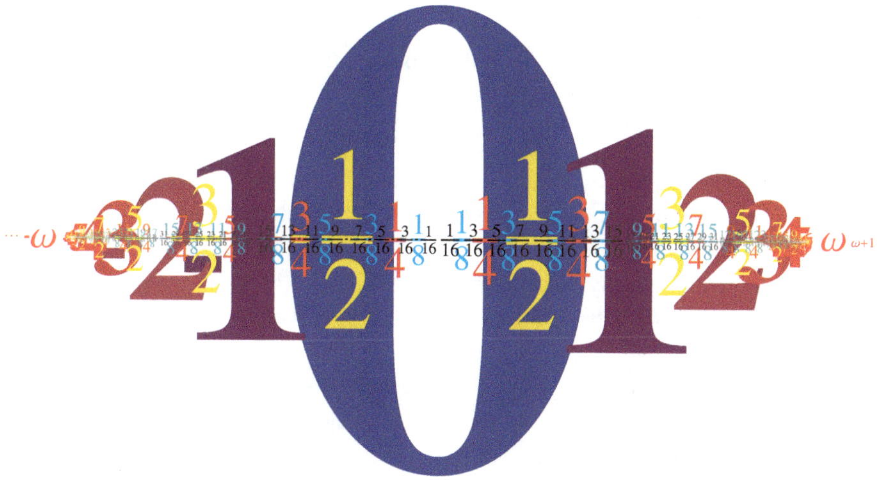

The class of all surreal numbers is denoted by $\mathbb{No}$, with the numbers born before ordinal birthday α denoted by $\mathbb{No}_\alpha$. Thus the surreal numbers accumulate as the days pass endlessly through the transfinite hourglass of time:

$$\mathbb{No} = \bigcup_{\alpha \in \mathrm{Ord}} \mathbb{No}_\alpha.$$

The Surreal Numbers Form a Proper Class

Like the ordinals and the class of all sets, the surreal numbers form what is called a *proper class*—the class of surreal numbers $\mathbb{No}$ is a definable collection of mathematical objects, but it is not a set. In a sense, it is too large to be a set. One can think of the surreal numbers in a potentialist manner as an unfinished project, for which more and more numbers are created, but the process is never completed.

In particular, we carried out the surreal number generation rule through all the ordinals, but there are no stages after all the ordinal stages are complete. We form the surreal number $\{L \mid R\}$ only when L and R are *sets* of surreal numbers. We do not form any surreal number, for example, to fill the gap between 0 and the class of all positive surreal numbers. This would require a proper class upper set—no set of surreal numbers can specify that gap. One can begin to see on these grounds why $\mathbb{No}$ cannot be a set, for if it were, then we could form that gap and we would have the paradox of a positive surreal number that is smaller than all positive surreal numbers. This is a version of the Burali-Forti paradox for ordinals, which similarly shows that the class of ordinals is a proper class, for otherwise there would be ordinals larger than every ordinal.

In light of the proper-class nature of the class of surreal numbers $\mathbb{No}$, some foundational aspects of surreal numbers require delicate set-theoretic care. Meanwhile, in this elementary exploration, everything will be fine.

20.2 Surreal Numerals

Let us discuss a little more the surreal number notation that we have been using. We have said that the surreal number

$$x = \{L \mid R\}$$

is created at an ordinal stage α, when L and R partition the previously created numbers into the lower set L and the upper set R, meaning that every element of L is less than every element of R. I shall refer to $\{L \mid R\}$ as the *birthday* numeral of this newly created surreal number x, for it details exactly the circumstances of its birth, when we place x into the surreal order so it is above every element of L and below every element of R, situating it with respect to all previously created numbers.

One simplifying notational convention that we have already adopted is that we allow ourselves simply to list the numbers of L and R on each side of the surreal numeral notation without enclosing them within additional nested set-notation braces. That is, we write

$\{0\mid\ \}$ for the number 1 rather than $\{\{0\}\mid\varnothing\}$ to indicate the lower set $\{0\}$ and the empty upper set $\varnothing$, and the same applies for other surreal numbers. The meaning is perfectly clear without the extra braces, so let us drop them.

Another simplification is that it is often not necessary to partition all the previously created numbers, since the cut can often be specified uniquely by listing only some of the previous numbers on each side. For example, when L has a largest number, then that number alone determines the lower set. We had specified the number 2 with this convention simply as $\{1\mid\ \}$, for example, and not as $\{-1\quad 0\quad 1\mid\ \}$, although this would also be correct. Similarly, when L has no largest element, then any cofinal subset of L suffices to determine the cut.

In light of these conventions, a given surreal number will generally be represented by many numerals—different ways of writing the same surreal number. The situation is analogous to how we can write a given rational number in different but equivalent ways: $\frac{1}{2}=\frac{2}{4}=\frac{3}{6}$.

These conventions open the door to an even more general numeral usage, where we drop the requirement that L and R should mention only previously created numbers. Namely, let us henceforth allow all numerals of the form $\{L\mid R\}$, whenever L and R are sets of surreal numbers, regardless of the stages at which they are created, provided only that every number in L is less than every number in R. The meaning of this numeral, by definition, is that it represents the first-born surreal number filling this gap, the first-born surreal number that is larger than every number in L and smaller than every number in R. This usage is more general than what we had earlier because we now allow that L and R possibly have surreal numbers created only after the number represented by $\{L\mid R\}$ itself. But this will sometimes be convenient, so let us allow it.

With this generalized numeral notation, for example, we have many numerals to represent the number 1:

$$1 \quad=\quad \{0\mid\ \} \quad=\quad \{\tfrac{1}{2}\mid 2\} \quad=\quad \{-2\quad \tfrac{5}{8}\mid 5\} \quad=\quad \{\varepsilon\mid\omega\}.$$

In each case, the number 1, defined as $\{0\mid\ \}$, is the first-born surreal number filling the specified gap, so these numerals all denote the number 1.

20.3 The Surreal Order

According to the surreal number generation rule, as surreal numbers are created, we also systematically place them in order amongst each other. Namely, whenever a new number $x=\{L\mid R\}$ is created, occupying the gap between the lower set L and upper set R in the previously created numbers, then we place x above every element of L and below every element of R. This specifies how x relates to all the previously created numbers, but also, by transitivity, it tells us how x relates to the other numbers born on the same ordinal day; and the future numbers will find their order relation with x when they are born. In this way, the generation rule provides a total linear order on the surreal numbers.

Every Gap Overflows

I should like particularly to emphasize that although the number $x = \{L \mid R\}$ is defined to be the first-born surreal number filling the gap between L and R, it will never be the only surreal number occupying that gap. Immediately after x is created, we can subsequently create the numbers $\{L \mid x\}$ and $\{x \mid R\}$, which are distinct from x but find themselves also strictly between L and R:

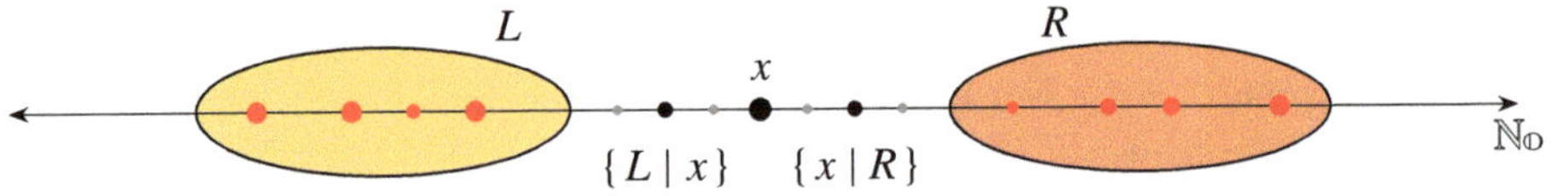

And the further gaps between those numbers will likewise immediately get filled, with new gaps created and filled afresh eternally. The gap between L and R ultimately overflows with surreal numbers. In particular, these observations show that the number $x = \{L \mid R\}$ is neither the supremum of L nor the infimum of R.

The Surreal Numbers Are Fundamentally Discontinuous

There is a certain heavy consequence of this for the foundational theory of the surreal numbers. The rude fact of the matter is that the surreal number line is fundamentally discontinuous—it is Dedekind incomplete and lacks the least upper bound property. The least upper bound property, of course, is the familiar completeness property of the real line, stating that every nonempty bounded set has a least upper bound. It is impossible to overstate the vital importance of the completeness property in the fundamental theory of the calculus and real analysis—completeness is a bedrock axiomatic principle pervasively used in every fundamental result of the subject.

And yet this completeness property fails quite badly in the surreal numbers—it fails for the surreal numbers in every nontrivial instance. To see this, observe that if L is any set of surreal numbers with no largest element and b is an upper bound of L, then the surreal number $\{L \mid b\}$ is a strictly smaller upper bound of L. So there can be no least upper bound:

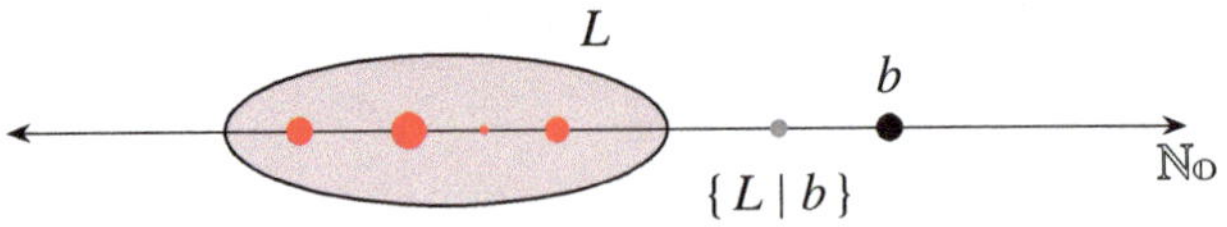

The greatest-lower-bound property similarly fails in the surreal numbers.

The same kind of reasoning shows that there are no nontrivial convergent sequences in the surreal numbers. This holds not just for ordinary sequences of length ω, but also for transfinite sequences of arbitrary limit ordinal length, as well as for the more general nets considered in topology—there are no nontrivial convergent instances (but there are nontrivial proper class sequences and nets that converge). Consider, for example, an increasing sequence x_n with candidate limit point ℓ. Let us simply wait until all those numbers are born, and then observe that the number $\{\, x_0\ x_1\ x_2\ \cdots \mid \ell \,\}$ sneaks between the sequence and the supposed limit, thereby preventing convergence:

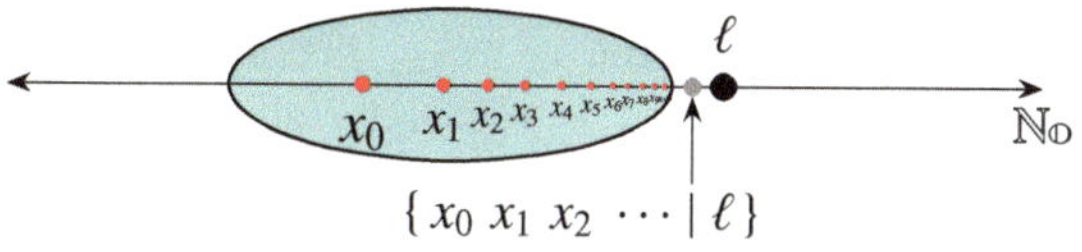

The reader is invited in the questions at the end of the chapter to show, without the monotone increasing assumption used here, that there are no nontrivial convergent sequences in the surreal numbers.

A Red Flag for Surreal Calculus?

So the surreal number line is at bottom discontinuous. This could be a red flag for those mathematicians who hope to use the surreal numbers at the core of a new foundational theory of the calculus—one based on the surreal numbers in place of the real numbers. The problem would be that most of the classical calculus theory is based fundamentally on continuity ideas—one is taking the limit of a sequence or finding a least upper bound in essentially every core calculus result, including the definitions of the derivative and the definite integral. An alternative surreal-numbers-based theory would therefore have to proceed in a very different manner, for in the surreal numbers, there simply are no nontrivial convergent sequences or least upper bounds. Nevertheless, the practitioners propose an alternative calculus based on the abundant surreal infinitesimals, for example, rather than limits. It is a work in progress by the experts.

I can imagine a future in which schoolchildren learn from a young age to count in the surreal numbers, perhaps mastering the ideas of this whole chapter in elementary school and studying surreal-numbers-based calculus soon after.

Alternative Account of the Surreal Order

It turns out to be possible to express the surreal order directly from generalized numerals, apart from how I defined the order as arising during the construction process of the surreal numbers via the surreal number generation rule. Namely, if we are given surreal numbers $x = \{\, X_L \mid X_R \,\}$ and $y = \{\, Y_L \mid Y_R \,\}$, each specified by a pair of lower and upper sets, which is not necessarily a partition of all the previous numbers, then what I claim is that $x \leq y$ in the surreal numbers if and only if these numerals exhibit no explicit obstacle to this. That is, $x \leq y$, provided that there is no $y_R \in Y_R$ with $y_R \leq x$, which of course would have forced x above y, and also no $x_L \in X_L$ with $y \leq x_L$, which would have forced y below x. The absence of these obstacles, I claim, is equivalent to $x \leq y$ as I defined it via the generation process.

To prove the forward implication is easy, for if $x \leq y$, then of course there can be no such obstacles since $\leq$ as we defined it from the generation rule is a linear order. Conversely, suppose that there are no such obstacles. In particular, we must have $x < y_R$ for every y_R in Y_R and $x_L < y$ for every x_L in X_L. If $x \leq y_L$ for some $y_L \in Y_L$, then of course $x < y$, and so $x \leq y$. Similarly, if $x_R \leq y$ for some $x_R \in X_R$, then again $x < y$, and so $x \leq y$. The remaining case is that $y_L < x < y_R$ and $x_L < y < x_R$ for all the points from the respective lower and upper sets. In other words, x and y are each in the gap specified by the other numeral, as well as their own. In this case, it follows that $x = y$ since whichever number is born first (or at the same time as the other) will be at least as good a candidate for being the first-born number in the other gap, and hence it will be equal to the other number. So $x = y$, and consequently $x \leq y$ as well, as desired.

The key point here is not merely that the no-obstacles criterion holds for the surreal number order as I have defined it, but rather that the no-obstacles criterion provides an alternative way to define the order relation $x \leq y$ in the first place. This would be a definition of the order by recursion on the numerals, without reference to the order as it is defined via the generation rule. The recursion is well founded because the no-obstacles condition makes reference only to simpler instances of the order, involving lower-rank numerals, since we are no longer comparing x with y but rather comparing either x or y with elements of the partitions of one numeral or the other, and these have lower rank in the hierarchy of numerals. The no-obstacles critierion therefore provides a stand-alone alternative numeral-based way to define the surreal order.

Alternative Numeral-Based Account of the Surreal Numbers

This observation leads to a useful alternative perspective on how we might understand what surreal numbers are. Namely, it is possible to develop the surreal numbers with a completely numeral-based approach, presenting them from the beginning on the basis of generalized numerals, taking the numerals rather than the numbers as the focal objects of study, and furthermore regarding the numerals to be recursively specified by lower and upper sets of numerals rather than sets of numbers. In this way of proceeding, the no-obstacle criterion is understood as the *definition* of the surreal order on the numerals, and not as a theorem that flows from the surreal number generation rule as in my approach. This account is also used to specify which forms $\{ L \mid R \}$ are allowed as numerals since we want every element of L to be less than every element of R.

In this numeral-based approach, the surreal numbers themselves arise by a process of abstraction from the numerals with respect to the accompanying equivalence relation

$$\{ X_L \mid X_R \} \equiv \{ Y_L \mid Y_R \} \quad \text{if and only if}$$

$$\{ X_L \mid X_R \} \leq \{ Y_L \mid Y_R \} \quad \text{and} \quad \{ Y_L \mid Y_R \} \leq \{ X_L \mid X_R \}.$$

In this approach, the surreal numbers thus arise in effect as metaphenomenal, amounting to the equivalence classes of the numerals with respect to this equivalence relation. One can then observe afterward, as a consequence of this development, that the generation rule is fulfilled.

The numeral-based conception of the surreal numbers arises quite naturally when one seeks to develop the theory of the surreal numbers as part of the general theory of Conway games, for surreal numerals indeed can be seen as special kinds of games. I greatly admire the theory of Conway games, with its own elegant development, and I am truly fascinated that one can view surreal numbers as games.

Nevertheless, I have sought in this chapter to give a stand-alone account of the surreal numbers alone, without delving into the theory of Conway games. And when considering the surreal numbers on their own, apart from the games, I find the surreal number generation rule deeply compelling and insightful as a fundamental principle, leading as it does to a clear and distinct conception of the surreal numbers. To my way of thinking, the generation rule expresses the central surreal conception, the elegant gap-filling idea, which leads ultimately to all the rest, to the surreal number order and all the further surreal number structure that we shall discuss presently. So I prefer to keep it at the center of the development. Surreal numbers are created individually and placed in order by the generation rule, with no need for an abstraction of numerals by an equivalence relation, and the numeral semantics are simply a matter of definition—a numeral represents the first-born number filling its gap.

Fortunately, it doesn't matter which presentation of the surreal numbers one adopts, since each way of proceeding leads on its own axiomatic principles to the defining properties of the other approach. So there is nothing at stake in the foundational dispute and the two conceptions give rise to exactly the same surreal number structure.

20.4 Universality

The surreal line $\mathbb{No}$ enjoys a remarkable feature, a beautiful mathematical property—the surreal numbers are universal for all linear orders. What this means is that every linear order $\langle L, \leq_L \rangle$, no matter how strange or unusual, will find a perfect copy of itself within the surreal number line. There will be an order-preserving embedding of L into the surreal numbers—indeed many such embeddings—each forming an isomorphic correspondence between L and the corresponding copy of it within $\mathbb{No}$. We already knew how to find copies of the natural numbers, the real numbers, and even the ordinals in the surreal numbers, but the universality claim is that every linear order has a copy in $\mathbb{No}$.

The proof of this statement is a transfinite generalization of Georg Cantor's argument that the rational line $\mathbb{Q}$ is universal for countable linear orders. In the surreal case, the phenomenon will flow directly from the gap-filling idea of the surreal number generation rule. The generation rule, after all, amounts to a transfinite generalization of the density criterion used in Cantor's theorem since a linear order is endless and dense if and only if it fills every gap specified by finitely many points. We shall define the embedding of L in $\mathbb{No}$ in stages, and the generation rule will enable us systematically to extend a partial embedding with any additional point of L, wherever it is, since every possible next point determines a cut in the previous points, so there will be a surreal number in the corresponding gap.

Let me explain. Suppose that $\langle L, \leq_L \rangle$ is a given linear order. By the well-order principle, there is an enumeration of the points of L in a well ordered sequence, indexed by ordinals:

$$p_0 \quad p_1 \quad p_2 \quad \cdots \quad p_\alpha \quad \cdots$$

This enumeration, of course, does not necessarily list the points of L in their $\leq_L$ order—the points may jump up and then down in the $\leq_L$ order as the enumeration proceeds. Rather, it provides a way to conceive the points of L as being born at ordinal stages—the point p_α of L is born on day α, for instance. We now map the points of L into the surreal numbers in stages, gradually defining the map $p_\alpha \mapsto x_\alpha$.

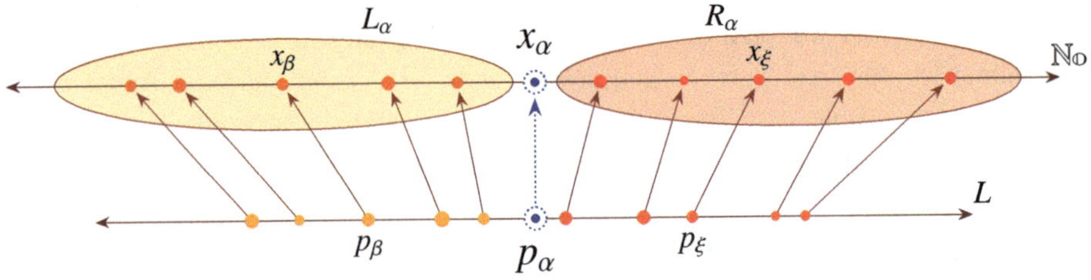

Suppose that at stage α, we have already specified the earlier part of the correspondence—we have specified the map $p_\beta \mapsto x_\beta$ for all earlier $\beta < \alpha$. It is time to consider the next point p_α itself. This point makes a certain cut in the L order amongst the earlier p_β, of course, being perhaps above some of them and below others. Below, let L_α be the set of the corresponding surreal numbers x_β for earlier $\beta < \alpha$ with $p_\beta < p_\alpha$; and above, let R_α be all those x_ξ for earlier $\xi < \alpha$ with $p_\alpha < p_\xi$. Every element in L_α is below every element of R_α, because inductively, our current partial map is order-preserving. So we can define $x_\alpha = \{ L_\alpha \mid R_\alpha \}$ as the surreal number born specifically to fill this gap, thereby extending the embedding one more step. In the end, after all steps are completed, the map $p_\alpha \mapsto x_\alpha$ provides the desired isomorphism of L with a copy of it in the surreal line $\mathbb{No}$.

In fact, there are many such embeddings of L into $\mathbb{No}$. One way to see this is to observe that at each stage, there were many alternative ways to choose x_α from the gap between L_α and R_α, since $\mathbb{No}$ has many points in that gap and every choice achieves a different embedding of L into $\mathbb{No}$. I claim more generally that for any interval $a < b$ in the surreal numbers, no matter how small, we can find a copy of L contained entirely within that interval. For example, we can find a copy of the real number line $\mathbb{R}$ between 0 and ε if we want, or indeed any linear order L, simply by putting 0 into all the lower sets and ε into all the upper sets as we proceed, so every x_α would be between them.

Let me also mention that the surreal numbers are universal not merely as an order, but also universal as an ordered field—that is, with respect also to the algebraic structure of addition and multiplication, which we shall begin to introduce on page 267. The surreal arithmetic turns the surreal numbers into what is called an *ordered field*, and the universality property is that every ordered field is isomorphic to a subfield of No. Without getting into any details, let me say only that this can be proved by building the desired embedding in a series of stages, similar to how we proved the order universality.

20.5 The Surreal Tree

Let us turn now to investigate another foundational feature of the surreal numbers, the underlying *surreal tree*, depicted in figure 20.1. The surreal tree is a binary branching tree, organized into ordinal levels, with a level for every ordinal—perhaps it helps to imagine some leveling of the fantastically drooping levels of the figure. Notice how every node in the tree splits into two child nodes, one on the left and one on the right, and at limit stages, we have nodes for all the possible branches traversing the tree to that level.

Every surreal number corresponds to a node in this tree and conversely. The child nodes branching to the right and left are the two surreal numbers born adjacent to it at the next stage, filling the gaps on each side between that number and the others. More generally, every branch in the tree indicates successively whether the intended number is above or below the previous number by branching right or left, respectively.

For example, we can see in the tree how the surreal number ω arises by branching to the right at every finite stage, and ε arises by branching in the positive direction at the first step (thereby getting above 0), but then to the left at all subsequent finite stages (thereby getting below all $1/2^n$). Similarly, we have branches proceeding to $\sqrt{2}$ or $-\omega$ and so forth.

Binary Sequence Representation

The surreal tree leads immediately to a binary-sequence representation of the surreal numbers, not a sequence of 0s and 1s but rather a sequence of + and – signs, corresponding to branching right or left in the tree, respectively. Every node in the tree is specified by the pattern of branching left or right so as to arrive at that node, so every surreal number is specified by a corresponding +– sequence of some ordinal length α, for numbers born on day α.

The surreal number 0, for example, is represented by the empty sequence $\langle\rangle$. From 0, we get to 1 by moving right and to -1 by moving left. In this way, we can realize the binary

Figure 20.1: The surreal tree, with apologies

sequence representation of the first few surreal numbers as follows:

$$\begin{aligned} 0 &= \langle\rangle \\ 1 &= \langle + \rangle \\ -1 &= \langle - \rangle \\ 2 &= \langle ++ \rangle \\ -2 &= \langle -- \rangle \\ \tfrac{1}{2} &= \langle +- \rangle \\ -\tfrac{1}{2} &= \langle -+ \rangle \\ \tfrac{3}{2} &= \langle ++- \rangle \\ -\tfrac{1}{4} &= \langle -++ \rangle \end{aligned}$$

Can you find the binary representation of the surreal numbers born on the next several birthdays? Readers with programming knowledge are encouraged in the questions at the end of the chapter to automate the task of converting a finite binary +− sequence to the corresponding dyadic rational number and conversely.

On the first infinite birthday, the number ω is represented by the sequence $\langle +++++\cdots \rangle$, and ε is $\langle +----\cdots \rangle$. Every ordinal α is represented by the sequence $\langle ++++\cdots \rangle$, consisting of exactly α many + signs. Positive infinitesimals begin with $\langle +-----\cdots \rangle$, with ω many − signs. Can you find the surreal binary representation of $\frac{1}{3}$, or $\frac{1}{3}+\varepsilon$? How about $\sqrt{\omega}$ or $\sqrt{\varepsilon}$? How does the binary sequence representation of x relate to that of $-x$?

20.6 Surreal Arithmetic

Finally, we come to the rich algebraic structure of the surreal numbers, moving beyond the order-theoretic features. Let us introduce and develop the theory of surreal addition and surreal multiplication, which together give the surreal numbers the structure of an ordered field. Mathematicians have also endowed the surreal field with further algebraic structure, augmenting it with versions of surreal exponentiation e^x, surreal trigonometic functions $\sin x$, and much more. But we shall focus here on the field structure.

Surreal Addition

For addition, we define the sum of two surreal numbers $x = \{X_L \mid X_R\}$ and $y = \{Y_L \mid Y_R\}$ by the following recursive rule:

$$x + y = \left\{ x + y_L \quad x_L + y \,\middle|\, x + y_R \quad x_R + y \right\}.$$

What we mean by this compact notation is that we are to include all possible choices of x_L from the lower set X_L and all possible $x_R \in X_R$, and similarly with $y_L \in Y_L$ and $y_R \in Y_R$. The lower and upper sets of $x + y$ here will have many entries if those of x and y are large;

but similarly, if X_L or another of the numeral sets is empty, then there is no x_L, so those corresponding terms simply do not appear in the numeral for $x + y$.

Considered carefully, the definition makes tremendous sense. Namely, since x and y fill the gaps in their own numerals, we know that $x_L < x < x_R$ and $y_L < y < y_R$, so if we are trying to make addition work with respect to the order—indeed, we aim for it to be what is called an *ordered group*—then we certainly want $x_L + y$ and $x + y_L$ to be less than $x + y$ and $x + y_R$ and $x_R + y$ to be above it. The definition of $x + y$ implements those requirements by placing those numbers into the lower and upper sets of the numeral for $x + y$, respectively. In short, the sum $x + y$ is defined to be the first-born surreal number that is larger than every $x_L + y$ and $x + y_L$ and smaller than every $x + y_R$ and $x_R + y$. Since furthermore this definition specifies $x + y$ in terms of simpler addition instances, it serves as a legitimate, well founded recursive definition of the operation on all surreal numbers.

Note that we defined $x + y$ in terms of the given numerals for x and y, and in this sense, we may fruitfully view surreal addition initially as an operation on surreal numerals rather than surreal numbers themselves. However, the operation is well defined with respect to the surreal equivalence of numerals—you always get the same surreal number for $x + y$ independent of which numerals for x and y that you use. In other words, the equivalence $\equiv$ of numerals is a congruence with respect to the surreal addition of numerals. And since the operation on numerals is well defined in this way, it induces a meaningful operation on the surreal numbers themselves, and this is what we think of as surreal addition.

Surreal Addition Is Commutative and Associative

It is not difficult to see by induction that surreal addition is commutative:

$$\begin{aligned} x + y &= \{x + y_L \quad x_L + y \mid x + y_R \quad x_R + y\} \\ &= \{y_L + x \quad y + x_L \mid y_R + x \quad y + x_R\} \\ &= y + x. \end{aligned}$$

The point here is that we assume inductively that the simpler instances are commutative, and thereby we can deduce commutativity for $x + y$. This is a form of strong induction on pairs of birthdays. Strong induction requires no anchor since the induction hypothesis is vacuous in the case of the simplest number $0 = \{\ \mid\ \}$, which has no lower or upper terms—we see that $x + 0 = 0 + x$ without any induction assumption because both these numbers are just x itself.

The reader is invited in the questions at the end of the chapter to check similarly that surreal addition is associative:

$$x + (y + z) = (x + y) + z.$$

Additive Identity

Let us verify next that 0 is the additive identity for surreal addition, meaning that $x + 0 = x$ for every surreal number x. Notice that since $0 = \{\ \mid\ \}$, there are no numbers of the form 0_L or 0_R, so we can calculate

$$x + 0 = x + \{\ \mid\ \} = \{\, x_L + 0 \mid x_R + 0 \,\} = \{\, x_L \mid x_R \,\} = x.$$

We assumed by induction that $x_L + 0 = x_L$ and $x_R + 0 = x_R$, since these are simpler instances of the additive identity property, and thereby we deduced that it also holds for $x + 0 = x$, as desired.

Several Sample Surreal Sums

Let us get some practice with surreal addition by computing some simple sums. Why not start with 1 + 1? We can simply unwind the definition of surreal addition to calculate

$$\begin{aligned} 1+1 &= \{0 \mid\ \} + \{0 \mid\ \} \\ &= \{\, 1+0 \quad 0+1 \mid\ \} \\ &= \{1 \mid\ \} \\ &= 2. \end{aligned}$$

Notice again that since there are no numbers as 0_L, 0_R, or 1_R, the expressions involving such terms simply do not appear in the numeral for 1 + 1.

Let us similarly calculate $\frac{1}{2} + \frac{1}{2}$ as follows:

$$\begin{aligned} \tfrac{1}{2} + \tfrac{1}{2} &= \{0 \mid 1\} + \{0 \mid 1\} \\ &= \{\, \tfrac{1}{2} + 0 \quad 0 + \tfrac{1}{2} \mid \tfrac{1}{2} + 1 \quad 1 + \tfrac{1}{2} \,\} \\ &= \{\, \tfrac{1}{2} \mid 1 + \tfrac{1}{2} \,\} \\ &= 1. \end{aligned}$$

For the last step, the sum is 1 because 1 is the first-born number in that final cut. Note that this is an instance where the definition of addition led to a generalized numeral since the lower and upper sets had numbers born after the resulting value 1.

Surreal Negation and Surreal Subtraction

We define the negative $-x$ of any surreal number x as follows:

$$-x = \{\, -X_R \mid -X_L \,\}.$$

By the notation $-X_R$ here, what I mean is the set of all $-x_R$ for $x_R \in X_R$, and similarly with $-X_L$. Using this, we can define subtraction simply as adding the negation:

$$y - x = y + (-x).$$

Let us make a few further observations about the nature of negation. First, observe that $-0 = 0$ since we still have the empty set on both sides. Also, we can prove inductively that $-(-x) = x$. By combining negation with addition, we can see inductively that

$$-(x + y) = (-x) + (-y).$$

It is also easy to see that $x \leq y$ if and only if $-y \leq -x$. More than this, I claim that $x \leq y$ if and only if $0 \leq y - x$. Using these observations, we can verify that $x + (-x) = 0$. To see this, observe first that since $x \leq x$, it follows that $0 \leq x - x$. And consequently, $-(x - x) \leq -0 = 0$; but $-(x - x) = -(x + (-x)) = -x + -(-x) = -x + x = x - x$. So we have seen that $0 \leq x - x \leq 0$, and thus $x - x = 0$, as desired.

To summarize, what we have shown so far is that surreal addition $x + y$ is an associative, commutative operation with an additive identity and an additive inverse $-x$ for every surreal number x. One can furthermore show that surreal addition interacts with the order in the sense that $x \leq y \leftrightarrow x + z \leq y + z$. These facts altogether mean that the surreal numbers form an ordered abelian group under addition.

The Dyadic Rationals as Surreal Numbers

Let us take a slight detour to justify the dyadic rational-number names that we had given to the surreal numbers arising at finite stages. We shall justify these names, such as $\frac{1}{2}$ and $-\frac{9}{4}$, by showing that the surreal arithmetic of these numbers agrees with the corresponding rational-number arithmetic of the names that we had assigned.

We have already observed that the surreal number $0 = \{ \mid \}$ is the additive identity in surreal arithmetic, which surely justifies this name. We take the surreal number 1 as $\{0 \mid \}$ to be a matter of definition, but this name will also be justified by the fact that it will turn out to be the multiplicative identity.

Let us show next that the natural number names that we had introduced respect the operation of adding 1. To clarify the discussion, let me introduce some subtle notation to distinguish between the natural number n and the surreal-number counterpart $\dot{n}$ to which we had assigned n. Specifically, we had defined an association of every natural number n with a surreal-number counterpart $\dot{n}$ as follows:

$$\begin{aligned} \dot{0} &= \{ \mid \} \\ \dot{1} &= \{\dot{0} \mid \} \\ n \dot{+} 1 &= \{\dot{n} \mid \}. \end{aligned}$$

To prove that this association is robust, let us simply verify that adding $\dot{1}$ to the surreal number $\dot{n}$ in surreal arithmetic gives the desired result $n \dot{+} 1$. This is true in the case $n = 0$

since $\dot{0} + \dot{1} = \dot{1}$. So let us suppose that it works for n and consider $n + 1$:

$$\begin{aligned} (n \dot{+} 1) + \dot{1} &= \{\dot{n} \mid \ \} + \{\dot{0} \mid \ \} \\ &= \{(n \dot{+} 1) + \dot{0} \quad \dot{n} + \dot{1} \mid \ \} \\ &= \{n \dot{+} 1 \mid \ \} \\ &= n \dot{+} 2. \end{aligned}$$

We see therefore that our correspondence respects the operation of adding 1, so by induction, all the natural number names are correct. (And thus we may also henceforth omit the dots.)

We have already verified that $\frac{1}{2} + \frac{1}{2} = 1$, which justifies the name $\frac{1}{2}$ for that surreal number. And we can similarly verify

$$\begin{aligned} \tfrac{1}{4} + \tfrac{1}{4} &= \{0 \mid \tfrac{1}{2}\} + \{0 \mid \tfrac{1}{2}\} \\ &= \left\{\tfrac{1}{4} + 0 \quad 0 + \tfrac{1}{4} \mid \tfrac{1}{4} + \tfrac{1}{2} \quad \tfrac{1}{2} + \tfrac{1}{4}\right\} \\ &= \left\{\tfrac{1}{4} \mid \tfrac{1}{2} + \tfrac{1}{4}\right\} \\ &= \tfrac{1}{2}. \end{aligned}$$

The final numeral represents $\frac{1}{2}$ because this is the first-born number in that gap.

More generally, suppose that q is the smallest positive number at some stage, so $0 < q$, with no numbers yet between them. Let us inductively assume also that there are no numbers yet between q and $q + q$, which amounts to assuming that $q = \{0 \mid q + q\}$. Let $p = \{0 \mid q\}$, filling the gap between 0 and q. Observe that

$$p + p = \{0 \mid q\} + \{0 \mid q\} = \{p + 0 \quad 0 + p \mid p + q \quad q + p\} = \{p \mid p + q\} = q.$$

For the last step, we have $\{p \mid p + q\} = q$ because our assumptions ensure that q is the only number currently in that gap. And notice that we fulfill the induction assumption with p since p is the only number now in the gap between 0 and $q = p + p$. Thus we may conclude that all the dyadic fractions $1/2^n$ are named correctly since we are indeed dividing by 2 each time. And by adding these together suitably, it follows that all the dyadic rationals are properly named.

More generally, one can argue that if $a < b$ at a finite stage, with nothing between them, then the number $x = \{a \mid b\}$ is exactly halfway between, in the sense that $x - a = b - x$. One can show that $x - a = \{0 \mid b - a\}$, which is half of the full difference $b - a$. And for these reasons, the dyadic rational number names that we associated with the finite-stage surreal numbers are correct.

Surreal Multiplication

We come now to surreal multiplication, a slightly more delicate matter than addition.

Let me first dispense with a false start, an incorrect, naive definition of multiplication that might initially seem attractive but which upon reflection we realize is incorrect. Namely, the initially attractive idea is that perhaps multiplication could be defined in direct analogy with the additive case as follows:

$$x \circledcirc y = \{\, x \circledcirc y_L \quad x_L \circledcirc y \mid x \circledcirc y_R \quad x_R \circledcirc y \,\}.$$

However, this definition will not work properly—it is not what we want. To see this, consider how we would calculate $1 \circledcirc 0$ with this definition, as follows:

$$1 \circledcirc 0 = \{\, 0 \circledcirc 0 \mid \ \} = \{\, 0 \mid \ \} = 1.$$

This is certainly not what we want! So $x \circledcirc y$ is not our desired multiplication concept.

The right definition of multiplication turns out to be the following, where, for simplicity, we often denote a product $x \cdot y$ simply by the juxtaposition of the factors, xy:

$$\begin{aligned} x \cdot y &= \{\, X_L \mid X_R \,\} \cdot \{\, Y_L \mid Y_R \,\} \\ &= \{\, x_L y + x y_L - x_L y_L \quad x_R y + x y_R - x_R y_R \mid x_L y + x y_R - x_L y_R \quad x y_L + x_R y - x_R y_L \,\}. \end{aligned}$$

This definition is motivated in part by the idea that if $x_L < x$ and $y_L < y$, then we should expect

$$0 < (x - x_L)(y - y_L) = xy - x_L y - x y_L + x_L y_L,$$

and consequently that

$$x_L y + x y_L - x_L y_L < xy.$$

This is why this term appears in the lower set of xy. Similarly we expect that

$$0 < (x_R - x)(y_R - y),$$

and consequently

$$x y_R + x_R y - x_R y_R < xy,$$

which is why this term also appears in the lower set of xy. Similar calculations lead to the expressions in the upper set, and thus to the definition of xy. Ultimately, therefore, we have defined xy to be the first-born surreal number that obeys these requirements.

Just as with addition, it is not difficult to prove that surreal multiplication is associative and commutative, and furthermore that surreal multiplication distributes over surreal addition $x(y + z) = xy + xz$.

Let's make some further sample calculations. To begin, it is easy to see that 0 is a multiplicative annihilator, meaning that $x \cdot 0 = 0$. We simply form the numeral for the product as follows. Notice that since there is no 0_L or 0_R, the terms with those expressions do not arise in the product numeral:

$$x \cdot 0 = x \cdot \{\ \mid\ \} = \{\ \mid\ \} = 0.$$

Let us next show that 1 is the multiplicative identity, calculating $x \cdot 1$. Since $1 = \{0 \mid \ \}$, there is no 1_R, so terms with that expression do not appear in the product:

$$\begin{aligned} x \cdot 1 &= x \cdot \{0 \mid \ \} \\ &= \left\{ x_L \cdot 1 + x \cdot 0 - x_L \cdot 0 \;\middle|\; x \cdot 0 + x_R \cdot 1 - x_R \cdot 0 \right\} \\ &= \{ x_L \mid x_R \} \\ &= x. \end{aligned}$$

Therefore, 1 is the multiplicative identity.

For our sanity, let us calculate $2 \cdot 2$ and $2 \cdot 3$ to see if this is really working. Note that there are no such numbers as 2_R or 3_R, so terms with these expressions do not arise when forming the product numerals:

$$\begin{aligned} 2 \cdot 2 &= \{1 \mid \ \} \cdot \{1 \mid \ \} \\ &= \{1 \cdot 2 + 2 \cdot 1 - 1 \cdot 1 \mid \ \} \\ &= \{3 \mid \ \} \\ &= 4. \end{aligned}$$

$$\begin{aligned} 2 \cdot 3 &= \{1 \mid \ \} \cdot \{2 \mid \ \} \\ &= \{1 \cdot 3 + 2 \cdot 2 - 1 \cdot 2 \mid \ \} \\ &= \{5 \mid \ \} \\ &= 6. \end{aligned}$$

Excellent! Just what we would want.

Let's also check what we had claimed earlier—namely, that $\varepsilon\omega = 1$. Notice that there is no ω_R, so those terms do not appear:

$$\begin{aligned} \varepsilon\omega &= \{0 \mid \cdots \tfrac{1}{4} \ \ \tfrac{1}{2} \ \ 1\}\{0 \ \ 1 \ \ 2 \cdots \mid \ \} \\ &= \left\{ 0y + \varepsilon n - 0n \;\middle|\; \varepsilon n + \tfrac{1}{2^n}\omega - n\tfrac{1}{2^n} \right\} \\ &= \left\{ \varepsilon n \;\middle|\; \varepsilon n + \tfrac{1}{2^n}\omega - n\tfrac{1}{2^n} \right\} \\ &= 1. \end{aligned}$$

The final value is 1 because one can show that εn is a positive infinitesimal for every natural number n, so these numbers are all less than 1, and the expression $\varepsilon n + \frac{1}{2^n}\omega - n\frac{1}{2^n}$ appearing in the upper set is infinite—definitely bigger than 1. So the first-born number in the gap is 1, and consequently, $\varepsilon\omega = 1$. This justifies our earlier claim that these two numbers are reciprocals.

In fact, one can prove that every nonzero number has a multiplicative inverse, so we have a notion of division x/y in the surreal numbers, provided that $y \neq 0$. Furthermore, the multiplicative structure respects the order in the sense that $x \leq y$ and $0 \leq c$ implies that $cx \leq cy$. Because of this, the surreal numbers become what is known as an *ordered field*. Indeed, we mentioned earlier that they are a universal ordered field—every ordered field embeds isomorphically as a subfield.

20.7 Surreal Structure

What is the core surreal number structure? Mathematicians often take the surreal numbers as an ordered field, with the basic operations of addition and multiplication. The order is definable from the algebraic structure since $x \leq y$ if and only if there is z such that $x + z^2 = y$. Viewing the surreal numbers as an ordered field, the core structure would be the field structure—addition and multiplication—and the order comes along with that.

The trouble with this perspective, however—the trouble with taking the surreal numbers at bottom as an ordered field—is that most of the other familiar structural features of the surreal numbers, including the birthday order and the surreal numeral $\{ L \mid R \}$ representations, are simply not structural with respect to the algebraic field structure. The fact of the matter is that the surreal number field has numerous automorphisms that simply do not respect this other structure. If one views the surreal numbers as an ordered field, therefore, then one cannot refer to the birthday order, or to left and right sets of the surreal numerals, because these features are not preserved by automorphisms of the field structure.

Let me explain. The surreal numbers form a real-closed field, which means that they form an ordered field in which every positive number has a square root and every odd degree polynomial has a root. In any real-closed field, one finds every rational number p/q, and indeed every real algebraic number, such as $\sqrt{2}$ or $\sqrt[3]{1 + \sqrt{2}}$. But meanwhile, in fact only the algebraic numbers are definable. A deep theorem of Alfred Tarski shows that in any real-closed field, every assertion is equivalent to a quantifier-free assertion in the language of ordered fields. It follows that the surreal number ω fulfills exactly the same field-theoretic assertions as any infinite surreal number. And since the surreal field is saturated and homogeneous, there will be a field automorphism moving ω to any other such infinite element. We can move ω to $\sqrt{\omega} + 5$, for example, or to $\frac{1}{2}\omega^2 - \sqrt{\omega + 17}$. Since $\omega = \{ 0, 1, 2, \ldots \mid \ \}$ and the automorphism fixes the positive integers, this shows that the surreal numeral notation is not structural. Similarly, every transcendental real number in the surreal numbers fulfills the same assertions as any surreal number in the same infinitesimal interval, so there are automorphisms moving π to $\pi + 1/\sqrt{\omega + 5}$. This shows again that the birthday order and the surreal numerals are not structural with respect to the field structure.

The situation is similar to that of the complex field $\mathbb{C}$, where we understand the complex numbers as $a + bi$, where $a, b \in \mathbb{R}$, but this representation amounts to extra structure not derivable from the complex field structure alone. Indeed, there are many distinct weird copies of $\mathbb{R}$ in $\mathbb{C}$, over which $\mathbb{C}$ arises as the algebraic closure, and one cannot define the real-part and imaginary-part operators purely from the complex field algebraic operations. The $a + bi$ notation is how we construct the complex field, in effect interpreting it in the real field with ordered pairs, but this is not derivable from the complex algebraic field operations. Similarly, the $\{ L \mid R \}$ notation for the surreal numbers is how we construct the surreal field, in effect interpreting it in set theory, but this is not derivable from the surreal algebraic field operations and order.

Questions for Further Thought

20.1 How many surreal numbers are there at the start of finite day n?

20.2 Can you describe exactly the birthday of $m/2^n$, assuming that this is in lowest terms? (Hint: The ceiling function $\lceil m/2^n \rceil$ might be part of your consideration.)

20.3 Compare and contrast the nature of filling gaps in the surreal numbers with the idea of a Dedekind cut in the rational numbers in the traditional account of the real numbers. What are the important differences in these two concepts of filling cuts in an order?

20.4 Describe exactly which surreal numbers are born on day $\omega + 1$, day $\omega + 2$, and so forth.

20.5 Show that there are no nontrivial convergent sequences in the surreal numbers.

20.6 If birthday α just occurred, show that all surreal numbers born on day $\alpha + 1$ are adjacent on that day to a previous surreal number—either the lower set has a largest number or the upper set has a smallest number. What is the situation at limit ordinal birthdays?

20.7 Summarize the foundational dispute in the two approaches to the surreal numbers—one founded on the surreal number generation rule and the other using the numeral-based, no-obstacles recursive definition of the order on numerals. Can you describe for each concept how we are to think about what a surreal number is?

20.8 Provide explicitly a copy of the real line $\langle \mathbb{R}, < \rangle$ between 0 and ε. By what ordinal stage does it appear?

20.9 What are the surreal binary $+-$ representations of $\frac{1}{3}$, $\frac{1}{3} + \varepsilon$, $\sqrt{\omega}$, and $\sqrt{\varepsilon}$?

20.10 For programmers: Write a program that converts a given finite $+-$ sequence to the rational number that it represents.

20.11 How does the surreal binary representation of a surreal number x relate to the sequence for x?

20.12 Argue that surreal addition is associative $x + (y + z) = (x + y) + z$ by reasoning inductively on the birthdays.

20.13 Is the surreal number $x + y$ born always after x and y, before, or at the same time? Give as full an answer as you can, exhibiting the range of behavior.

20.14 Calculate $3 \cdot 3$ in the surreal numbers.

20.15 Prove that the product of two positive surreal numbers is positive, the product of a positive and a negative is negative, and the product of two negative numbers is positive.

20.16 Is the surreal field a complete ordered field?

20.17 How would you develop the surreal-number analogue of the complex numbers, known as the *surcomplex field*?

Further Reading

Knuth (1974). A fanciful mathematical text relating the story of a young couple marooned on an island, who find a tablet and discover the surreal numbers. They explore and develop the theory from first principles and find themselves in the process. Reviewer John W. Dawson, Jr. (MR0472278) says of the book, "it is entertaining, and its plot does incorporate that sine qua non of contemporary literary success, an episode of illicit sex."

Credits

The surreal numbers were invented by John Conway in the early 1970s, and they were introduced by Donald Knuth (1974). The phrase "the unification of all numbers, great and small" is borrowed from the article of Philip Ehrlich (2012) on the surreal numbers, which adapts the title "All numbers great and small" of chapter 0 in Conway (2000), which is itself presumably borrowed from James Herriot's wonderful book, *All Creatures Great and Small.*

21 The Axiom of Choice

Abstract. The axiom of choice is seen by many mathematicians as expressing an enduring logical truth, yet by others as an unjustified, perhaps arrogant, overreach. Let us explore the nuances of this fundamental, yet controversial principle.

At the candy shop as a child, your fingers on the counter's edge, you peer up at the big jars of candy, filled with sweets. You place your quarter firmly on the counter, and the kind owner nods—you choose exactly one candy from each jar and are soon proudly carrying your treats out of the shop in a crisp white bag.

In mathematics, one often finds oneself in just such a situation, with a collection of nonempty sets and an urgent desire, for some mathematical purpose, to select a single element from each set in the family. Is it always possible? Can we always choose a mathematical sweet from each set? Perhaps it seems obvious—just pick an element arbitrarily, at random, from each set. Can we expect to choose like a kid in a candy shop?

And yet this principle of being able to choose an element from each set in a family of sets is sometimes viewed in mathematics as controversial. The objection is that in some cases, we might not be able to describe any rule or constructive procedure by which we shall make the choices, and ultimately the worry is that in those kinds of cases, there may actually somehow be no coherent way of choosing. The perspective is that mathematical existence claims should be deeply connected with the constructions by which we may realize them. And so some mathematicians therefore hesitate to agree that the principle holds universally, and many keep careful track of exactly which mathematical arguments use the choice principle and which do not. Meanwhile, other mathematicians find the principle unproblematic and make free use of it—they take the axiom of choice as asserting a fundamental feature concerning the nature of sets, a logical truth about which kinds of sets there are.

21.1 The Axiom of Choice

To state the principle precisely, the *axiom of choice* is the assertion that every family $\mathcal{A}$ of nonempty sets admits a *choice function*, a function f, for which $f(A) \in A$ for every set A in the family. The function f in effect "chooses" the element $f(A)$ from the set A, systematically for each set in the family, just like choosing a sweet from each candy jar.

The axiom is not about making a single choice from a single nonempty set, but rather about the possibility of having a system of choices for an entire family of nonempty sets. When the family is enumerated by an index set I like this,

$$\mathcal{A} = \{ A_i \mid i \in I \},$$

then it is often attractive to view the choice function equivalently as defined on I, so we require $f(i) \in A_i$ for every $i \in I$. And when the family $\mathcal{A}$ consists of disjoint sets—that is, where the various sets in the family have no elements in common—then a *choice set* is a set B such that $A \cap B$ has exactly one element for each set A in the family. The reader will prove in the questions at the end of the chapter that the choice-set formulation of the axiom of choice is equivalent to the choice-function formulation.

A Different Kind of Axiom

The axiom of choice, introduced by Ernst Zermelo first in 1904 and then again in 1908 along with the other axioms of set theory, has a different character than the other axioms. The other axioms assert that certain sets exist, such as the power set of a set or a definable subset of a given set, and in these cases, the sets asserted to exist are uniquely determined by the features that the axiom asserts of them. With the axiom of choice, in contrast, we assert that there is a choice function, but choice functions are seldom unique; in nontrivial instances, we can easily change a few values and thereby produce many additional choice functions.

Shoes and Socks

Bertrand Russell relates an allegory about the axiom of choice like this:

> A millionaire possesses an infinite number of pairs of shoes, and an infinite number of pairs of socks. One day, in a fit of eccentricity, the millionaire summons his valet and asks him to select one shoe from each pair. When the valet, accustomed to receiving precise instructions, asks for details as to how to perform the selection, the millionaire suggests that the left shoe be chosen from each pair. Next day the millionaire proposes to the valet that he select one sock from each pair. When asked as to how this operation is to be carried out, the millionaire is at a loss for a reply, since, unlike shoes, there is no intrinsic way of distinguishing one sock of a pair from the other. In other words, the selection of the socks must be truly arbitrary. Bell (2021, fn 4); see also Russell (1907, §III, 1959, p. 78).

The lesson is that we do not need the axiom of choice to make choices when we can describe a definite procedure for making the choices, as with the shoes. The power of the axiom, rather, arises in the cases in which we seem unable to describe any definite way of choosing, as with the socks.

Nevertheless, perhaps one has a view of the nature of set theory that it is very rich, including all manner of sets having quite arbitrary natures, including amongst them all manner of choice sets and choice functions, even if those choices do not follow any constructive procedure that we might hope to describe. Such is the expansive, abundant nature of set theory that might lead one to the axiom of choice.

Inadvertent Use of the Axiom of Choice

The axiom of choice comes so naturally to human thinking that mathematicians often use it inadvertently, without realizing. Indeed, Zermelo had responded to some of his most strident critics by finding instances in the research literature where those same mathematicians had used the axiom implicitly in their own work without remarking upon it (see Moore, 1982, p. 149). He had thereby caught his critics red-handed—they were critical of the axiom, but they were nevertheless using it in their mathematical proofs as though it were perfectly valid. Some critics continued to use the axiom implicitly in this way for years afterward.

Therefore, if you find yourself tempted to deny the axiom of choice, then to avoid this embarrassing hypocrisy, it would seem best for you to ensure that you are not also using the axiom in your mathematical proofs.

21.2 The Well-Order Principle

We say that a set A is *well ordered* by a relation $\leq$ when this is a linear order on the set with the further property that every nonempty subset of A has a least element. If there are any elements at all, therefore, then there will be a globally least element, and then a next element, and a next, and so on for as long as elements remain. We may thus begin counting through any well-order as though counting in the natural numbers, and on into the transfinite ordinals—recall how we learned to count in the ordinals in chapter 14. There is a fundamental connection between well-orders on a set and the ordinals, for the ordinals indeed are well ordered by their natural order, and conversely every well ordered set is isomorphic to an ordinal. A well ordered set can thus be enumerated by ordinals in the order $a_0, a_1, a_2, \ldots, a_\alpha, \ldots$, where a_α is the αth member, the least element above all the previous elements. So we have an abundance of well-orderings, counting ever higher in the ordinals to staggering heights:

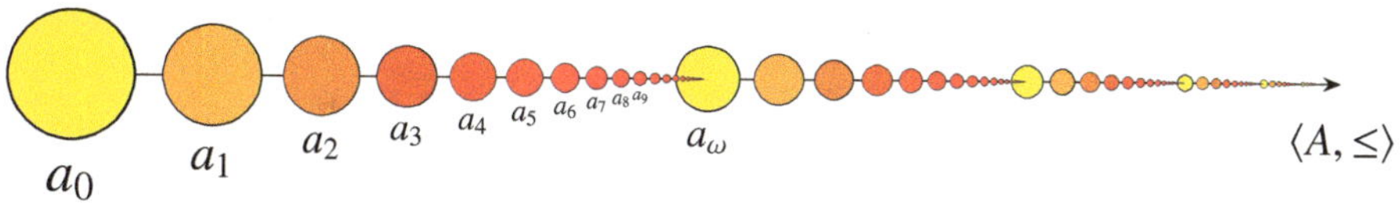

The *well-order principle* is the assertion that every set can be well ordered.

Wait,...what? *Every* set? Sure, it is easy to see that many familiar sets can be well ordered. Every countable set, for example, can be well ordered by the correspondence that it has with a set of natural numbers. And the set of countable ordinals is well ordered by its natural order, which therefore makes an uncountable well-order. With other uncountable sets, however, it is not always clear how to construct a well-order. What would be a well-ordering of the set $\mathbb{R}$ of all real numbers?

The usual order is certainly not a well-order since the positive real numbers, for example, have no least element. For a well-order, we would need a totally different ordering idea. You might try to build one, perhaps starting with zero at the bottom, then the positive integers, then the negative integers after that; then π and e; and then $\sqrt{2}$, and then...well? It just isn't fully clear how to proceed in any definite way—there seem to be numerous haphazard choices regarding what comes next. This is a deep hint about the involvement of the axiom of choice.

Zermelo's Shocking Result

Despite the seemingly hopeless situation, Zermelo proves in 1904 using the axiom of choice that indeed every set admits a well-order. The response to this remarkable result is an immediate outcry—a vigorous mathematical and philosophical debate in the international mathematical community. Mathematical historican Gregory Moore describes it like this:

> When in 1904 Zermelo published his proof that every set can be well ordered, many questions lay unresolved.... What methods were permissible in mathematics? Must such methods be constructive? If so, what constituted a construction? What did it mean to say that a mathematical object existed? Normally mathematicians avoided such quasi-philosophical questions, and addressed them only when they felt their discipline to face a crisis. That Zermelo's proof precipitated such a crisis was shown by the extent of the resulting controversy. From 1905 to 1908 eminent mathematicians in England, France, Germany, Holland, Hungary, Italy, and the United States debated the validity of his demonstration. Never in modern times have mathematicians argued so publicly and so vehemently over a proof.

The difficulty of imagining a well-order on the set of real numbers is a particular source of objection. In response to the criticisms, Zermelo articulates in 1908 the fundamental principles of set theory that his argument uses, formulating the theory that we now know as *Zermelo set theory*, which includes the axiom of choice from the start. The theory was later expanded by a few additional axioms, the foundation and replacement axioms, to form the theory now known as *Zermelo-Fraenkel set theory* (ZFC), widely recognized today as the standard default theory of sets, hugely successful as a foundation of mathematics.

Zermelo's Proof That the Axiom of Choice Implies the Well-Order Principle

I should like to convince you that indeed every set can be well ordered, just as Zermelo claims. Let us go through Zermelo's elegant argument, following a simple idea using the axiom of choice. Given any set A, we shall construct a well-ordering of it in a transfinite recursive process. To begin, by the axiom of choice, we may fix some way of choosing elements from any given nonempty subset of A. Zermelo's keen idea was simply to iteratively choose elements from A, choosing successively from amongst the elements not yet chosen. That is, we choose at the start an element of A as the bottom element of the order. Next, amongst the elements remaining, we choose another element to come next; and so on. At every stage, if there are any elements remaining, we choose one of the remaining elements as the next element in the order. In this way, we have recursively defined a well-ordering of A, completely determined by the specific way of choosing elements from nonempty subsets of A.

Stated a bit more formally, one should fix a particular choice function f on the set $\mathcal{A}$ consisting of all the nonempty subsets of A, and then consider all well ordered sequences from A that conform with that choice function in the manner that I just described, so the next point on the sequence is always the element of A that the choice function picks from the elements not yet on that sequence. All such well ordered sequences cohere with one another since they can have no least place of disagreement. The union of all the sequences, therefore, is a maximal well ordered sequence from A, which cannot leave any elements of A remaining, for if there were points remaining, then we could make a longer sequence by using f to choose an element from those remaining elements. So A is well orderable.

The Axiom of Choice Is Equivalent to the Well-Order Principle

Zermelo thus proves that the axiom of choice implies the well-order principle. To prove the converse, suppose that $\mathcal{A}$ is a family of nonempty sets, and let U be the union of all the sets in the family. By the well-order principle, there is a well-ordering $\leq$ of this set U. For a choice function, we can simply define $f(A)$ to be the $\leq$-least member of A. This chooses a particular element from each set, so we have got the desired choice function. Thus, the two choice principles are equivalent.

21.3 Zorn's Lemma

A third choice principle, known as *Zorn's lemma*, is introduced by Max Zorn (1906–1993) in 1935, although it was already proved independently in 1922 by Kazimierz Kuratowski (1896–1980), so it is also sometimes known as the *Kuratowski-Zorn lemma*. The principle asserts that if you have a partially ordered set $\langle P, \leq\rangle$, and every linearly ordered subset of P has an upper bound in P, then there is at least one maximal element.

Zorn's lemma is especially useful when undertaking constructions of a kind often seen in algebra, for we can take P as the set of partially completed constructions ordered naturally by extension. To find a basis in a vast infinite-dimensional vector space, for example, we consider the partial order of all linearly independent sets, ordered by extension, and observe that a basis for the whole space is the same as a maximal linearly independent set. These algebraic-construction applications generally satisfy the upper bound requirement because the union of a chain of partial constructions is itself a larger partial construction extending each of them—for example, the union of a chain of linearly independent sets in a vector space is also linearly independent. So, by Zorn's lemma, there will be a maximal partial construction. The main point, then, is that any maximal partial construction must be totally finished since every strictly partial construction can be continued a bit more.

We can prove that Zorn's lemma implies the well-order principle in just this way. Fix any set A, and let P be the collection of well ordered sequences of distinct elements of A, ordered by extension. The union of any chain of such well ordered sequences is itself a well ordered sequence extending them, so we satisfy the upper bound requirement. By Zorn's lemma, therefore, there will be a maximal well ordered sequence of distinct elements of A. Being maximal, this sequence must already contain every element of A, however, since otherwise we could make a strictly longer sequence by adding any particular missing element to the top. So A is well orderable, as desired.

Conversely, let us prove Zorn's lemma from the axiom of choice. Given a partial order $\langle P, \leq \rangle$ satisfying the upper-bound requirement, fix a way of choosing for each nonmaximal element p an element p^+ strictly above it, and also a way of choosing an upper bound for any linearly ordered chain. We now simply iterate these choices by transfinite recursion. Start from any point p_0, and if it is not maximal, we choose a point $p_1 = p_0^+$ strictly above it, and then another point p_2 above that, and so forth, with a point p_α for each ordinal α as the process continues. At limit stages, these form a linear chain, and we choose an upper bound to continue. The iteration cannot continue forever, however, since we cannot inject every ordinal into P, an observation due to Friedrich Hartogs, so it must eventually stop. But the only way that it stops is by reaching a maximal element.

Zorn's lemma encapsulates many useful consequences of this kind of transfinite recursion, while also conveniently sublimating the underlying details. Indeed, this was part of Zorn's purpose in introducing his lemma—to help his readers avoid transfinite recursion. Should we thus view Zorn's lemma as concealing this illuminating underlying method? Or is it rather helpfully abstracting away from those details to reach the important main conclusions? To be sure, Zorn's lemma continues to be used often today by mathematicians who are less familiar with the ordinals and transfinite recursion.

But for those of us, perhaps like many of my readers, who may know the ordinals well and who enjoy a proper transfinite recursion like a stroll in the park, the use of Zorn's lemma can sometimes seem to miss the main show—one climbs to a maximal point by iteratively picking something bigger. It is totally clear.

Choice Equivalents

So we have three equivalent principles—the axiom of choice, the well-order principle, and Zorn's lemma—but one must admit that our prereflective intuitions about these principles might not have been all the same. Zermelo's proof may seem frankly like a bit of mathematical magic, for we started with the axiom of choice, a principle that many find initially to be totally unobjectionable, even banal, and we ended with the conclusion that every set can be well ordered, a principle that many mathematicians find mysterious.

Jerry Bona tells it as a joke:

> The axiom of choice is obviously true, the well-order theorem is obviously false, and about Zorn's lemma, who can tell? (see Krantz, 2002, p.121)

But of course, we have proved that they are all equivalent. Personally, I do not care much for that joke if it is taken to suggest that we aren't able to understand in a deep way how and why the principles are equivalent. We can understand this deeply—one constructs a well-order on a set simply by iteratively choosing a next point at each stage from those points that remain. And one climbs to a maximal point in Zorn's order by successively choosing larger points until none remain.

There are many further fruitful mathematical statements equivalent to the axiom of choice. Herman and Jean Rubin (1985) fill a book with well over 100 instances, including numerous weakenings of the principle. For example, the axiom of choice is equivalent to the linearity of cardinality, the assertion that for any two sets, one of them injects into another; it is also equivalent to the assertion that every vector space has a basis, as well as the assertion that every nonempty set carries a group structure.

21.4 Weakened Forms of Choice

The axiom of choice admits various natural weakenings, which furthermore enable some of the key consequences of the axiom from a strictly weaker principle.

Countable Choice

The axiom of *countable choice*, for example, makes the choice-function assertion for all countable families $\mathcal{A}$. If there are only countably many sets in the family, then the choice function will be making only countably many choices altogether. Countable choice is enough to prove, for example, that a countable union of countable sets is countable, that every Dedekind-finite set is numerically finite, and that sequential convergence for functions on the real numbers, which holds of f when every convergent sequence $x_n \to x$ has a convergent image $f(x_n) \to f(x)$, is equivalent to the $\epsilon\delta$ notion of convergence. None of these is provable without any choice principle.

In chapter 11, we proved that if the axiom of choice holds, then Dedekind finite is the same as numerically finite. In fact, one needs only the countable axiom of choice to make this argument. To see this, suppose that X is a numerically infinite set. For each natural number n, let Y_n be the collection of n-tuples of distinct elements from X. Since X is not finite, we can prove by induction that every Y_n is nonempty. By the countable axiom of choice, we can pick specific $s_n \in Y_n$. From these, we can build up a countably infinite subset of X as a_0, a_1, and so forth. If we have defined this below n, then s_{n+1} must have an element that does not appear amongst a_k for $k < n$, since it has $n + 1$ elements, so let a_{n+1} be the first such new element in s_{n+1}. Therefore, we have found a countably infinite subset of X, which implies that X is not Dedekind finite. So under countable choice, the finite sets and the Dedekind-finite sets coincide.

Dependent Choice

The principle of *dependent choice* asserts that one can make countably many choices in succession. That is, if R is a serial binary relation on a set A, which means that for every a, there is at least one b for which $a \ R \ b$, then the principle says that, indeed, there is an infinite sequence $a_0, a_1, a_2, \ldots$, such that $a_n \ R \ a_{n+1}$ for every n. This principle is strictly stronger than countable choice, and it is useful when undertaking countable-length constructions, where the later options may depend on the earlier choices that were made. Such constructions often arise in measure theory and topology.

Please take care with the subtle difference between countable choice and dependent choice, since one can find many instances where mathematicians wrongly state that they are using only countable choice when their argument actually uses dependent choice.

Pair Choice and Finite Choice

The principle of *pair choice* makes the choice-function assertion for families of pairs. This is Russell's classic case of choosing one sock from each pair. More generally, *finite choice* gives us a choice function for any family of nonempty finite sets. Both of these principles follow from the *linear-order principle*, asserting that every set admits some linear order since if you place a linear order on the union of the family, then every finite set will have a least element, which defines a choice function. Meanwhile, none of these principles are equivalent to each other and none are provable without any choice principle.

Choice for Finite Families

Finally, consider the axiom of *choice for finite families*, which asserts that every finite set of nonempty sets admits a choice function. This principle actually comes for free, for it is provable without making any choice assumption, which the reader is invited to do in the questions at the end of the chapter. In this sense, the axiom of choice is problematic only when it comes to making infinitely many arbitrary choices.

21.5 Vitali and Nonmeasurable Sets

The axiom of choice has a remarkable consequence for the theory of Lebesgue measure, which also emerged at the dawn of the twentieth century, aiming to provide a foundational account of probability and randomness for the real numbers. Introduced by Henri Lebesgue (1875–1941), the Lebesgue measure $\mu(A)$ of a set of real numbers $A \subseteq \mathbb{R}$ is a nonnegative real number (or ∞) that tells us in a sense how massive the set is. The measure of an interval $[a, b]$ is simply the length $b - a$, and from this, it follows that individual points get measure zero $\mu(\{x\}) = 0$. The measure is additive, which means that $\mu(A \cup B) = \mu(A) + \mu(B)$ when A and B are disjoint, and not only that, it is countably additive, which means that $\mu(\bigcup_n A_n) = \Sigma_n \mu(A_n)$ whenever the sets A_n are disjoint. The measure is consequently *monotone*, which means that if $A \subseteq B$, then $\mu(A) \leq \mu(B)$, and it is *translation invariant*, which means that $\mu(A) = \mu(A+q)$, where $A+q$ is obtained from A by shifting everything by q. The theory altogether has been hugely successful, clarifying much of our understanding of random processes.

Giuseppe Vitali proves in 1905, however, that the axiom of choice implies a certain fundamental limitation of the theory—namely, there must be nonmeasurable sets, sets that cannot coherently be assigned any particular value as their measure. To do this, he defines that two real numbers are equivalent $x \sim y$ if they differ by a rational number. Let V be a *Vitali* set, one choosing exactly one element from each equivalence class, and we may assume that V is contained in the unit interval by always choosing from that part of each equivalence class. The translation set $V + q$ is disjoint from $V + r$ for distinct rational numbers since otherwise, $u + q = v + r$ for some $u, v \in V$, but this implies that $u - v$ is rational and hence $u \sim v$; but V has only one member of each equivalence class, so $u = v$, and thus $q = r$, contrary to assumption. Since every number is equivalent to some number in V, it follows that the rational translations $V + q$ are disjoint and cover $\mathbb{R}$. So V cannot have measure 0, since then the whole real line would be a countable union of measure zero sets and hence measure zero itself. But V cannot have a positive measure because it will have infinitely many disjoint translations $V+q$ contained in $[0, 2]$, moving V by some small amount $q < 1$, and this would force the measure of $[0, 2]$ to be larger than 2. So the Vitali set cannot have a measure at all.

Banach-Tarski Paradox

An extreme form of nonmeasurability arises with the Banach-Tarski theorem, due to Alfred Tarski and Stefan Banach (1892–1945). What the theorem asserts is that it is possible to decompose the unit sphere into finitely many pieces, such that by applying certain rigid motions to these pieces, we can put them together again to form *two* spheres, each the same size as the original.

What? Yes, let me say it again. We can cut the sphere into finitely many pieces and by rearranging them we can make two spheres, each identical to the original sphere. By repeating the process, we could have four spheres, then eight, and so forth, all made ultimately from the same substance of the original sphere. How is this possible?!

One way of reconciling the theorem with our intuitive ideas about how objects behave under rigid rearrangement is to recognize that the partition pieces arising in the theorem are quite bizarre—one should not imagine carving them with a knife; they are each dense in the unit ball, mixed intricately amongst each other, more complicated than fractals. The sets in the partition are necessarily nonmeasurable since rigid motion preserves measure.

The existence of the partition is deduced using the axiom of choice in a manner that builds on the Vitali construction. In rough outline, one first finds two independent rotations of the sphere—they generate what is called the *free group* on two generators, which means that distinct nontrivial patterns of applying those rotations and their inverses never result in the same overall symmetry. Next, we say that points in the sphere are equivalent if they are mapped to each other by any such pattern of these rotations. Applying the axiom of choice, we select representatives from each equivalence class. Finally, a simple, clever decomposition of the free group itself applied to those representative points results in the desired decomposition of the sphere.

The Self-Replenishing Gumball Machine

Since the full proof of the Banach-Tarski paradox is perhaps just a little too difficult for the focus of this book, let me discuss instead what I view as a toy version of the phenomenon, which does not require the axiom of choice but does exhibit what I see as a core feature of it. Namely, I shall exhibit a certain finite geometric figure for which a rigid motion transforms it into itself, but with something extra left over—a gumball gift.

Allow me therefore to introduce the *self-replenishing gumball machine.* This is a gumball machine, fully stocked with infinitely many gumballs, but with a certain very curious aspect of its behavior. Namely, if you should turn the knob of the machine, then you will get a gumball out, as with any gumball machine, but the special feature of this machine is that after the rotation, as a consequence of the geometry of the device, the internal arrangement of the gumball inventory remains identical to what it had been before the rotation—there remains a full stock of gumballs available, despite the fact that a gumball has been dispensed. In short, the machine in effect replenishes itself immediately and automatically upon dispensing any gumball, without requiring any additional gumballs. I find this pleasantly magical in the same way as the Banach-Tarski result—a little mathematical miracle, but completely understandable geometrically.

To assemble the machine, we start with infinitely many identical gumballs placed on the number line at the natural-number positions, as pictured here. Let us assume that the gumballs are truly identical, with the same size, shape, color, and taste—indeed, they are identical in every way that might matter to us. The numbers shown here and slight color variations are just for our reference in the discussion and not actually part of the gumballs:

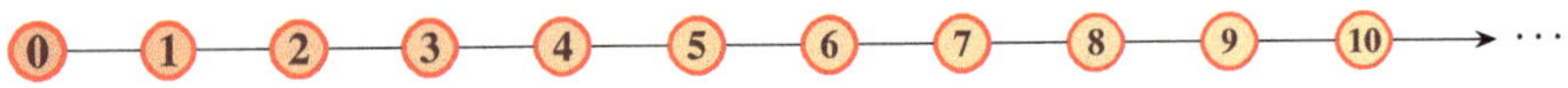

Next, we place the machine's main internal gear, a unit circle at position 0, and proceed to load the gumballs into the machine by wrapping the line around it—every gumball from the number line will find its corresponding place on the circle:

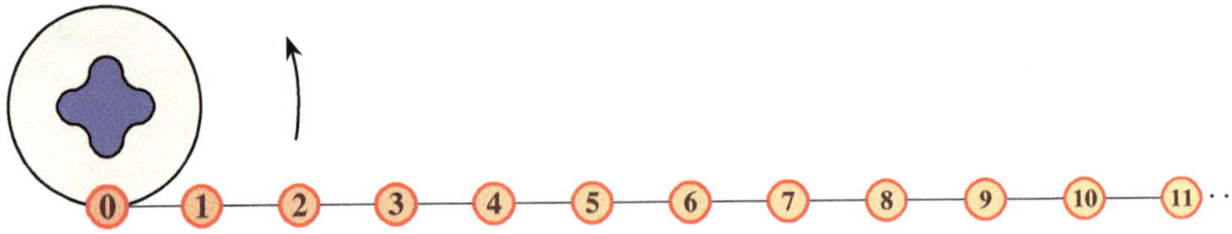

The result is our gumball machine, fully stocked with gumballs:

Notice that no two gumballs occupy exactly the same place on the circle. The reason is that ball n will land at radian angle n, measured from the bottom position, so the angular difference between ball n and another distinct ball m will be the integer angle of $n - m$ radians, and this is never equivalent to $2\pi k$ for any integer k since π is irrational, and therefore the positions on the circle are different. So we have infinitely many gumballs wrapped around the circle, with no two of them in the same place.

If you would like a gumball, as I assume you would, simply turn the knob of the machine—no quarter required. The whole circle turns through one radian angle, and out pops a gumball from the bottom, gumball 0:

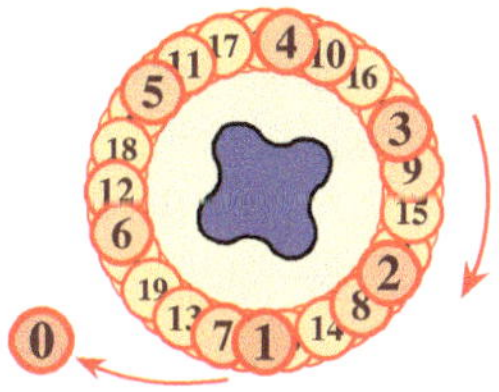

Because of the rotation, gumball 1 has moved into the bottom position, ready to be dispensed next, and gumball 2 has moved to where 1 had been, and so forth. In general, gumball $n + 1$ has moved to where gumball n had been. Consequently, after the rotation, the machine remains fully stocked! It has a gumball in every position, just as before, even though a gumball has been dispensed. Remember that since we can't tell the gumballs apart, the internal state of the machines is identical before as after the rotation. But now we have the free extra gumball in our hand:

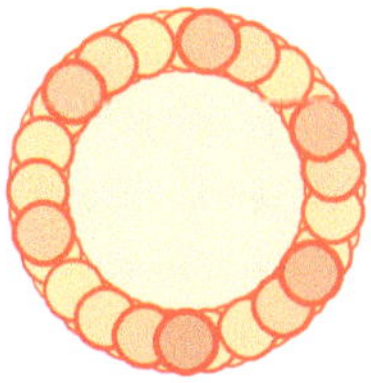

So what has happened here? With a rigid motion, a turn by one radian, a little over 57 degrees, we got a gumball, and the internal state of the machine afterward is identical to what it had been before the turn. We could make two turns of the knob to get two gumballs, or three, or any finite number, and afterward the internal state of the machine would still be identical to what it had been at the start, fully stocked with gumballs. In short, this is the self-replenishing gumball machine.

A more sophisticated machine uses nested circles of gumballs—each turn of the machine gives you a free gumball from each gear, even while the internal state of the machine remains unchanged, fully stocked, the same as before the gumballs were dispensed. By using continuum many nested circle gears, one will get uncountably many free gumballs with each turn, while the machine remains fully stocked after each turn. Amazing!

I find this to be a simple toy form of the Banach-Tarski paradox, even though it does not use the axiom of choice, because it shows how a finite geometric figure can undergo a rigid motion, rotation in this case, and become identical to itself as it was, but with something extra as well.

21.6 Consistency of the Axiom of Choice

Addressing what had become decades of contention over the axiom of choice, Kurt Gödel proves in 1938 that the axiom of choice is consistent with the other axioms of set theory if those axioms are themselves consistent. This is a *relative* consistency result, showing the consistency of one theory assuming the consistency of another, and in light of the Gödel incompleteness theorem, we can never prove outright that our foundational theories are consistent. The relative consistency result, however, shows that we needn't be worried about the axiom of choice on the grounds of consistency. It is perfectly safe to assume that the axiom of choice holds with all the other axioms of set theory—as safe as it is to assume those other axioms without the axiom of choice in the first place.

Gödel's Constructible Universe

To prove the relative consistency result, Gödel mounts an ingenious inner-model construction. He proves that every model of the ZF axioms of set theory has inside it a certain inner model, the *constructible universe*, where both the axiom of choice and the continuum hypothesis hold, along with all the other ZF axioms of set theory.

The constructible universe L is built in a grand transfinite recursive process, where at each stage we have a partial universe L_α, which is extended to the next stage $L_{\alpha+1}$ by adding all subsets of L_α that are definable in L_α by an expression in the first-order language of set theory with parameters from L_α. This is a thinner analogue of the cumulative hierarchy discussed in section 15.2, since the constructible hierarchy adds the fewest sets possible at each stage, keeping only those sets that would absolutely have to be there to form a model of the ZF axioms. And indeed Gödel proved that L is a model of all the ZFC axioms, including the axiom of choice, and also fulfills the generalized continuum hypothesis.

Global Choice

Gödel observes in 1938 that the axiom of choice holds in the constructible universe L in a particularly strong manner. Namely, in L, there is a definable global choice function, a definable manner of picking an element from absolutely every set. In this sense, in L, one can have choice functions that universally agree in their choices, always choosing the same element of a given nonempty set, regardless of the family $\mathcal{A}$ in which these sets reside.

Gödel proves this by showing that there is a definable global well-ordering of the entire set-theoretic universe L. The order is easy to understand once one has understood how the constructible universe L is built. Namely, every set in L is added at some ordinal stage α for a specific reason—it is defined as a subset of L_α by a specific formula, with specific parameters. We can therefore simply define that $x <_L y$ in the global L-order, if either x was added at an earlier stage than y; or they were added at the same stage but x was added because it was defined by a formula with a smaller numerical code; or they were added at the same stage and with the same formula, but x was added with parameters appearing lexically earlier in this order. This is a definable, global well-order, from which one can define a global choice function in L, so the axiom of choice holds in L.

Definable Well-Orderings of the Real Numbers

Now I would like to highlight a certain consequence of Gödel's argument for some common ways of talking about the axiom of choice and well-orders. Namely, one often hears mathematicians express the view that uses of the axiom of choice in mathematics are inherently about nondefinable choices. For example, one sometimes heard it said plainly that there is no definable well-ordering of the real numbers.

Nevertheless, those remarks are not actually correct inside the constructible universe, which does have a fully definable well-ordering of the real numbers, and indeed it has a definable well-ordering of the entire set-theoretic universe. The specific well-order of the real numbers in L has a complexity known as Δ^1_2, which is relatively low in the complexity hierarchy of descriptive set theory, the subject in which one analyzes the exact complexity of sets of real numbers and relations on the real numbers.

So ultimately, it is incorrect to think that using the axiom of choice is necessarily about making nondefinable choices. Rather, it is possible that there are fully definable global choice functions, so it is consistent with the axioms of set theory that the axiom of choice holds with fully definable choice functions in all cases. Meanwhile, to be sure, it is also relatively consistent that the axiom of choice holds, but there are no definable global well-orders or choice functions. So these extra features that Gödel had observed are something special about the constructible universe.

Axiom of Choice Is Independent

We have mentioned the relative consistency of the axiom of choice with the other axioms of set theory, but the question about its possible failure remained open for several decades more. This was finally resolved in 1963 by Paul Cohen, who proved that it is relatively consistent with the ZF axioms of set theory that the axiom of choice can fail. In his extremely influential work, Cohen (1963), he introduces the celebrated forcing method and proves that every model of set theory has a forcing extension with an inner model of ZF in which the axiom of choice fails. Variations of the method are subsequently used to show diverse ways in which the axiom of choice can fail, often quite spectacularly. For example, it is consistent with ZF that the axiom of choice fails for sets of pairs, even pairs of sets of real numbers, thereby fulfilling Russell's allegory about choosing from an infinite collection of pairs of socks. In some set-theoretic universes, you just can't choose one sock from each pair.

21.7 A Counterpoint

I would like to make a change in what seems to have become a common way of discussing the axiom of choice. Namely, one commonly hears the axiom of choice described as useful for certain mathematical arguments, but problematic in light of the Banach-Tarski paradox and other counterintuitive consequences. To my way of thinking, however, a more balanced discussion arises when we also highlight the counterintuitive situations that can occur when the axiom of choice fails. For example:

- Without the axiom of choice, a real number can be in the closure of a set of real numbers $X \subset \mathbb{R}$, but not the limit of any sequence from X.
- Without the axiom of choice, a function $f : \mathbb{R} \to \mathbb{R}$ can be sequentially continuous at a point x, meaning that every convergent sequence $x_n \to x$ has a convergent image $f(x_n) \to f(x)$, but the function is not continuous there in the ϵ, δ sense.
- Without the axiom of choice, a set can be infinite but have no countably infinite subset. There can be an infinite set with all subsets either finite or the complement of a finite set. Thus, it can be incorrect to say that $\aleph_0$ is the smallest infinite cardinality since these sets would have an infinite size that is incomparable with $\aleph_0$.
- Without the axiom of choice, there can be a nonempty tree T with no leaves, but which has no infinite path. That is, every finite path in the tree can be extended to further steps, but there is no path that goes forever.
- Without the axiom of choice, there can be an equivalence relation on $\mathbb{R}$ such that the number of equivalence classes is strictly greater than the size of $\mathbb{R}$. That is, you can place the various real numbers into buckets, with every bucket having at least one real number, such that the number of buckets is strictly greater than the number of real numbers. It is totally bizarre! This situation is a consequence of the axiom of determinacy and is relatively consistent with the principle of dependent choice and the countable axiom of choice.
- Without the axiom of choice, there can be a field with no algebraic closure. Without the axiom of choice, the rational field $\mathbb{Q}$ can have different nonisomorphic algebraic closures–it can have an uncountable algebraic closure as well as a countable one!
- Without the axiom of choice, there can be a vector space with no basis, and there can be a vector space with multiple bases of different cardinalities.
- Without the axiom of choice, the set of real numbers can be a countable union of countable sets (yet still uncountable). In this case, the theory of Lebesgue measure is a complete failure—it is much worse than just having Vitali's nonmeasurable sets.

I take all these examples to support a call for balance in our discussions about the axiom of choice. Let's not hear just one side of the story, heaping doubt on the axiom via the Banach-Tarski result and perhaps another counterintuitive consequence. Rather, let's also hear how the axiom of choice rules out some downright bizarre phenomena, in many instances more bizarre than the Banach-Tarski-type results.

Intrinsic Versus Extrinsic Justification

The axiom of choice is used throughout mathematics, and many mathematicians find it indispensable as a tool of reasoning. Is this a reason to believe in it? If so, it would be an *extrinsic* justification, supporting an axiom because it provides the consequences that you want. Perhaps this would be something like scientific practice, where we judge a scientific theory in light of its ability to predict and agree with experimental observations.

In mathematics, however, we also often find *intrinsic* justification of a principle, when we take it to express a core truth of a mathematical subject matter. This is the attitude in play when people refer to self-evident axioms, such as in geometry. Most of the axioms of set theory are taken to express fundamental truths of set theory in this intrinsically justified manner. The axiom of extensionality, by which sets are equal when they have the same elements, and the axiom of separation, by which we can form a subset of those individuals in a set exhibiting a certain property, are seen as expressing core features of set theory.

What I should like to emphasize is that despite the adamant objections to the axiom of choice that one may hear from some corners of mathematics, nevertheless many other mathematicians regard the axiom of choice to have this kind of strong, intrinsic support. They take the axiom to express in part a certain sweeping vision of the nature of the set-theoretic universe, one in which sets of all kinds and patterns exist, where choices are made in all possible manners and ways that are quite arbitrary, without necessarily conforming with an easily described rule or procedure. The sets exist, including the choice sets, even if we cannot describe them. On this view, the axiom of choice is seen as something like a logical truth of mathematics—those are the kinds of sets that we want the theory to be about.

To deny the axiom of choice on the grounds that we lack descriptions of how to choose is seen on that view as a change in the subject matter, a change that would make our existence claims in mathematics not about what there is in the pure platonic realm of ideal forms, but rather something about us, about our ability or inability to know the truth or to effectively carry out the processes that might lead us to the truth. It is a different focus.

But meanwhile, many mathematicians do care deeply about and adopt that focus, taking a constructive approach to mathematical claims, emphasizing as a core value of mathematics the constructive procedures from which our mathematical insights arise, the means by which we know the mathematical truths that we claim to know.

What is your view?

Questions for Further Thought

21.1 Suppose that we have a family of sets $\mathcal{A} = \{A_i \mid i \in I\}$. What does the existence of choice functions for this family have to do with the product $\prod_{i \in I} A_i$?

21.2 Prove that the choice-function formulation of the axiom of choice (asserting that every family of nonempty sets has a choice function) is equivalent to the choice-set formulation (asserting that every family of disjoint nonempty sets admits a choice set). (Hint: For the converse implication, given a family of nonempty sets, you will have to disjointify it to produce a family to which the choice-set formulation applies.)

21.3 Prove using countable choice only that every Dedekind-finite set is numerically finite. (Hint: Consider an infinite Dedekind-finite set A. For each positive integer n, let A_n consist of the length-n sequences of distinct elements of A. Apply countable choice and then construct a countably infinite subset of A.)

21.4 Prove, without assuming any form of the axiom of choice, that the axiom of choice for finite families holds. (Hint: Prove this by induction on the size of the family.)

21.5 Prove, without assuming any choice principle, that all instances of the pair choice axiom hold for families of pairs of real numbers.

21.6 Is it correct to speak of *the* Vitali set? If not, is there some sense in which this is sensible?

21.7 Consider the various transfinite recursive constructions that were undertaken in section 15.3. Which of them used the axiom of choice? Can you prove any of the decomposition results in that section without using the axiom of choice?

Further Reading

Moore (1982). A sweeping account of the history of the axiom of choice.

Bell (2021). A solid summary of the axiom of choice in the *Stanford Encyclopedia of Philosophy*.

Hamkins (2022). A nice discussion on MathOverflow concerning perspectives on Zorn's lemma as an "old friend" versus a "historical relic."

Credits

Parts of this chapter are adapted from my discussion of the axiom of choice in Hamkins (2021b). Many of the historical claims about Zermelo and the history of the axiom of choice are to be found in Moore (1982). A complete account of the Banach-Tarski paradox can be found in Tomkowicz and Wagon (2016).

22 Infinitary Hat Puzzles and the Aftermath

Abstract. What color is your hat? The genre of logic conundrums known as *hat puzzles* grows from some easy finite puzzles ultimately to absurd, spectacular settings with infinitely many people and infinitely many hats, whose solutions often require sophisticated mathematics, including whimsical applications of the axiom of choice.

22.1 The Midnight Ball

We are all at a festive ball with infinitely many guests. The music is divine, and the formal dances involve such elegant intricate choreography—we undergo graceful whirls in a nested interlocking fractal pattern. The evening is infused with excitement.

Midnight approaches, and a murmur of giddy anticipation goes through the crowd. In preparation for the culminating special event, every guest has had a natural number written on their forehead with black lipstick. Perhaps some numbers are repeated, others absent—we can't be sure. We can all plainly see everyone else's number, however, but nobody knows their own number or how the numbers were chosen. Nevertheless, at the stroke of midnight, one of us will be designated as Queen of the Ball, who will be allowed to erase her number and replace it with any number of her choosing. And then, immediately after this, we shall all simultaneously announce our best predictions of our own numbers, as well as we are able to do so. We had had time at the afternoon garden party earlier in the day to plan and agree upon a strategy, but naturally once the numbers were imprinted, all communication about them is forbidden.

The queen, of course, will guess her own number correctly, but can we somehow guarantee that others also will guess correctly? Surely it would be hopeless for any significant number of us to guess correctly, right? What if it had been real numbers on our foreheads? Or ordinals? Would it matter if the number of guests had been a much larger uncountable infinity?

Think about it...

We shall obtain a solution to that puzzle by the end of this chapter, but let us warm up first with some easier finite puzzles, to get a sense of the style of reasoning, before moving more fully to the infinitary puzzles, including the midnight ball.

All these conundrums exist in the genre of logic puzzles known as *hat puzzles*, in which invariably you find yourself placed in a curious situation where everyone is wearing a hat of a certain color—or a number on their forehead, you get the idea—but nobody knows their own hat color, and everyone must guess their own hat color or to try to deduce it somehow. A distinctive common feature of these puzzles is that by clever mathematical strategizing, people can guess correctly far more often than one might naively expect. At the midnight ball, for example, we shall describe a mathematical strategy ensuring that absolutely everyone guesses correctly—yes, absolutely all the guests will correctly predict their own numbers. It is incredible.

The original hat puzzles featured just a few people, perhaps standing in a circle, and we shall see some of these original puzzles presently, but eventually there were variations with dozens of people, and then hundreds. Ultimately, of course, we find ourselves with infinitary puzzle variations like the midnight ball—puzzles with infinitely many people trying to guess their number or hat color. These infinitary hat puzzles often become mathematically subtle and sophisticated, and it is a core part of the genre to assume that everyone inhabiting the puzzle is capable of perfect logical reasoning, and furthermore able to implement any mathematically possible strategy, such as announcing a real number with perfect accuracy and precision or mounting a collective application of the axiom of choice. Meanwhile, to attempt hand signals or other communication outside the stated channels of a puzzle is totally against the spirit and style of the genre.

22.2 Finite Hat Puzzles

Let's have some fun! We shall begin with several finite hat puzzles and then return to the midnight ball later in the chapter.

Raising Hands for Red Hats

You and two friends stand in a circle facing each other, and each of you is given a colored hat, either red or blue. You can all see each other's hats, but not your own. You are told to raise your hand if you see a red hat, and whoever can deduce their own hat color first will receive a prize. At the start, all three of you raise your hands, and then stand in awkward silence for a good long moment. Eventually, your friend Savannah says, "Red!" She is right, but how did she know? What were you and your other friend about to say?

Give it some thought before continuing to my account of the solution. Can you figure out how she knew? What would be your thinking as the awkward moments proceed while you stand with your friends in that situation?

Think about it...

Since all three of you raised your hands, each of you must have seen at least one red hat, and so, clearly, there is at least one red hat amongst the three of you. But actually, upon reflection, there must be at least two red hats since otherwise, the isolated red-hat wearer would not have seen any red hats and thus wouldn't have raised a hand. So all of you know that there are at least two red hats. Your friend Savannah, who you admit (with a mixture of admiration and envy) is perhaps slightly the quickest of the three of you, reasoned like this: "If my hat weren't red, then the others would both see only one red hat, but since they know that there must be two red hats, and since they are both very sharp, they would have deduced already by now that they must have the other, second red hat. Since instead we have the awkward silence with no announcement, that must not be the case, so I must be wearing a red hat!" In fact, both you and your other friend had reasoned just the same and were both also about to say "Red!" just like Savannah, and for exactly the same reason.

Blindfolded Hat Guessing

You and two of your friends are shown five hats—two green and three gold—and then you are blindfolded and each of you has one of the hats placed on your head (and the two extra hats are removed unseen). You are each to deduce the color of your own hat. Your first friend removes his blindfold, observes your hats, and explains that he cannot on that basis deduce the color of his own hat. He walks out, still wearing his hat. Your second friend then removes her blindfold, looks at your hat, and says that she also cannot deduce the color of her hat. Without removing your blindfold, can you deduce the color of your hat?

Think about it...

The only way that your first friend could have known the color of his hat would have been if he saw you both wearing green hats, since then he would have known he had gold, as there were only two green in the original collection. So he must not have seen two green hats on you, and thus either you or your friend must be wearing gold. When your second friend says that she doesn't know her hat color, it must mean that you have gold, since otherwise she would have known that she was the one amongst you two that had gold. So you must have gold. And you know this without removing your blindfold.

Coin-Flip Hat Colors

You and two friends are each given a gray or tan hat, determined in each case by a coin flip. Nobody knows their own hat color, but everyone can see the other two. At the bell, everyone will either announce a guess as to their own color or remain silent. The three of you win, as a team, if at least one person guesses correctly and nobody guesses incorrectly. So you will lose if everyone is silent or if someone makes a wrong guess. What is your best chance of winning? You may arrange a strategy before the coins are flipped and the hats given out, but of course, no communication is allowed after that.

Think about it...

Of course, you can win with at least 50 percent probability just by agreeing that a designated person will say “gray” regardless, and nobody else will say anything. This fixed strategy will be correct half the time. Perhaps it seems difficult to do better than half—won’t we have to be wrong as often as we are right? If so, winning half the time may seem optimal.

Actually, it turns out to be possible for you to do much better than half. The three of you should agree to follow this rule: If you see two hats of the same color on the others, then guess the opposite color for yourself, but otherwise remain silent. What is the chance that this strategy works?

Well, there are eight possible patterns of colors that the three of you might be allocated. There is the all-tan pattern, the all-gray pattern, three patterns with exactly one tan hat and two gray, and three patterns with exactly one gray hat and two tan. In the all-tan and all-gray situations, of course, the strategy that we described here will lead everyone to guess wrongly, since everyone will see two hats of the same color and guess the opposite color for themselves, but this will be incorrect in these monochromatic cases. In the six latter cases, however, where there are two hats of one color and one of the other, then the odd-color person will be the only one seeing two same-color hats, and they will correctly guess the opposite color for themselves. So the strategy will succeed in six of the eight cases, which makes for a 75 percent chance of winning.

This solution is optimal—there is no strategy to achieve higher than 75 percent. The reason is that as we had suggested earlier, for every circumstance in which a person individually has guessed correctly, there must be another circumstance (the one where their hat color is different) for which they make the same guess on the same basis, but incorrectly. Therefore, if the strategy wins in seven out of the eight cases, there will be at least seven correct guesses, so there must be other possible circumstances in which there are seven incorrect guesses. But there are only three remaining opportunities for incorrect guesses—the three guesses in the one losing instance—so it is impossible to do better. The best that is possible is to spread six winning guesses out in six separate cases, while concentrating the six corresponding losing guesses together in the two losing cases. And this is just what our solution did. So it is optimal.

Beetles in Boxes

You stand with your fellow entomologists in a group of six, each of you having just been given an exquisite black lacquer box with a beetle inside—either a checkered beetle or a long-horned harlequin beetle. Everybody knows the situation, but nobody knows which type of beetle is in which box, including their own. Each person is allowed, however, to peek briefly inside the box of exactly one other person of their choice, and then at the count of three, everyone predicts which type of beetle they themselves have. Is there a strategy to ensure that someone is correct? How many correct guesses can you guarantee?

Think about it…

We can guarantee that exactly half will be correct. To achieve this, we can simply divide ourselves into pairs and agree within each pair that a particular one of us will assume that we have the same type of beetle and the other will assume that they are different. When the time comes, we each look at our partner's beetle and make our predictions accordingly. Exactly half of us will be correct every time, since within each pair, either the beetles are indeed the same or they are different.

This solution is optimal, in the sense that there can be no strategy to achieve better than half on average. To see this, consider any one person who, having looked at another person's beetle, decides on that basis what their own beetle type is. So their individual strategies will be to guess that it is the same, that it is different, that it is a checkered beetle, or that it is a harlequin beetle. For each such deterministic way of proceeding, they will be right exactly half the time over all possible circumstances, since for every configuration in which that particular guess is correct, there is another in which the beetle is opposite, in which that guess is incorrect. So over all the logically possible configurations, exactly half of all the guesses will be correct and half incorrect. So we cannot achieve more than half correct on average. And this remains true for probabilistic mixtures of strategies.

22.3 The Philosophical Society

You stand amongst 100 fellow ambitious philosopher candidates at the annual initiation test for the philosophical society, held on the society's elegant formal grounds. You all stand in a line facing the society hall, aiming to enter into membership as a class, if you pass the collective test that will presently take place. Each of you is wearing either a black or white hat that has been placed upon your head, and everyone can see the hats of those in line ahead of them, but not their own or those behind; the colors of those hats are not known. Starting from the back, each candidate will shout out a guess for their own hat color, either white or black, and then the next candidate, and so forth, proceeding to the front. To avoid a humiliation suffered once years ago, a sorry incident in which every single candidate guessed wrongly, the tradition now is that the candidate at the very back of the line may inspect her hat, and even select another color if desired, before the process starts.

According to the strict society rules, the class is admitted to membership only in the extraordinary event, seldom achieved, that everyone guesses their hat color correctly, in which case the class is accepted as a whole. If there is even a single error, alas, the entire class is sent down, rejected as unsuitable.

How well can you expect to do? There was time earlier in the day for you all to agree on a strategy, if that would help, but communication at the initiation test itself, such as by modulating the voice in attempt to convey extra information beyond the guess itself, is strictly forbidden and absolutely disqualifying.

Think about it…

We can easily guarantee that at least half will be correct. Simply divide the line into successive pairs, the first two, the next two, and so forth, and within each pair, the first candidate makes as their own guess the color of the hat that they observe on their partner in front of them, who then correctly makes that same guess as their own color. This achieves at least half correct, and if the person at the back swaps her hat color to match her partner's, then we will get at least 51 correct—a strict majority. The odds are good to achieve much more than half if the colors were chosen randomly, since we can expect about half of the remaining first guesses in each pair will also be correct by chance, achieving around 75 correct on average. However, we are very unlikely to be universally correct since we'd have to get lucky 49 times. Furthermore, we do not actually know that the hat colors were selected randomly—perhaps they were chosen instead specifically to thwart an anticipated plan. In the worst case, this pairing-off plan doesn't ensure any more than 51 correct.

Can we do better? Yes—in fact, we can fully meet the membership requirement! Everyone will guess correctly, I claim, if we agree on the following plan. The philosopher at the back looks ahead and counts the number of white hats in the line ahead. If this is even, then that philosopher will pick a white hat, otherwise pick black, and make her guess accordingly. The second philosopher sees the whole line ahead, except herself, but knowing what the first philosopher had announced, she can deduce whether her own hat is white or black. For example, if she observes an even number of white hats ahead, and the first philosopher had also announced white, then she knows that she must be black. The third philosopher can now reason similarly. Knowing the observed parity of white hats ahead and the announcements of the previous philosophers, he can work out whether he has white or black. And so on down the line. Each subsequent philosopher deduces their hat color correctly based on what they observe ahead and what they heard from the philosophers behind. Everyone will be correct! The entire candidate class is thus admitted to membership and marches directly into the society hall, where lively philosophical discussions and celebratory drinking ensues.

The Physics Society

Across the lane is the physics society, which holds its annual initiation test on the same day. The physicists use 10 hat colors, however, rather then merely 2—the governing body has mainly applied physicists—so they carry out an enormous basket of hats. The procedure otherwise is the same. All 100 candidate members line up toward the society building, and each is given one of the hats. Everybody can see all the hats ahead of them, but not their own or those behind. After lining up, the candidate at the back is allowed to inspect and change hats if desired, and then, starting from the back, each person announces successively a prediction of their own hat color. If everyone is correct, then the entire class is admitted, and otherwise everyone will be sent down. How well can the physicists do?

Think about it...

Indeed, the physicists will succeed, being all correct in their predictions. Their strategy will simply be to use arithmetic modulo 10 in place of the previous even/odd parity argument of the philosophers, which amounts to arithmetic modulo 2, after all. Namely, the physicists will all agree in advance on a scheme of numerical values of the hat colors from 0 to 9. The first physicist will observe all those ahead and compute the total sum, taking the result modulo 10 (the remainder when dividing by 10)—this is just the final decimal digit of the sum in decimal notation—and then adopting that color hat as his own and announcing it. The second physicist can observe those ahead of her, compute the sum, and then compute what color her hat must be in order for the first physicist to have made the announcement that he made. For example, if she observes a residue of 7 in the total sum ahead of her, but the first physicist had announced 3, then she knows that she must have color 6, since $7 + 6 \equiv 3 \mod 10$. And so forth down the line. Each next physicist can compute what color they must have to square what they see with the previous announcements.

A similar solution works at the chemistry society around the corner, where they wear special helmets mounted atop with a pure sample of one of the chemical elements in the periodic table (118 currently). The more dangerous and difficult elements are officially allowed by the society rules, but these are seldom used now in practice after the disaster of '87. The chemists simply compute modulo 118, rather than modulo 10 as with the physicists, but the solution works exactly the same in principle.

At the astronomical society, meanwhile, each candidate wears a crown mounted with a bulb emitting light exactly in the spectrum of a known star. Since the database of stars is finite, the modular solution works just as well for them.

The reader is invited in the questions at the end of the chapter to describe the procedure and solution at the musical society, where each music candidate has a tuning fork tuned to a particular frequency.

The Mathematical Society

At the mathematical society, instead of hats the candidates wear large slates on their backs, upon which a natural number is written in chalk. There is in principle no official upper bound for the numbers that may be used, and the actual numbers used seem to get arbitrarily larger each year—an attempt by the governing body to thwart the modular arithmetic strategy that had taken over most of the other societies.

The 100 mathematics candidates are lined up facing the society entrance. Each mathematician can see only the numbers of those in front but does not know their own number or the numbers of those behind. After lining up, as a courtesy the candidate at the back is allowed to change her number before the process starts, but then starting at the back with her, each mathematician shouts out a prediction for their own number in turn, and admission is won only if absolutely everyone is correct without error.

Do the mathematicians have a strategy to enter into membership?

Think about it...

Indeed they do, and this case is actually easier than the others since no modular arithmetic is required. The first mathematician simply adopts as her own the total sum of all the numbers ahead and announces this as her guess. And then the next can easily work out what their number is by subtracting from that number the sum that they observe. And then the next, and so forth, all the way down the line.

A contentious argument arose at the governing body meeting following the first successful use of this strategy, with some fellows claiming that the entrance exam had become an embarrassing farce, being now solvable so easily—they look like fools—and a proposal was made to replace the use of natural numbers in the puzzle with arbitrary real numbers. Would this help?

No—a new young mathematical fellow points out, the very same total-sum strategy works just as well with real numbers. The first candidate at the back can simply adopt and announce as his own number the real-number sum of the subsequent numbers, and then the next candidate, and each subsequent candidate in turn can work out their own number knowing the previous numbers.

Indeed, this method works in any group (actually, in any cancellation semigroup). And since every nonempty set admits a group structure, provided the axiom of choice holds, this method provides a solution to the puzzle with any nonempty set of allowed labels, using any set known in advance. The mathematical candidates would simply have to agree on the group structure in advance and then implement the total-sum-observed strategy. The top agenda item on the upcoming society meeting, therefore, is to debate whether officially to adopt the use of proper classes...

The Garden Party—A Simple Infinitary Puzzle

At an afternoon garden party, we find ourselves amongst a countable infinity of other guests, and the game will soon begin. Each of us has been given an envelope containing an engraved card on which a natural number is elegantly written. We may not look at our own cards, but each of us is allowed discreetly to look at the cards of at most finitely many other people (nobody except the game masters will know who has looked at whose card). When everyone is ready, then simultaneously each of us announces a finite prediction set—a set of numbers that we hope will include our own actual number, the number on our card.

For example, perhaps after looking at a few other numbers, you guess that your number is one of them or close to one of them, within 10. Perhaps someone else doesn't look at all and simply guesses all the numbers up to googol plex bang.

How well can we do? Can we ensure that someone is correct? Can we ensure that infinitely many people are correct? What proportion of people might we hope to be correct?

Think about it . . .

We can easily get half correct by dividing ourselves into pairs, each looking at our partner's number and making our guess be all the numbers up to that number. In each pair, after all, one of the numbers will be at least as large as the other, so at least one person will be right in every pair (they'll both be right if they happen to have the same number). We can similarly divide ourselves into groups of 100, and within each group, we look at all the other people's numbers, and put forth as our prediction set all the numbers up to the largest of them. This will be correct for at least 99 out of the 100 since only the person with the strictly largest number amongst us will be wrong. So this strategy achieves a density of 0.99 correct.

We can make asymptotic density 1 in the limit by using larger and larger groups in succession. That is, let us have first a group of 10, then a group of 100, then 1000, and so forth, with everyone announcing all the numbers up to the largest amongst the others in that group. Since all but one person at most in each group will be correct, and the groups become arbitrarily large, the asymptotic density of correctness will be fully 1 in the limit. Furthermore, by taking the groups sufficiently large sufficiently quickly, we can make the occurrences of incorrectness as sparse as desired, below any specified nontrivial scarcity function.

22.4 The Aftermath

Gathered for the newly efficient afterlife batch processing at the Pearly Gates, we find ourselves deceased but nevertheless strangely refreshed, standing amongst infinitely many philosophers and mathematicians awaiting instructions in the Prejudgement Antechamber. We gradually realize our newfound powers of thought and perception, able now to process even infinite quantities of information in a snap, as well as an enhanced capacity for moral discernment. We will soon enter the Judgement Chamber, God explains, where we shall have a chance at redemption. There, God will reveal to each of us His final judgement of all the others, and we shall each then silently pronounce to the creator our own self-judgement, whether of *sufficient* virtue or *insufficient*. If infinitely many of us are in error—that is, in disagreement with the judgement of God, who is all-knowing—then regrettably we shall all together be sent down. That amounts to batch processing. But meanwhile, if only finitely many of us are in error, then indeed the merciful Lord will allow all of us happily to enter Heaven as a group. We have 10 minutes, God announces, to decide on a plan. After this, all further communication is strictly forbidden, upon the worst penalty imaginable.

It is immediate, anxious chaos. What are we to do? How can we possibly ensure that at most, finitely many of us will be in error?

Think about it...

One brilliant young mathematician proposes a plan. "Let us use the axiom of choice," he says, and indeed they can all observe the truth of it. Consider all the possible patterns of judgement that could conceivably be made upon us, he explains, and let us say that two such patterns are equivalent if they differ at most finitely from one another. This is an equivalence relation, which thus divides the possible judgements into classes of equivalent judgement patterns. By the axiom of choice, we can select a representative pattern from each equivalence class, a particular individual pattern in each equivalence class, which therefore differs at most finitely from the others in that class but stands as an agreed representative of that class. It is important, he emphasizes, that we must all agree in advance on the particular choices of representatives. Upon learning God's judgements, each of us will see the whole pattern of judgement except for our own. This is enough to determine the equivalence class, and our plan will simply be for each of us to pronounce our self-judgement in accordance with the chosen representative of the observed equivalence class.

We enter the Judgement Chamber, and sure enough, each of us observes that all the others have either a halo over their heads or a small ball of fire, but none of us perceives our own judgement yet. We duly pronounce our self-judgement in accordance with the chosen representative of that class. Does it work? Yes, indeed it does, since that chosen representative differs only finitely from the true pattern.

The Midnight Ball (Revisited)

Consider again the midnight ball puzzle from the beginning of the chapter. We have each had a natural number written in black lipstick on our foreheads, and everyone can see all the others' numbers, but we do not know our own. Nevertheless, at the stroke of midnight, the Queen of the Ball is allowed to erase her number and adopt a new one of her choosing, and then we all announce together our predictions of our numbers.

Incredible as it may seem, we can ensure that absolutely everyone guesses correctly. The strategy is this. We consider the space of possible patterns of numbers that we might receive, and say that two such patterns are equivalent if there are at most finitely many differences between them. By the axiom of choice, let us agree on a chosen representative in each equivalence class. Just before midnight, we each observe all the other numbers and thereby come to know the equivalence class. Thus we also all know the chosen representative of that class, which differs only finitely from the true pattern of numbers. The queen will erase her number and replace it with the total difference in value between what she observes and what she would have observed with the chosen representative (ignoring her own value in both cases). The key observation is that with this information on the total value of the difference as she sees it, each of us can now simply compute our own number. Namely, we see all the numbers that the queen sees except for our own, and therefore by adjusting for the known difference from the chosen representative, we would know what number we must have. So we will all be correct!

The solution works with real numbers just the same, and indeed it works in any group, of any size. Since every nonempty set carries a group structure (using the axiom of choice), it follows that we can implement the argument with any set of labels and any size infinity for the number of people. No matter how many guests are at the midnight ball, and no matter what kind of labels are used on the foreheads or how many, whether finite or some vast uncountable infinity, nevertheless there will be a strategy by which we can absolutely all guess correctly.

In the special case of countably many people and real number labels, the reader is invited in the questions at the end of the chapter to figure out how to solve this version of the puzzle without using the axiom of choice, and in particular without choosing representatives of equivalence classes.

22.5 Set Theory Prisoners

An evil demon is holding uncountably many set theorists captive. He explains to us how he will presently arrange us into a well ordered sequence, with everyone facing the same direction upward in the order. On the back of everyone's head, he intends to inscribe an ordinal, visible only to those behind. All the ordinals will be less than a certain private ordinal θ that he is thinking of, but he won't tell us what it is. And then, at the designated time, we must all shout out simultaneously an ordinal that we predict is the ordinal on the back of our head. These ordinals must all be less than θ, or we shall all be put to death immediately. Otherwise, if all the predictions are correct except for at most finitely many errors, then we shall all be released. We have a few moments to plan before all this will happen. Will we be able to succeed?

Think about it...

Yes, in fact we shall have a strategy to succeed. One of us had tentatively proposed that if we are arranged into a well-order of order type γ, an uncountable ordinal, then we could attempt to implement our previous methods by defining the relation of eventual agreement on the sequences of length γ, by which sequences are equivalent when they are identical after some index point. Since each of us will see the tail of the sequence of actual labels, we will each know the eventual-agreement equivalence class, and we can make our guesses in accordance with the agreed upon representative of that class. These guesses will agree eventually from some point on with the true pattern of labels.

But this is not good enough! Eventual agreement for sequences of length γ, after all, still allows for infinitely many differences, so long as the sequences agree after some point. It will not help to use the finite-difference equivalence relation in place of eventual agreement, by which two γ-sequences are equivalent when they disagree on only finitely many coordinates, since the set theorists are not all able to observe the finite-difference equivalence class of their pattern—at positions ω and beyond in the well-order, they are missing infinitely much information. We need another method.

To succeed, let us proceed according to the following strategy. First, we should all agree on a well-ordering $\lhd$ of all the possible γ-sequences of sets that might arise as a labeling pattern. This would be possible by using the axiom of choice if we knew the ordinal θ from which the labels would be drawn (or even merely an upper bound on θ), for we would simply fix a well-ordering of θ^γ. But since we do not know what θ is, and we do not know γ in advance either, we shall instead appeal to the global choice principle, a stronger uniform version of the axiom of choice, to fix a well-ordering of the proper class consisting of all well ordered sequences of ordinals of any ordinal length.

While we set theorist prisoners are being assembled into the γ-sequence, we each have a chance to observe γ, as well as our own place in the sequence. Next, having agreed on the well-order $\lhd$ of the possible patterns of labels of that length, each of us can survey the pattern of ordinal labels on those ahead of us, and we could make our prediction in accordance with the $\lhd$-least pattern that agrees with our observed data. Our provisional strategy is thus simply to guess in accordance with the least pattern that agrees with what we see ahead.

Except—and this is how we deal with the deathly θ issue—we wouldn't want to risk death by having anyone guess an ordinal that is not below θ, and to ensure that this doesn't happen, we shall adopt the uniform policy that if someone's predicted ordinal is larger than all those they see ahead, then they should instead simply guess the ordinal 0. In short, none of us will ever state an ordinal that is larger than all the ordinals that they observe ahead, and consequently, all our announcements will be safe in light of the θ rule. I shall argue at the end that this policy will affect altogether only finitely many people, and therefore ultimately it will have no effect on whether our strategy is successful.

The key observation to make is that whenever one of us stands somewhere behind another, then the first sees all the same observation data as the second (and more), so any labeling pattern that stands in agreement with what the first sees will also be in agreement with what the later set theorist sees. It follows that as one moves from the back to the front in our ordering, the labeling patterns being used for prediction at each stage are descending in the $\lhd$ order or staying the same. Since $\lhd$ is a well-order, these can descend only finitely many times altogether since there is no infinite descending sequence in a well-order. For this reason, there will be altogether only finitely many overall patterns that are used by any of us in our predictions.

Next, notice that in any well ordered sequence of ordinals, there can be only finitely many instances where an ordinal exceeds all the later ordinals in the sequence. The reason is that for any such instance, the next instance will involve a strictly smaller ordinal, and these ordinals can descend only finitely many times. Since we have observed that only finitely many patterns will be used for prediction, and these each have only finitely many instances of an ordinal exceeding all later ordinals, it follows that only finitely many people will be affected by our θ-policy.

Finally, we observe that whenever someone's predicted ordinal differs from their actual label, then indeed those behind will be using a different prediction pattern, one with the correct value for that person, since this is part of the observation data for those behind. Thus, an incorrect prediction value is exactly one of the cases where the prediction pattern has dropped in the $\lhd$ order. And since this occurs only finitely many times, only finitely many of us will be wrong! We set theorists shall be released by the evil demon.

Questions for Further Thought

22.1 Explain in full detail the solution at the mathematical society initiation test for the case of real numbers.

22.2 Design a version of the puzzle for the musical society, where each candidate has a tuning fork tuned to a particular (real-number) frequency. How will you ensure that the candidates will know only the tones of those in front, but not their own or those behind? Is there a winning strategy by which all guesses will be correct?

22.3 Explain in full detail how and why it is that we all are able to get into Heaven in the Aftermath parable.

22.4 Suppose that there is a vast infinity of us, each wearing a black or white hat. We can all see each other's hats, but not our own. One amongst us is chosen and allowed to look at her hat and change it to the other color if she desires. Now, we all simultaneously announce a guess as to our hat color. Show that there is a strategy that will ensure that we are all perfectly correct. Show the same with any finite number of hat colors.

22.5 Without using the axiom of choice, explain how we nevertheless have a strategy by which we shall all be correct at the midnight ball in the version of the puzzle where there are countably infinitely many people and we each have a real number on our foreheads.

22.6 Infinitely many philosophers stand in a line, like the natural numbers, facing toward infinity. Each wears a black or white hat, seeing only those hats ahead of them, and is unaware of their own hat color or those behind. The philosopher at the back is allowed to inspect and change her hat color if desired. Then, starting from the back, the philosophers shout out in turn a prediction for the color of their hats. Is there a strategy by which they can ensure that they are all correct?

22.7 In the previous question, suppose now that the hat colors are chosen from a continuum of possibilities, like the real numbers. Show that there is a strategy for them all to be correct, except possibly philosopher 0. Can you prove this without using the axiom of choice?

22.8 Which of the various puzzles admit the possibility of a nefarious person causing disaster? And how bad does it get? For example, no one person can affect the success of the result in the Aftermath. But at the midnight ball, what happens if the Queen of the Ball doesn't follow the plan? What happens exactly at the mathematical society initiation if the 37th candidate makes a mistake?

22.9 Imagine an alternative mode of release for the set-theory captives. They are arranged as before in a well ordered sequence, labeled with arbitrary sets, but now they sound off their predictions successively, starting at the back. They are released only if all are correct except at most one error. Can they succeed? In fact, after the first announcement, all the rest can announce simultaneously, and all correct. (Hint: We do not even need the axiom of choice for this version of the puzzle, if the first set theorist makes a good announcement.)

Further Reading

Horsten and Welch (2013). Highly recommended, the full original, erudite parable of the Aftermath.

Hardin and Taylor (2008b). A nice introduction to infinite hat problems.

Credits

The coin-flip hat puzzle was reportedly created by Todd Ebert in 1998. The argument for the optimality of the solution is based on answers to my questions on MathOverflow at https://mathoverflow.net/q/475606/1946. Hardin and Taylor (2008b) credit the origin of the first infinite hat problems to Yuval Gabay and Michael O'Connor in spring 2004. The form of hat puzzle in the garden party is due to Jeroen Winkel, and extremely general uncountable forms of this puzzle are considered by Jeroen Winkel and Andreas Lietz in their article, "Infinite Hat Problems and Large Cardinals." The parable of the Aftermath is due to Leon Horsten and Philip Welch (2013), but my account is a simplification, presenting it essentially as a version of the now classic two-color infinite hat puzzle, whereas Horsten and Welch focus on a subtle prisoner's-dilemma conflict that arises when the rules are: If infinitely many err in their self-judgement, then all are sent down, but if only finitely many err, then only those are sent down. This change causes a rift in the incentives, since everyone wants not only for the group to be mostly correct, but also for themselves individually to be correct. The solution to the set theory prisoners puzzle is a generalization of the method due to Hardin and Taylor (2008a), who consider only the two-color case with ω_1 many people.

23 The Guessing-Box Puzzle

Abstract. Which number is in your box? Using the axiom of choice, we shall find a strategy for the guessing-box puzzle that will enable us, with a shocking degree of reliability, to predict the contents of various boxes not yet opened.

23.1 The Grand Guessing-Box Hall

We stand in nervous excitement amongst 100 fellow mathematicians assembled in the foyer of the Grand Guessing-Box Hall, which houses infinitely many distinctive wooden boxes, each containing a real number. Perhaps one box has π, another $\sqrt{3}$, and so forth, and perhaps the numbers might appear multiple times or not at all; we can't be sure. The numbers were not necessarily chosen randomly, but they might have been chosen to thwart an anticipated guessing plan. In the guessing-box puzzle, each of us shall indeed aim to guess the contents of many of the boxes without opening them. A cheer goes up at the announcement that the games are imminent.

According to the established procedure, each mathematician successively will enter the main hall alone, and during the allotted time inspect whichever and as many boxes as desired, opening them and learning the number inside each one, and proceeding to further boxes, in a process perhaps determined by the particular numbers observed, except that ultimately, one box must be left unopened. For this final box, the mathematician will discreetly make a prediction of the real number it contains. After the accuracy of the prediction is checked and recorded, the mathematician—whether elated or crushed—will exit to the rear garden, where refreshments will be available, while the boxes are resealed and the room reset to the original state for the next mathematician to enter. We win, as a team, if all our predictions are correct except for at most 1 of us guessing incorrectly. That is, 99 of us must guess correctly.

Can we do it? We had all seen the room briefly, and we have had a few minutes to plan, but naturally, no communication will be allowed once the play begins.

Think about it...

Incredible as it may seem, we can win! There is a strategy to ensure that we shall all predict correctly, except at most one of us guessing in error. The solution will use the axiom of choice, much like the infinitary hat puzzle solutions of chapter 22.

Our winning guessing strategy proceeds like this. Having seen how the boxes are laid out in the hall, we shall agree on a way of conceiving of them as 100 disjoint infinite sequences of boxes, one sequence associated with each of us. If there had been uncountably many boxes available, it is no matter, since we shall use only countably many. Each of us has been assigned in effect an infinite sequence of real numbers, the as-yet-unknown contents of the boxes in our assigned sequence. We shall consider the relation of *eventual agreement* on such sequences, by which one sequence $\{x_n\}$ is said to be equivalent to another $\{y_n\}$ if they eventually agree; that is, if $x_n = y_n$ for all $n \geq N$ for some number N. Let us call the smallest such N the *agreement point* of these two sequences, the point where they first come into an enduring agreement. Eventual agreement is an equivalence relation, which therefore partitions the collection of sequences into the various equivalence classes of sequences. By the axiom of choice, we may choose a distinguished representative from each equivalence class, an agreed scheme of choices.

When mathematician i out of the 100 enters the hall, she begins by immediately opening up all the boxes on everyone else's sequences, reserving her own sequence for a gentler subsequent treatment. For each $j \neq i$, she can thus know the jth sequence in its entirety and compare it with its chosen representative to learn the agreement point N_j, the point from which the jth sequence begins to agree with the numbers on the corresponding agreed representative of its equivalence class.

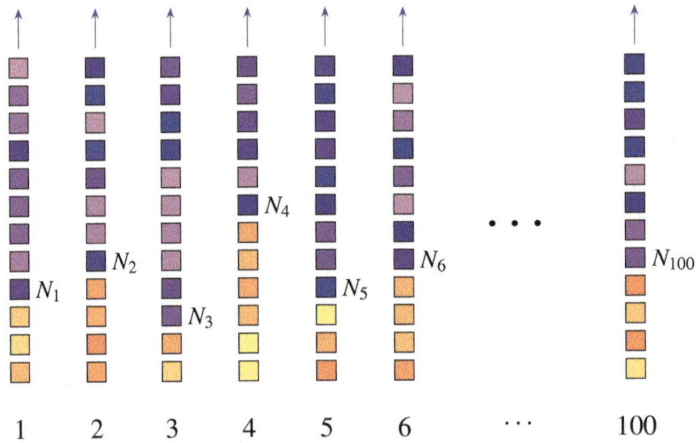

Mathematician i now computes $N = \max\{ N_j \mid j \neq i \}$, which she will take as a proxy for the (as yet unknown) agreement point N_i of her own sequence. She next opens all the boxes on her own sequence past this number N, and thereby comes to know the eventual-agreement equivalence class of her own sequence. She then predicts that box N on her sequence, which has not yet been opened, has the same number as the corresponding number on the agreed representative of the equivalence class of her sequence.

The key point is that her actual sequence will begin to agree with the representative of its equivalence class beginning at the actual agreement point N_i, so if she is lucky and it happens to be that $N_i \leq N$, then her guess at box N of her sequence will be correct. But in particular, this means that if $N_i \leq N_j$ for some $j \neq i$, then indeed she will be lucky in this way since this will cause $N_i \leq N$, and thus she will be right. It follows from this that all but one of us will be lucky in just this way—there will be at most one mathematician who fails: the mathematician with the largest value of N_i, strictly larger than all the others. Only one of the numbers, after all, can be strictly larger than all the others. So at least 99 of us will guess correctly. And if 2 mathematicians happen to have the same largest agreement point, with none strictly larger than all the others, then in fact all 100 of us will be correct.

Other Things in the Boxes

It was not important for the solution that the boxes had real numbers inside, and the same strategy will work regardless of what kind of things might be in the boxes. For any set B of possible contents, if every box has some element of B, repetitions allowed (it makes the puzzle only harder to allow additional possibilities), then we can still allocate the infinite sequences of boxes to the mathematicians and agree on a choice of representatives for the eventual-agreement equivalence classes for sequences over the set B. Each mathematician i will enter the room as before and open all the boxes for the other mathematicians' sequences and thereby learn N_j for $j \neq i$, and take $N = \max\{N_j \mid j \neq i\}$ as a proxy for her unknown N_i. Then she opens her own boxes beyond N, learns her equivalence class, and then predicts that box N on her sequence agrees with the representative of that class. She will be right if $N_i \leq N_j$ for some $j \neq i$, and this will be true for everyone except possibly the one mathematician with a value of N_i that is larger than all other N_js. So it doesn't matter what kind of things will be in the boxes, so long as the set B of possible objects is known in advance.

Using Global Choice When *B* Is Not Known in Advance

If the set B is not known in advance, then there will still be a solution by appealing to the global choice principle, a strong uniform version of the axiom of choice, by which there is a proper class well-order of the entire mathematical universe, including all possible sequences of mathematical objects of any kind. This is enough to choose representatives of the equivalence classes, regardless of the particular set from which the objects were drawn.

Getting Infinitely Many Guesses Right

Next door is the Grander Guessing-Box Hall, with rather stricter requirements for success. At this guessing club, each mathematician must leave infinitely many boxes unopened rather than just one, and make a guess about each of them, all at once. Nevertheless, even for this more difficult game, there is a strategy to ensure that 99 of the 100 mathematicians can be entirely correct about every single one of their guesses.

For the solution, we allocate to every mathematician not just an infinite sequence of boxes, but rather an $\omega \times \omega$ matrix of boxes. We may conceive of each matrix as a sequence of columns. Let us say that two such matrices are equivalent if from some column index onward, all the columns are totally identical. And let us agree on a choice of representative from each matrix equivalence class and then proceed similarly as before.

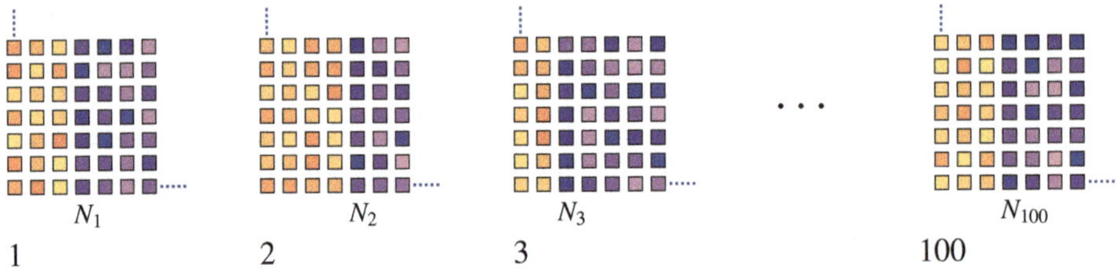

Namely, mathematician i enters the room and opens all the boxes in all the other mathematicians' matrices. She thereby learns the agreement points N_j, the point in matrix j at which the columns begin to be in perfect agreement with the chosen representative. Taking $N = \max\{ N_j \mid j \neq i\}$ as a proxy for her own N_i, she then opens all the boxes in her columns beyond N and thereby learns the equivalence class of her matrix. She then uses the unopened boxes in column N for her predictions, guessing that they agree with column N of the representative of her class. If $N_i \leq N$, then she will be entirely right. But this will be true for at least 99 of the mathematicians since the only way that it can fail is if N_i is strictly larger than all the other N_j, and this can be true for at most one mathematician, the one with the strictly largest N_i. So 99 of us will make all our infinitely many guesses with perfect correctness.

23.2 Infinitely Many Mathematicians

Clearly the solution works not just with 100 mathematicians, but with any finite number. With 1000 of us, we can ensure that 999 are correct. What about infinitely many mathematicians? At the Grandest Guessing-Box Hall, we stand with infinitely many mathematicians packed tightly into the foyer, ready to enter the main hall and begin the guessing game. Can we all guess correctly except at most one of us?

Let's consider first the case of countably infinitely many mathematicians and countably infinitely many boxes, with each mathematician making a prediction about one unopened box. The original solution strategy with 100 mathematicians cannot be implemented directly in this context because the supremum $\max\{ N_j \mid j \neq i\}$ might be infinite, and therefore it could not serve as a useful proxy for N_i.

Nevertheless, there is a strategy by which all infinitely many of us will guess correctly except at most one of us. For this solution, again we imagine that the boxes are partitioned into disjoint infinite sequences of boxes, with one allocated to each mathematician. Each mathematician i will again immediately open all the boxes on the other sequences and thereby come to know all the other mathematicians' sequences, as well as their corresponding agreement points N_j, the place where those sequences come into agreement with the agreed-upon representatives of their equivalence classes. Let us also agree on a choice of representative sequence of natural numbers $\langle n_0, n_1, n_2, \ldots \rangle$ for the equivalence class of the sequence $\langle N_0, N_1, N_2, \ldots \rangle$, considered under eventual agreement. Since each mathematician i knows all the N_j values except N_i, we will thus all come to know the same representative n_j sequence. So there is a finite set I of exceptional indices, such that $n_j = N_j$ for all j outside this exceptional set I, and mathematician i can observe this set correctly for all $j \neq i$ but doesn't know whether $i \in I$. Mathematician i defines N to be the maximum of n_i and all the N_j for $j \in I$ with $j \neq i$. She will take this N as a proxy for her unknown N_i and open the boxes on her sequence beyond N, observe the resulting equivalence class, and then make her prediction about the unopened box N on her sequence in accordance with the representative of that class. If i is not actually in I, then $n_i = N_i$ and so $N_i \leq N$, which makes N good enough for guessing, since she will be making her prediction in the agreement part of her sequence. And if i is in I, then she will have $N_i \leq N$, provided that $N_i \leq N_j$ for some $j \neq i$ with $j \in I$, again making her guess correct. But this property will be true for all but one of them in I, since it can fail only for the $i \in I$ that happens to have an N_i that is strictly larger than all the other N_j for $j \in I$. So all but one of us will guess correctly, and the strategy succeeds. It is incredible.

Continuum Many Mathematicians

Consider now an uncountable case of the puzzle. Suppose that we have continuum many mathematicians, one for each real number, and continuum many boxes in the hall, each with a real number inside. Each mathematician will enter in turn and have a chance to open some of the boxes but must leave continuum many boxes unopened and make predictions about their contents. In short, there are continuum many mathematicians, each making continuum many guesses.

What I claim is that there is a strategy by which every single mathematician is perfectly correct in all their guesses except for one unfortunate mathematician at most, who may make some incorrect guesses.

The solution is this. There are continuum many boxes, but since $\mathbb{R}$ is equinumerous with $\mathbb{R} \times \mathbb{R}$, we may think of them as a continuum of continuums. Indeed, since $\mathbb{R}$ is equinumerous with $\mathbb{R} \times (\omega \times \mathbb{R})$, we can allocate to each mathematician a countable sequence of such continuums of boxes, in effect a matrix of size $\omega \times \mathbb{R}$ with ω many columns, each of size continuum. We define as before the eventual column-agreement relation on such matrices and agree on a choice of representative for each equivalence class. Each mathematician i from the continuum will enter the hall on their appointed turn and open all the boxes on the other mathematicians' matrices and thereby learn those equivalence classes and their agreement points N_j. But let us also assume that we had all also agreed on representatives for the finite-difference relation on continuum-indexed functions $\langle N_j \mid j \in \mathbb{R}\rangle$ of natural numbers. So we have a representative $\langle n_j \mid j \in \mathbb{R}\rangle$, where each n_j is a natural number and $n_j = N_j$ for all but finitely many j, those in a finite set I. Mathematician i knows all the n_j and knows N_j for all $j \neq i$, but she doesn't yet know her own N_i, and she doesn't know whether $i \in I$. Nevertheless, she defines N to be the maximum of n_i and all the N_j for $j \in I$ with $j \neq i$ and takes this as a proxy for her own N_i. So she opens all the boxes beyond the Nth column of her matrix and thereby learns the equivalence class, and then guesses all the continuum many boxes in her column N in accordance with the representative of that class. If i is not I, then $n_i = N_i$, and so $N_i \leq N$, which will make all her guesses at or past the agreement point (hence all correct). Furthermore, if $N_i \leq N_j$ for some other $j \in I$, then indeed $N_i \leq N$, placing her in this lucky case. The only way that mathematician i can be wrong is if $i \in I$, but the true value of N_i is strictly larger than all the other N_j for $j \in I$. And this can be true for at most one mathematician. So amongst the continuum many mathematicians, all of them will be fully correct in all their continuum many guesses, except possibly one unlucky mathematician who may make incorrect guesses.

The Higher Infinite

The argument generalizes to any infinite cardinal κ. For κ many mathematicians and at least κ many boxes in the hall, and any nonempty set B of possible box contents, then in the puzzle where each mathematician will make predictions about κ many boxes, then there is a strategy for all of them to be fully correct in all their guesses, except possibly one poor mathematician, whose answers, I admit, might be completely wrong.

More Mathematicians than Boxes

For certain variations of the puzzle, we shall be unable to ensure so much correctness in our guessing. For example, if we have uncountably many mathematicians, but only countably many boxes, then I claim that there will some arrangements where uncountably many of the mathematicians guess incorrectly. Consider the trivial arrangement at first, where every box has 0 inside. The mathematicians will make some predictions, but since there are more mathematicians than boxes, it follows by the pigeonhole principle that there must be some box chosen as a prediction box by uncountably many of them. By changing the value of this box, we can make uncountably many of them wrong. More generally, whenever there are more mathematicians than boxes, then it is impossible to win, even in the formulation where each person guesses about only one box.

23.3 No Solution with Finitely Many Boxes

What about the case with only finitely many boxes? Suppose that we have finitely many mathematicians and finitely many boxes, each containing a real number, and they aim to enter the hall one at a time in turn according to the procedure, each one opening some boxes and making a prediction about some unopened box.

It strikes me as hopeless for them to succeed to any great extent—shouldn't we generally expect them all to be wrong? And yet this seems to be one of those intriguing mathematical cases, where something that seems intuitively obvious turns out to be fussy and delicate to prove. I had wanted to find an elementary, transparent argument showing that for any given strategy, there would be a robust sense in which they are almost always all wrong. But alas, I have found no such sweeping, easy argument. Worse, it isn't clear what sense of "almost always" should be meant here since if the strategies do not arise from measurable functions, then the corresponding correctness and incorrectness sets might not be measurable—they could be paradoxical sets like the Vitali set.

Nevertheless, I do have an argument showing that there will be infinitely many instances where they are all wrong. Unfortunately, it is not completely elementary since it will make an appeal to Ramsey's theorem, a certain difficult result in combinatorics, which underlies the subject of Ramsey theory, named for the brilliant philosopher-mathematician Frank P. Ramsey. But kindly allow me to sketch out the details for those who are interested, while others may skip this part.

We have finitely many mathematicians and finitely many boxes in the Guessing-Box Hall, each with a real number, and each mathematician will ultimately venture a guess about one box. What I claim is that for any specific procedure by which they will make their predictions, there will be infinitely many states of the boxes for which the predictions are all incorrect. Indeed, I will find such incorrect configurations of the boxes where the boxes contain natural numbers. Consider what I call the *increasing* box arrangements, where the boxes are laid out in a line and contain natural numbers that are increasing in order. For any such configuration, consider how the given strategy unfolds. Ultimately, some of the mathematicians will make their guess about certain boxes and others about other boxes, and there will be a pattern of correctness and incorrectness in these predictions. There are only finitely many such patterns of which-boxes-are-guessed and which-mathematicians-are-correct (I do not take the numbers in the guesses themselves as part of this data). Every box configuration leads to one of these patterns. It follows as an immediate consequence of the infinitary Ramsey's theorem that there is an infinite set H of natural numbers, such that any increasing box configuration arising from numbers in H will lead to the same pattern of boxes-guessed and mathematicians-correct—the set H is homogeneous in this sense (and this kind of thing is what Ramsey theory is all about). I claim that for these configurations, none of them can be correct. The reason is that I could lay out an increasing pattern from H using numbers spaced well apart within H. If the homogeneous pattern was that a certain mathematician was correct, then by changing his box in this configuration to the next largest element of H, it would still be an increasing arrangement from H, but now his prediction would be wrong, contradicting homogeneity. So the homogeneous pattern must be that all increasing configurations from H lead to everyone guessing incorrectly. So I have found infinitely many configurations for which everyone is wrong. In light of the finite Ramsey theorem, the argument actually shows that for any sufficiently large finite set of possible box contents, there will be a configuration using those contents in which everybody is wrong.

Meanwhile, when there are very few possible box contents, then there are strategies to ensure some success. For example, if every box contains either 0 or 1, then this is an instance of the beetles-in-boxes puzzle described in section 22.2, where we could ensure that always half would be correct by pairing them off, with one person in each pair assuming that the two boxes are the same and the other assuming that they are different. This strategy requires each mathematician to look at only one other box, but more elaborate strategies work with more looking and greater success. There seems to be a transition in the finitely-many-boxes case between success when there are very few kinds of things to place in the boxes to complete failure via the Ramsey theorem argument when the number of possible things to place in the boxes becomes sufficiently large. Ultimately, it becomes a subtle problem in combinatorics.

Questions for Further Thought

23.1 In the original guessing-box puzzle, with 100 mathematicians and infinitely many boxes, following the strategy described in this chapter, will the mathematician who guesses wrongly know before making her prediction that it will be she, if anyone, who guesses wrongly? If this is not true in all circumstances, what are the circumstances in which it is true—that is, in which a person would know in the midst of the process that everyone else will guess correctly?

23.2 Show that if there are more mathematicians than boxes, then there are many states of the boxes that are losing (in the sense that not all-but-at-most-one were correct), and indeed, every state of the boxes is very close to many losing states.

23.3 Prove that there can be no winning strategy in the guessing-box puzzle in which the mathematicians end up always make their guesses about the same boxes.

Further Reading

Kuperberg (2013) and Naslund (2013). This is further discussion of the guessing-box puzzle on MathOverflow.

Credits

I learned of the numbers-in-boxes puzzle on MathOverflow in a question by Denis Kuperberg (2013), who reportedly learned of the puzzle from a blog post of David Madore (2009), and a version appeared on a 2009 M2 exam of Boban Velickovic in Paris. But the puzzle reportedly originated in 1997–1998 with Christophe Chalons, appearing briefly in his PhD dissertation at Paris VII, written under the supervision of Velickovic, but also in Chalons (2007). The solution with countably infinitely many mathematicians guessing, all but one at most being correct, is due to Eric Naslund (2013). The improvement to allow uncountably many mathematicians, each making uncountably many guesses with all being correct except for one mathematician, is my own observation. I asked on MathOverflow whether anyone has a truly elementary proof of failure in the finite-box case (see https://mathoverflow.net/q/476227/1946), and the reader may find further discussion there, including an answer by Elliot Glazer.

24 We Can Predict the Future

Abstract. According to the nearly perfect prediction theorem, there is a prediction strategy for the current and future values of a given function f on the real numbers based on the history of prior values—one that predicts every function correctly at almost every point. The proof is a pure-existence mathematical result, establishing the existence of the nearly perfect prediction strategy from the axiom of choice without revealing the specific details of any particular prediction.

Suppose that, like stock-market traders trying to guess today's opening price and the subsequent price movements of a certain favored stock, we aim to predict the current and future values of an unknown function based on the observed history of previous values. The true function $f : \mathbb{R} \to \mathbb{R}$ will be revealed only gradually, and we seek to predict the current value $f(t)$ at the current time t in light of the historical data, as well as how the function will continue beyond t into the immediate future.

24.1 Predicting the Present

Let us begin with the seemingly easier task of predicting the present. We want to predict the current value $f(t)$ of the function f on the basis of its historical values before time t. Of course, if we knew somehow that the function were continuous, then we could easily predict the current value simply as the limit of the previous values:

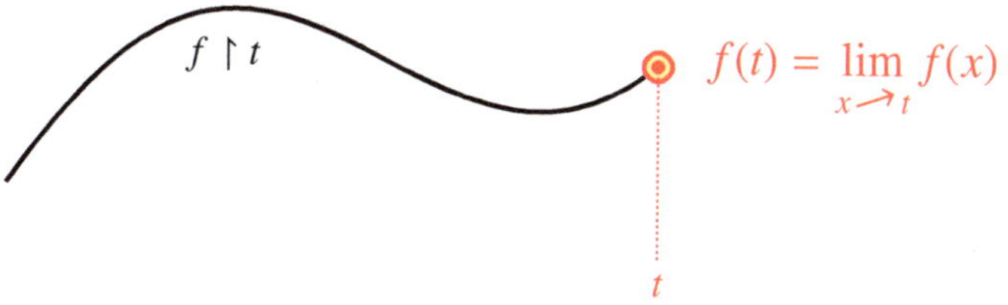

The continuous case is thus a trivial kind of case for predicting the present—the limit value will be correct in every continuous instance.

The true function, however, might not be continuous—it could be very badly behaved, perhaps discontinuous at every point. Nevertheless, we hope to predict the current value $f(t)$ based on the history of the function until now. Let me emphasize that we aim to predict the values of the function exactly, not merely approximately. Close doesn't count.

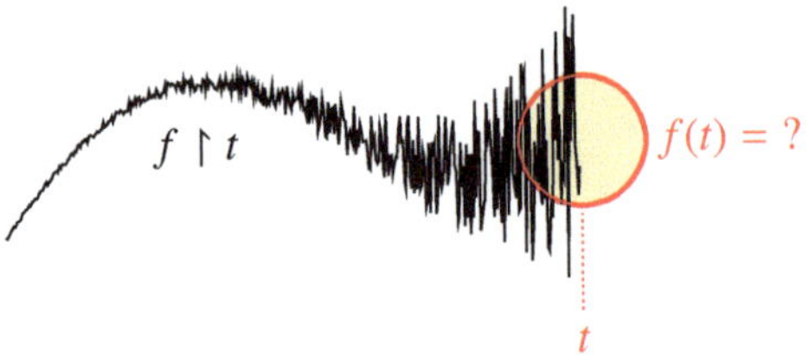

In the general case, it may seem that no amount of information about the prior values $f(x)$ for $x < t$ would have any bearing on the current value—after all, the true current value $f(t)$ could be anything, to say nothing of the future values beyond t. Right? How could we possibly predict the correct value of $f(t)$ with any reliability? Shouldn't this be completely hopeless? We might hit a few lucky guesses, sure, but for any particular prediction procedure, wouldn't there be numerous functions that evade the predictions entirely?

Actually, no. The amazing mathematical fact, proved using the axiom of choice, is that there is a nearly perfect prediction strategy—one for which every function is predicted correctly at almost every point.

The proof casts us floundering in the deep philosophical waters of mathematical ontology. For although the proof establishes that indeed there is such a nearly perfect prediction strategy, unfortunately it does not reveal any useful details about the actual predictions that are to be made. This will be a pure-existence proof, showing that the strategy exists in the platonic realm of ideal forms, but without revealing sufficient information for us actually to implement it in practice. You had best keep your day job.

To my way of thinking, nevertheless, even the pure mathematical existence of such a nearly perfect prediction method is astounding, so let me tell you the story of this remarkable theorem, the nearly perfect prediction theorem, proved by Christopher Hardin and Alan Taylor (2008a). To begin, for any function $f : \mathbb{R} \to \mathbb{R}$ on the real numbers, let us write $f \restriction t$ to refer to the historical part of the function prior to time t; that is, restricting the function to the domain $(-\infty, t)$ with all the observed previous values $f(x)$ for $x < t$. A current-value prediction strategy is an operation P that looks at this historical data of any given observed prior values $f \restriction t$ up to some point t, and on that basis makes a numerical prediction $P(f \restriction t)$ about what the current value of the function will be:

Historical data		Predicted value at t
$f \restriction t$	$\longmapsto$	$P(f \restriction t)$.

The prediction is correct at time t for the function f if indeed $f(t) = P(f \restriction t)$.

We may state the present-prediction form of the theorem as follows.

Nearly perfect prediction theorem. *There is a nearly perfect prediction strategy P—one for which every function* $f : \mathbb{R} \to \mathbb{R}$ *is predicted correctly at almost every time t:*

$$f(t) = P(f \upharpoonright t).$$

To clarify the sense of "almost every" here, the proof shows that the predictions will be correct except on a countable, nowhere dense set of prediction-failure points t. What is more, the prediction-failure points will be well ordered as a set of real numbers. It follows that every real number s will be followed by an interval $(s, s + \varepsilon)$ on which the predictions are correct—they are correct up to the next failure point, if any—and indeed, there will be arbitrarily long intervals on which the predictions are perfectly correct. It is incredible.

I am astounded by this theorem since as we said, we had thought perhaps any given prediction strategy might admit functions f evading the predictions entirely, but here, we have a strategy that predicts every single function correctly almost always. I remember the exhilaration in the audience at our CUNY Logic Workshop in New York when Taylor spoke on his theorem shortly after having proved it with Hardin. What a joy it was to experience.

Let me sketch the proof, which surprisingly is not terribly difficult. Consider the class of all functions $f : \mathbb{R} \to \mathbb{R}$. By the well-order theorem, we may fix a well-ordering $\trianglelefteq$ of this collection of functions. Thus, we have placed them in a linear hierarchy with the property that every nonempty set of functions has a least member. If we imagine the hierarchy as a priority preference relation, where we prefer functions appearing lower in the hierarchy to those appearing higher, then what the well-order property amounts to is that every nonempty set of functions has one that is strictly preferred to all the others.

Using this well ordered preference hierarchy, we can now define the prediction strategy. Whatever observed historical information $f \upharpoonright t$ we are given, we should consider all possible functions that agree with that data to this moment and let f^* be the most-preferred function amongst them—the function appearing least in the $\trianglelefteq$ hierarchy while standing in agreement with the given history $f^* \upharpoonright t = f \upharpoonright t$. We simply predict that the function f agrees with this most-preferred function f^*:

$$\text{Let} \quad P(f \upharpoonright t) = f^*(t).$$

In short, we predict that the current value of the function agrees with the current value of the $\trianglelefteq$-most preferred function f^* standing in agreement with the observed data $f \upharpoonright t$ of prior values.

Let us now make the key observation. Suppose that the true function is f and consider the predictions made at times s and some later time t. Since the available history $f \upharpoonright s$ is part of the historical data available at the later time $f \upharpoonright t$, the prediction made at the later time t will be considering a more restrictive class of functions. Therefore, whichever function f_s^* is used at time s will be less than (or equal to) the function f_t^* used at time t, with respect to the $\trianglelefteq$ order. In short,

$$s < t \quad \Longrightarrow \quad f_s^* \trianglelefteq f_t^*.$$

As time passes, we are either climbing in the well ordered $\trianglelefteq$-hierarchy of functions or staying the same.

If the prediction made at s was incorrect, however, it means that $f_s^*(s) \neq f(s)$, and consequently f_s^* is not the same as f_t^*, which agrees with f up to t. Therefore,

$$\text{prediction is incorrect at } s \text{ and } s < t \quad \Longrightarrow \quad f_s^* \triangleleft f_t^*.$$

In other words, as time passes, the predicted continuation functions at the prediction-failure points climb strictly higher in the $\trianglelefteq$-hierarchy of functions. Since that hierarchy is well ordered, it means that these prediction-failure points themselves must be well ordered as real numbers.

From this, all the amazing consequences follow. Since the prediction-failure points are well ordered, above every point s there is a first next point at which the prediction is incorrect, if there are any such points above s at all, and consequently, on that whole open interval $(s, s + \varepsilon)$, the predictions are totally correct, without exception. The prediction-failure points are thus a countable, nowhere dense set. It is countable since we can associate each failure point s with a rational number in this interval. Since there must also be an absolutely smallest prediction-failure point r, if any, all the predictions made at times $t \in (-\infty, r)$ are correct, so there are arbitrarily long intervals on which all the predictions are correct. The real line is thus exhausted by a countable well ordered sequence of intervals, which abut each other and their limits, with the predictions fully correct in the interior of each interval. Amazing!

Continuous Functions

The prediction strategy given here does not necessarily respect continuity since perhaps the $\trianglelefteq$-least function f^* in agreement with $f \upharpoonright t$ is not continuous at t, even when $f \upharpoonright t$ is continuous. Nevertheless, we can ensure that the prediction process will always guess in accordance with continuity simply by placing all the continuous functions at the bottom of the $\trianglelefteq$ order. That is, first well-order the continuous functions, and then well-order all the rest of the functions and place them on top of this hierarchy. This is still a well-ordering of the functions, and it will have the property that if the historical data $f \upharpoonright t$ has a continuous extension, then a continuous function will be chosen as the most preferred one. So this strategy will guess completely correctly all continuous functions, while also guessing every function almost always correctly.

24.2 Predicting the Future

Let us now look at the harder task of predicting not only the present values, but also the immediate future values of an arbitrary unknown function, based on the observed prior history. We shall make predictions about how the function will continue.

Specifically, a *continuation prediction strategy* is a function F that takes as input the historical values $f \upharpoonright t$ of a real-valued function up to some time t, and on that basis makes a prediction $F(f \upharpoonright t) : \mathbb{R} \to \mathbb{R}$, a function on the real numbers that agrees with all the observed prior values before t, predicting how the function will continue to the current and future values beyond t:

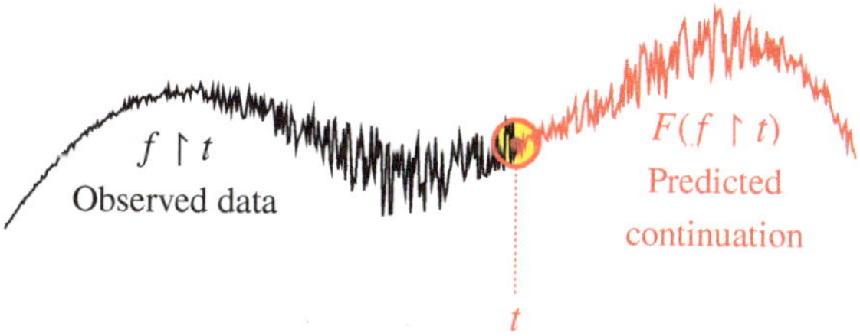

Such predictions can often be incorrect, of course, especially in the distant future. Nevertheless, we shall prove that there is a continuation prediction strategy that is almost always correct, at least in the immediate future. What I mean by this is that the predicted continuation function $F(f \upharpoonright t)$ almost always agrees with f for at least a short time beyond t. Specifically, for almost every t, there will be $\epsilon > 0$ such that $f(x) = F(f \upharpoonright t)(x)$ for all $x < t + \epsilon$. The continuation prediction thus gets the right value not only up to and at t itself but also has the right values in the immediate future of t.

Nearly perfect prediction theorem (future). *There is a nearly perfect continuation prediction strategy F, one predicting the present and immediate future values of every function $f : \mathbb{R} \to \mathbb{R}$ correctly almost always. That is, for every f and almost every t, the predicted continuation agrees with f in an interval extending strictly beyond t:*

$$\exists \epsilon > 0 \quad \forall x < t + \epsilon \qquad f(x) = F(f \upharpoonright t)(x).$$

Again, the sense of "almost always" here is that the predictions are correct except on a countable, nowhere dense set of exceptional points, which furthermore are well ordered as a set of real numbers. Consequently, every point s is followed by an interval $(s, s + \epsilon)$ on which all the predictions are fully correct for this entire interval, and indeed there will be arbitrarily long intervals on which the predictions are fully correct.

To prove the theorem, we again let $\trianglelefteq$ well-order the collection of all functions, and let $F(f \upharpoonright t) = f^*$ be the $\trianglelefteq$-least function $f^* : \mathbb{R} \to \mathbb{R}$ that stands in agreement with the observed historical data $f \upharpoonright t = f^* \upharpoonright t$. In short, we predict that the function will continue in accordance with the $\trianglelefteq$-most-preferred function that agrees with the prior data.

We have already seen that this prediction method correctly predicts the current value $f(t)$ for almost every t. I claim further that the continuation prediction will almost always remain correct into the immediate future extending strictly beyond t. Consider two points in time $s < t$. The prediction made at time s is based on the prior history $f \upharpoonright s$ available at that time, and this data is included in the data $f \upharpoonright t$ that is used at the later time t. Therefore, the later prediction will be finding the most-preferred function (least in the $\trianglelefteq$-hierarchy) amongst a more restrictive class of functions. For this reason, the earlier prediction must be at least as preferred as the later prediction:

$$s < t \quad \Longrightarrow \quad F(f \upharpoonright s) \trianglelefteq F(f \upharpoonright t).$$

Furthermore, if the continuation prediction made at time s is not fully correct between s and t, particularly if the prediction was not correct for the immediate future of s, then at time t, this deviation will be observable from $f \upharpoonright t$, so we will not be selecting the same function at that time. Consequently, we will have $F(f \upharpoonright s) \triangleleft F(f \upharpoonright t)$, which means that we have climbed strictly higher in the $\trianglelefteq$-hierarchy. In short, at every prediction-failure point, we climb higher in the hierarchy of functions, so the collection of exceptional points, the points at which the prediction is not correct for the immediate future, will be well ordered in exactly the same way as a set of real numbers:

$$\text{For all prediction-failure points } s, t \quad (\, s < t \quad \Longleftrightarrow \quad F(f \upharpoonright s) \triangleleft F(f \upharpoonright t)\,).$$

This implies as before that the prediction-failure points, being well ordered in the real line, must form a countable set, measure zero, and nowhere dense. Further, every point s will be followed by an interval $(s, s + \epsilon)$ on which all predicted continuations are fully correct on this interval. And further, there will be an absolutely least failure point r, if there are any at all, so the predicted continuations are correct for the entire interval $(-\infty, r)$, which includes arbitrarily long intervals on which the predictions are totally correct.

I find it interesting to notice a certain robust stability feature in the predictions—whenever the continuation prediction is correct for the immediate future of time t, extending to some $t + \epsilon$, then the continuation predictions made at times u in that interval will also be correct for the duration of this same interval. The reason is that they will be selecting the very same $\trianglelefteq$-least function f^* as at time t since this function remains in agreement with the extra observed data available at time u and will therefore continue to be the $\trianglelefteq$-least function with that property.

Predicting the Future with Continuous Functions

Let us revisit the case of continuous functions. We had observed earlier that predicting the present value $f(t)$ of a continuous function f is trivial since it will just be the limit of $f(x)$ as $x \to t$ from below. Predicting the *future* values of a continuous function, in contrast, does not seem trivial in this way at all. Nevertheless, the nearly perfect prediction procedure provides a method to predict how the continuous function will continue. I find this amazing.

Nearly perfect prediction theorem (continuous). *In the context of continuous functions $f : \mathbb{R} \to \mathbb{R}$, there is a nearly perfect continuation prediction strategy C—one for which for every continuous function f and for almost every point t, the predicted continuation $C(f \restriction t)$ is correct for the immediate future:*

$$\exists \epsilon > 0 \quad f(x) = C(f \restriction t)(x) \quad \textit{for all } x < t + \epsilon.$$

This follows immediately as a special case of the nearly perfect prediction theorem, if we well-order only the continuous functions, or alternatively simply place the continuous functions at the bottom of the $\trianglelefteq$ order. If the true function is continuous, then at every point, the data $f \restriction t$ will have a continuous extension, so the predicted continuation will be continuous. Isn't this special case still amazing? We can predict the immediate future almost always with continuous functions.

24.3 Pure Existence

So we have proved that there is a nearly perfect prediction strategy. The reader will have noticed, however, that the proof is not constructive. The proof does not provide us with an explicit prediction strategy—one that we can actually use to make predictions, but rather proceeds from the axiom of choice via the well-order theorem. So we have at bottom a pure-existence proof—we have proved as a mathematical fact that there is a nearly perfect prediction strategy, but we cannot say much concretely about what the predictions are that it makes. The full prediction procedure will remain fundamentally unknown, and perhaps unknowable. For this reason, we find ourselves unable to implement the nearly perfect prediction strategy in practice; stock traders will not be able to use it. As we said, you had best keep your day job.

What is this strangely weak kind of existence, where the mathematical object is said to "exist" but we can't compute it or know much about it? Well, the nearly perfect prediction strategy does exist in the same way that other abstract, nonconstructive mathematical objects exist—it enjoys a pure-existence ontology in the platonic realm of ideal forms. We might imagine the perfect prediction strategy standing aside other such ideal nonconstructive objects, such as the duplicated cube, the true solutions to computably undecidable decision problems, the busy beaver function, and so forth. Although we cannot calculate the actual predictions, nevertheless I am confident that the iridescent mathematical stock traders of the platonic realm are truly making bank.

Questions for Further Thought

24.1 Show that there is no prediction process that can be guaranteed correct at any particular fixed time t. Even with the nearly perfect predictor, at any particular moment, it could be wrong. How do you square this with the fact that it is indeed nearly perfect?

24.2 Is there a nearly perfect prediction theorem for functions $f : \mathbb{N} \to \mathbb{N}$ on the natural numbers? The prediction would be made for the current value $f(n)$ based on the prior values $f \upharpoonright n$.

24.3 Is there a nearly perfect prediction theorem for functions $f : \mathbb{Q} \to \mathbb{Q}$? Or for $f : \mathbb{Z} \to \mathbb{Z}$?

24.4 What is the connection between the nearly perfect prediction theorem and the set-theorist prisoner puzzle in section 22.5?

24.5 In the proof of the present+future version of the nearly perfect prediction theorem, will it be the case that whenever the prediction is correct for the present value, it is also correct for a short time into the immediate future?

Further Reading

Hardin and Taylor (2008a). The original account of the perfect prediction theorem, also with some hat games.

Credits

The nearly perfect prediction theorem is due to Christopher Hardin and Alan Taylor (2008a, 2008b), who also analyze many further variations, in which the predictions depend only on the recent past instead of the full history, and furthermore, where the predictions obey a certain translation invariance—if the recent past at one point follows the same pattern as the recent past of another point, the pattern of the predictions will also match. The image of stock traders in the platonic realm was generated by the author via DALL·E. A shorter version of this chapter appeared as a "Short Stories" column in the AMS Notices as Hamkins (2025c).

25 Infinite Liars

Abstract. Regrettably, this chapter consists entirely of lies—every sentence in it, including this one, is false.

When my daughter Hypatia was very young, she liked to greet visitors to our home at the door with a friendly challenge:

> Answer *yes* or *no*: Will you answer *no*?

The guest would return her cheerful smile but stand flustered in the doorway with ponderous uncertainty. They were stymied by her question, for if they should answer *yes*, they would be saying that they should instead have said *no*, and if they should answer *no*, then it would mean that they should have said *yes*. My daughter had trapped them in an impossible logical labyrinth.

25.1 Epimenides Paradox

A much earlier logical labyrinth occurred on the island of Crete in the sixth century BC, when Epimenides announced:

> All Cretans are liars.

Being a Cretan himself, you see, Epimenides was calling himself a liar. Does this mean that the statement itself is a lie?

Well, for someone to be a liar, must *everything* they say be a lie? I do not think so. If someone tells lies only some of the time, then I would call them a liar, even if at other times they tell the truth. (I do not feel similarly about describing someone as honest or as a truth-teller—they must be reliably honest for this designation.) So it seems that Epimenides could consistently be telling the truth with this particular statement, even if, as seems dubious, every Cretan including Epimenides himself has told a lie at some point.

Some accounts of Epimenides have him saying:

> Cretans are always liars.

Does this change things?

I do not think so—not much anyway. After all, a musician is still a musician when the song has ended; an artist remains an artist when the brush is put down. And so also with liars. If all Cretans have told a lie, then they will always be liars from now on, even if they sometimes assert truths.

Consider, in contrast, the more extreme statement:

All Cretans always lie.

This is a totally different situation. Epimenides cannot truthfully make this statement, for if he is truthful here, then in virtue of the meaning, since he is Cretan, his statement must also be a lie. But can this statement be a lie? Well, it certainly seems so, and in fact I find it very likely that this statement is not true since I believe that most Cretans are good honest people, who seldom lie. This statement could merely be one regrettable case of a Cretan untruth, with the truth being that only some Cretans lie, and even then only sometimes.

25.2 The Liar Paradox

The *liar paradox* refers to the troubling conundrum that we face when trying to assign a coherent truth value for the following sentence, known as the *liar sentence* or sometimes personified simply as *the liar*:

This sentence is false.

The sentence asserts its own falsity. The central problem is that if the sentence were true, then precisely because of what it asserts, it would also have to be false, contrary to the assumption; but if it were false, then again precisely because of what it asserts, it would have to be true. It seems that the sentence, therefore, can be neither true nor false, posing a fundamental challenge for a complete and successful theory of truth.

The No-Truth-Value Solution Strategy

Perhaps one might hope to resolve the liar paradox by saying simply that the liar sentence has no truth value. It expresses a proposition that is neither true nor false, a gap in the truth values. Can one dissolve the liar paradox this way?

The Liar's Revenge

Some philosophers rebut that response by introducing the following sentence, known as the *liar's revenge*:

Either this sentence is false or it has no truth value.

If the liar's revenge were true, then in light of what it asserts, it would either be false or have no truth value, but both of those contradict the assumption that it is true. If it were false, in contrast, then in light of the first clause, it would have to be true, which again is a contradiction. Finally, if the liar's revenge had no truth value, then precisely because of what it asserts in the second clause, what it says would be the case, so it would be true, which would contradict the assumption that it had no truth value. So it seems wrong to say of the liar's revenge that it is true, wrong to say that it is false, and also wrong to say that it has no truth value. Crazy!

The liar's revenge thus seems directly to undermine the no-truth-value strategy of answering the original liar paradox. If it were a robust strategy to assert that the liar simply carries no truth value, then we should expect this strategy also to work for the liar's revenge, which doesn't seem to be the case.

False Versus Not True

Some philosophers prefer to state the liar sentence as the following assertion:

> This sentence is not true.

That is, they prefer to take the liar to assert its own untruth, as opposed to its own falsity. They do so to avoid a possible objection that could be made by someone who finds a distinction between "not true" and "false."

A straightforward approach to truth and falsity would take "false" simply to mean "not true," especially when applied to sentences with a definite meaning, and with this straightforward account, the need for a modified statement of the paradox may seem less urgent. But meanwhile, if there will be a question of meaningfulness, then perhaps it will be better to use "not true" in place of "false."

Would you say, for example, that the following sentence is false?

> 'Twas brillig, and the slithy toves
> Did gyre and gimble in the wabe.

Or how about this one—would you say it is a false sentence?

> Shillyhumph ghghx88shhs.

It would seem reasonable to say instead that these assertions are nonsense, with no meaning—they are neither true nor false because they do not express any proposition. But in this case, the sentences would reveal a distinction between falsity and untruth, for we have said that these sentences are not false, and yet they are not true. The sentences thus show that being false is not the same as being untrue.

But notice how the not-true formulation of the liar paradox thus seems to take on board much of the liar's revenge reasoning. Namely, if the sentence were true, then it should not be true; and if it were not true, then it should be true. This latter reasoning seems to work whether or not the sentence has a truth value since if the sentence had no truth value, then in particular, it wouldn't be true, so what it says would be the case. So it would be true after all. For this reason, the not-true formulation of the liar sentence can be seen already as an efficient form of the liar's revenge.

Obvious Truth

Meanwhile, if you should reflect on the matter, the sentence

> This sentence is not obviously true.

does seem indeed to be true. If it were obviously true, then in particular, it would also be true, which in virtue of what it says would mean that it is not obviously true—a contradiction. So it isn't obviously true, which is just what it says. So it is true, but not obviously true.

Consider in contrast the following sentence:

> This sentence is obviously not true.

It cannot be true, since then it wouldn't be. That is obvious. So it is obviously not true. But that is just what it asserts!

I find it interesting to consider the *obvious* operator and how it interacts with logic. For example, it seems clear that "obviously not φ" implies "not obviously φ," but not always conversely, since perhaps the question of φ or not φ is confusing—not obvious either way—in which case we would have both "not obviously φ" and "not obviously not φ." It seems reasonable to claim that obviously (A and B) is equivalent to (obviously A) and (obviously B), but the similar claim for the disjunction "or" seems wrong. That is, obviously (A or B) can be true even when neither A nor B is obvious. What is the logic of obviousness?

The Both-True-and-False Solution Strategy

Some philosophers embrace a solution strategy for the liar paradox by declaring that it is both true and false. In paraconsistent logic, for example, there can be true contradictions, statements that are both true and false. These would be truth-value gluts, as opposed to gaps, as in the no-truth-value solution strategy considered earlier. In the paraconsistent approach to a theory of truth, one develops not only the theory of truth, but also a simultaneous theory of falsity, with the idea in mind that these designations might overlap. And indeed they do overlap in the liar sentence, which is declared both true and false. Does this resolve the paradox? To my way of thinking, no, it does not bring us insight, but can be seen rather merely as pointing at the paradox, underlining it and taking it fully on board by incorporating it into the very foundations of the logic but not resolving it in any way. Paraconsistent logic itself, after all, is as paradoxical as the liar paradox—it does not seem to help us find our way out of paradox.

25.3 Avoiding Self-Reference

People often observe that the liar sentences exhibit a self-referential nature, with each sentence making an assertion about itself. And perhaps it is natural to become suspicious of self-reference when it seems to be featured prominently in such paradoxes. Is self-reference itself the source of paradoxicality?

The Twin Liars

Some philosophers attempt to eliminate the self-reference from the liar paradox by various means. Consider, for example, the following pair of sentences, which I dub the *twin liars.*

> The next sentence is true.
>
> The previous sentence is not true.

The thing to notice here is that each sentence refers only to the other sentence and not directly to itself, so in this way, they might plausibly be described as having avoided self-reference. Meanwhile, these sentences together exhibit the liar paradox logic. Namely, if the first sentence is true, then the second must be true, which means the first sentence is not true after all. And if the first sentence is not true, then in light of what it asserts, the second sentence must not be true, in which case the first sentence is true. We have thus reached a contradiction in any case.

Indirect Self-Reference

Another attempt to avoid self-reference, which I have heard from philosopher Hartry Field, is to describe a situation where the liar agent does not refer directly to himself. For example, imagine that someone is walking on the street and sees a live television broadcast in a store window of a person talking, and he says:

> What that person on the television is saying right now is not true.

But actually the person being shown on the television is himself right in that moment, and what he was saying just then is the very same utterance. Does this method actually avoid self-reference?

My own considered view is that self-reference is not especially worrisome or problematic, and there is no urgent need to avoid it. Indeed, we shall discuss later in this chapter a remarkable mathematical theorem showing that any sufficiently strong mathematical language admits the capacity for self-referential assertions, even just in the formal language of arithmetic (see page 338). The basic idea is that formal mathematical assertions can be represented with numbers (by encoding their syntactic symbols into the digits of the number, for example) and since arithmetic assertions can make assertions about these numbers, they can thus refer to themselves and even express their own provability or consistency. The fact of the matter is that self-reference is inextricably a part of every minimally sufficient mathematical theory. So not only do we have no need to avoid self-reference, this will be impossible anyway for any sufficient theory.

Meanwhile, my view also is that the attempts that we have mentioned here to avoid self-reference in the liar are not actually successful—self-reference remains—and yet self-reference is not ultimately the source of paradox. I do not place the blame on self-reference, but rather on mistaken ideas concerning the nature of truth.

25.4 The Infinite Liar

Let us now consider another attempt to avoid self-reference in the liar paradox by constructing an infinite list of sentences, a system of assertions that I call the *infinite liar*. This scheme of sentences is also known as the Visser-Yablo paradox, discovered by Albert Visser and Stephen Yablo. The infinite liar proceeds in an infinite list of sentences, each asserting the falsity of all the later sentences on the list. Notice that none of the sentences directly refers to itself, but rather only to the later sentences on the list:

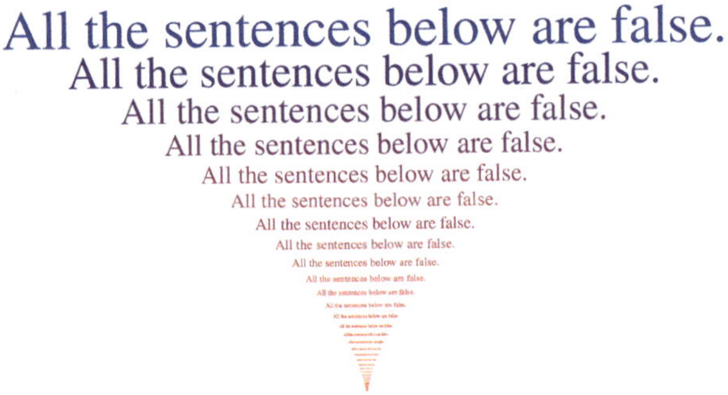

If any of the sentences in the list were true, then all the subsequent sentences below it would be false; but in this case, in light of what they assert, they should all be true, contrary to what we just said. So none of the sentences can be true, and therefore, they are all false. But given this, in light of what they assert, they should all be true. We reach a contradiction in any case.

Here is a variant dual form of the infinite liar:

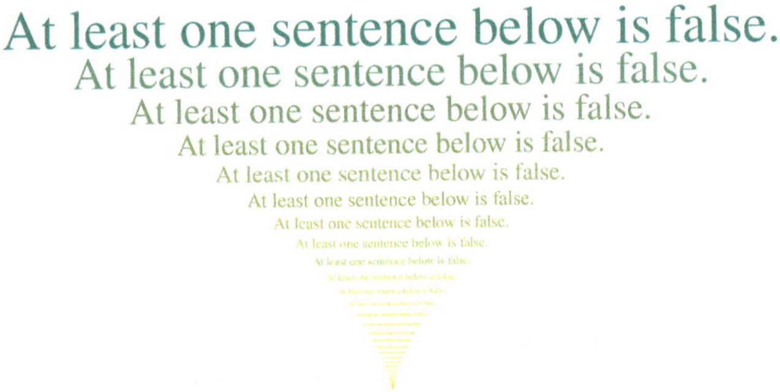

If any of the sentences were to be false, then in virtue of what that sentence asserts, there would be no false sentences subsequently. But then all the subsequent sentences, in virtue of what they assert, would be false, contrary to our conclusion that none of them are. So none of the sentences can be false. And therefore, in virtue of what they assert, all of them are false. Contradiction.

Part of the explicit motivation for constructing the infinite liar was to find a version of the liar paradox that avoided self-reference. Although each sentence in the infinite liar refers only to the later sentences and not to itself, in my view the infinite liar still partakes of self-reference. Each sentence in the infinite liar refers in effect to its own place in the system, since the sentences that it refers to directly are "below" the given sentence, later than the given sentence on the list of sentences in which it appears, and this is a form of self-reference. For example, we cannot freely rearrange the sentences and expect to preserve the referents and meaning.

The infinite liar is often presented as an indexed family of sentences s_0, s_1, s_2, and so on, with a sentence s_n for each natural number n:

(s_n) Every sentence s_k for $k > n$ is false.

Here, the sentence s_n refers most directly to the sentences s_k for $k > n$. But to my way of thinking, the mention of n here in the sentence s_n is already a form of self-reference. Furthermore, one naturally regards the scheme of sentences in parametric form $s(n)$, with the sentence $s(x)$ in general form having a free variable x not yet specified, asserting that $\forall y > x\ s(y)$. And this formula s is clearly self-referential.

We can also adapt the revenge idea to the infinite liar as follows:

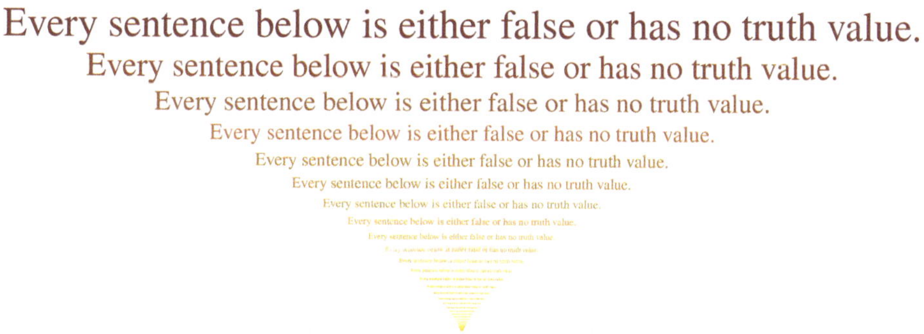

Notice that every sentence on the list implies all the later sentences since the collection of sentences referred to by "Every sentence" is getting smaller. So if any of them is true, then all the later ones must also be true. But this would contradict what the initial true sentence asserts. So none of them can be true, and therefore all of them must either be false or have no truth value. But that is precisely what they all assert, so they would all be true. So there is no coherent way for them to be true or false or to lack a truth value.

Or perhaps one prefers the "not true" formulation, which as we argued previously is a form of the liar's revenge:

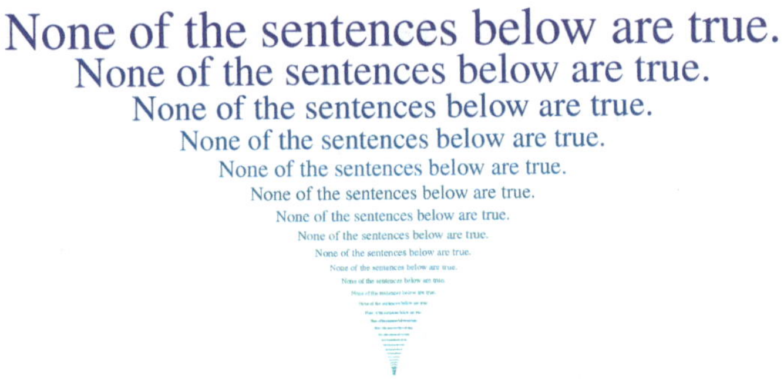

If any of them are true, then all subsequent sentences are not true; but in this case, in light of what they assert, they should all be true. Contradiction. Thus, none of them are true. But in this case, they should all be true. This reasoning seems to work even when one entertains the possibility that the sentences might be neither true nor false.

25.5 The Disquotational Theory of Truth

The disquotational theory of truth consists of the idea that what it means to assert that a sentence is true is simply to assert the sentence itself:

"Snow is white" is true if and only if snow is white.

More generally,

"________" is true if and only if ________.

According to this theory, to make a truth assertion simply amounts to removing the quotation marks—it is *disquotational.*

The disquotational theory of truth is sometimes described as a deflationary account of truth in that it deflates what might otherwise be an overblown complicated approach to truth—a sentence is true, in the disquotational theory, simply if the proposition that it asserts is the case. In this sense, every particular use of the truth concept for a specific assertion is eliminable since we can instead assert the content of that sentence.

The Use/Mention Distinction

The disquotational theory of truth builds a bridge between *use* and *mention*, crossing the use/mention dichotomy, the distinction between using a word and merely mentioning it. In section 8.1, I had related the story of my Oxford colleague, who at high table had announced that he generally uses pants as trousers, or perhaps merely "pants" as "trousers," which would be much less shocking; and we discussed my daughter's question concerning whether everything rhymes with itself—or was it whether "everything" rhymes with itself? Or whether everything rhymes with "itself"? By convention, we typically indicate mentions as opposed to uses of a word in writing by enclosing the word within quotation marks.

According to the disquotational theory, truth assertions amount to a traversal from mention to use. On the left of the disquotational biconditional, we are merely mentioning the sentence "Snow is white" because we are talking about this sentence, asserting of it that it has the property of being true, while on the right, we are using the sentence, asserting the propositional content that the sentence expresses. According to this theory of truth, to make a truth assertion—in effect to remove the quotation marks—is to transition from mention to use with the same expression.

Not every mention of a sentence need explicitly involve quotation marks, of course, since I can refer to the first sentence of *Moby Dick*, as I am doing just now, without actually asking you to call me Ishmael.

The Grelling-Nelson Paradox

A word is *autological* if the word describes itself, if the word itself is or has the property that the word expresses. The standard example is "polysyllabic," which does indeed have several syllables. Other examples include "adjectival," which is indeed an adjective; "unhyphenated" is unhyphenated; and "pronounceable" is pronounceable. All these words are therefore autological.

A word is *heterological*, in contrast, if it fails to describe or hold of itself. The word "monosyllabic" is not monosyllabic; "unwritable" is writable, and therefore not unwritable; and the word "explosive" is not literally explosive. So these words are heterological.

Is "red" red? Well, that particular copy of the word, written here in this chapter, is black, not red, but we could have written the word with red ink—would that make it red? Which of the following words is autological and which heterological?

RED ORANGE PURPLE BLACK BLUE GREEN

Or is there a category error here? When we talk about a certain word, do we mean a particular instance of the word, written on a piece of paper or carved into the brick wall under a bridge. Or is the word instead an abstraction of all those particular instances of it? The word "combustible" might seem combustible if you are reading this book in a format printed on paper, since you could ignite the book (but of course . . . uh . . . you would never do that . . . right?). If you were reading an electronic copy of the book, however, or perhaps as I expect in the near future (ha!), you are reading the copy of my book that is engraved in stone at the Smithsonian for the eternal benefit of all future generations, then you would say that it is not combustible.

Perhaps all this confuses type with token, for a given word has many instantiations. There is arguably just one word "tenacious" in the English language, even though it could be written many times. If words are indeed abstractions in this sense, then it would seem that they have no color at all, so every color name word is colorless. On this view, despite naive appearances, even "**Red**" is heterological, while "**Colorless**" is autological, since the word referred to is an abstraction beyond the color used in representing it.

Incidentally, the mental confusion that arises from confusing the meaning of a word with the manner of writing it, such as in the color examples here, is called the *Stroop effect*, discovered by John Ridley Stroop in the 1930s. Here are other kinds of examples:

Italic *Bold* capitalized LOWER CASE Small Large

But let me finally get to the point with heterologicality. Consider the question of whether "heterological" might be autological or heterological:

Is "heterological" heterological?

The question is asking in effect whether the word "heterological" applies to itself, and therefore this question, perhaps confusingly, seems to be the same as asking:

Is "heterological" autological?

The paradox here is that if "heterological" is heterological, then what this means is that it does not apply to itself, so it should not be heterological. And if it isn't heterological, then "heterological" does not apply to itself and so it should be heterological. In either case, we reach a contradiction. This is known as the *Grelling-Nelson paradox*, introduced in 1908 by Kurt Grelling and Leonard Nelson.

Iterated Truth Predicates

Suppose that we are considering a mathematical structure M of some kind, perhaps a group, a graph, or an order of interest. The essential structural features of M can be expressed in a formal language associated with the atomic structure of M, a language allowing reference to that structure and the quantifiers $\forall x$ and $\exists x$ ranging over the domain of individuals of the structure M. For example, the features of an order structure, such as transitivity and reflexivity, can be expressed in terms of the order relation $\leq$ employed in it; the features of a group are expressed in terms of the algebraic group operation and the identity element and inverses; and the features of a graph are expressed in terms of the adjacency edge relation.

In any such mathematical context, it is straightforward and unproblematic for us to add a truth predicate $T(\varphi)$ for making truth claims about such an assertion φ in that original language, asserting that it holds in M, even allowing the reference of arbitrary individuals from the domain of M. That is, we do not make nested application of the truth predicate T, but rather apply it only to assertions in the original language of the structure.

This is merely the first layer of what will become a very tall hierarchy of iterated truth predicates. Let us denote the original language of the structure by $\mathcal{L}_0$, and let us use T_0 for the truth predicate that we just mentioned, fulfilling the disquotational truth scheme $T_0(\varphi) \leftrightarrow \varphi$, but only for assertions φ in the original language $\mathcal{L}_0$. We now form a new, larger language $\mathcal{L}_1$ by expanding the original language $\mathcal{L}_0$ of the structure with this truth predicate T_0, which is interpreted in the corresponding expansion of M.

But the main point is that we can simply do it again, adding another truth predicate T_1 for making truth claims in the language $\mathcal{L}_1$. That is, $T_1(\psi)$ will hold of any $\mathcal{L}_1$ assertion ψ if and only if ψ holds in M. The predicate T_1 in effect enables us to make truth-about-truth assertions, but perhaps it is more accurate to say that T_1 makes T_1-truth assertions about assertions involving T_0-truth. Adding T_1 to the language forms a new, larger language $\mathcal{L}_2$.

In this way, we iteratively expand the language with additional truth predicates T_n in a sequence of layers. At the bottom level, we have the truth-free assertions of the original language, and each next layer is formed by adding a truth predicate for assertions of the previous layers. The construction proceeds infinitely, even transfinitely, achieving a coherent system of truth predicates T_α for every ordinal α, each referring to the truth of assertions at lower levels of the hierarchy. Every assertion in the accumulating language falls under the truth predicates of higher levels. Such is the unproblematic nature of the theory of well founded iterated truth, the completely clear, gold-standard instance of a theory of iterated truth.

Although the theory of well founded iterated truth is precisely designed for making truth assertions about truth, and truth-about-truth-about-truth and so on, nevertheless the liar sentence will not arise in this account of iterated truth. At no stage do we achieve a self-applicable truth predicate, a predicate T_α that applies to assertions in a language that includes the predicate T_α itself. Rather, what makes this account well founded is that each level of the truth predicate T_α applies only to assertions in the language $\mathcal{L}_\alpha$, which includes all earlier truth predicates T_β for $\beta < \alpha$, but not T_α itself.

The liar sentence, in contrast, is making a claim about the truth not of an assertion using only truth predicates from an earlier level of the hierarchy, but rather of an assertion employing the very same truth predicate about which the claim is made.

The No-Proposition Solution Strategy

These ideas lead to another resolution strategy for the liar paradox. Namely, because the liar makes a truth assertion about itself (the very same assertion), the truth predicate in play is self-applicable in a way that does not and cannot occur with any well founded truth predicate. The sentence is therefore not meaningful on the basis of the well founded theory of truth. Indeed, because of this self-nesting of the truth predicate, the liar sentence is not even syntactically well formed for well-founded iterated truth predicates. What is the theory of truth, then, to which it does refer?

One way to answer this question is that the paradox itself can be taken as an argument that there is no satisfactory theory of truth available for it—the contradiction at the heart of the liar paradox shows that there is no coherent theory of truth in play. And if no relevant theory of truth is being employed in the liar sentence, then the liar sentence's use of the concept of truth and falsity is groundless, without meaning. In this sense, the liar sentence is not meaningful—it does not express a proposition.

The proposal here is that the liar sentence offers only an illusion of meaning. Perhaps because we are used to making truth assertions and even truth-about-truth assertions in the unproblematic case of the well founded iterated theory of truth, the liar sentence conveys a superficial appearance of being meaningful. But since ultimately, there is no coherent theory of truth underlying the liar usage, the sentence is not actually meaningful. In particular, despite appearances, it does not assert its own falsity.

Is this account subject to the liar's revenge? If we take the liar as not meaningful, then the revenge strategy might call for us to form the following sentence:

> This sentence is either false or not meaningful.

The revenge line of reasoning then proceeds as follows: If we say that this sentence like the liar is not meaningful, then precisely in light of what the sentence asserts, it would be true since we would fulfill the second clause of the disjunction. So it would be meaningful after all.

But is that analysis correct? Consider the following sentence:

> This sentence is either globharbash or not meaningful.

Is this sentence meaningful? Does it express a coherent proposition? I do not think it does. It seems to express a disjunction, but the first clause is gibberish, and we would seem to be unable to say exactly what meaning the disjunction has that the sentence asserts. Because of this, I find no coherent proposition expressed here. Nevertheless, one can seem to carry out the revenge analysis here by saying that if the sentence were not meaningful, then the second disjunctive clause would be fulfilled and therefore the sentence would be true (and hence meaningful). One can also take that alternatively as an indictment of the revenge analysis itself. Since we said that this sentence does not have a coherent meaning, the fact that the revenge analysis leads us to a wrong conclusion tends to pour cold water on the revenge analysis itself.

Ultimately, the no-proposition response to the liar regards both the liar sentence and the liar's revenge sentence as not expressing propositions. There is no meaning to be found in them, despite their superficial appearance as expressing one.

And yet, a critic of the position will point out that this solution attempt of the liar amounts to the claim that the liar sentence is not meaningful, and therefore, in particular, the claim that it is not true. In short, on this view, we seem to ascent to the following proposition:

> The liar sentence is not true.

But now the critics shout in unison: *This is exactly what the liar sentence itself asserts!*

This is irritating trouble for the no-proposition position, which seems to be in the ridiculous position of holding that "This sentence is not true" is not true, even though that is what the sentence itself appears to assert. On the no-proposition account, one must bite the bullet, insisting that the liar sentence "This sentence is not true" does not express a proposition, so despite appearances, it is wrong to say that it expresses its own untruth. On this view, the liar sentence doesn't express anything at all:

Lia This sentence is not meaningful.

Tru What twaddle nonsense! That sentence has no meaning.

Lia But darling, that is what I just said.

Was Lia's statement meaningless or not? The trouble for Tru is that to assert that Lia's sentence is meaningless appears to be the very same sentence that she asserted. He will reply:

Tru No, your sentence did not have that meaning, because it had no meaning at all.

The Liar Paradox Is a Form of the Burali-Forti Paradox

Consider the following sentence, which I shall call the *well founded liar's revenge*:

This sentence is not true according to any well founded theory of truth.

What are we saying here? The sentence seems to be saying that it will not be declared true by any particular truth predicate T_α arising in the well founded hierarchy of iterated truth. Thus, the sentence in effect quanties over all the levels T_α of the hierarchy of well founded truth. For this reason, the sentence is not expressible in the language $\mathcal{L}_\alpha$ at any ordinal level of the hierarchy, since those languages have the predicates T_β only for $\beta < \alpha$. Thus, the sentence is not of the form that could be declared true by any predicate T_α in the hierarchy. In this sense, we might therefore judge that the proposition asserted by the sentence indeed seems to be the case. This makes the sentence rather similar to the "This sentence is not obviously true" case.

Perhaps one is tempted to say that the sentence is true, meaning that it is true according to a continuation of the hierarchy beyond all the ordinals. We can naturally form the truth predicate T_{Ord}, after all the ordinals, for assertions in the union of the languages $\mathcal{L}_\alpha$. In this way, the liar paradox is revealed as a form of the Burali-Forti paradox, consisting of the observation that the class of ordinals is a transitive, well ordered class of ordinals, but cannot itself be an ordinal, for then it would be an ordinal that is larger than every ordinal. Indeed, the class of ordinals is a proper class and not a set. The point is that the concept of truth in play for saying that the well founded liar's revenge is true would similarly be a proper-class continuation of the hierarchy of truth beyond every ordinal stage.

25.6 The Russell Paradox

We saw in section 13.6 how Bertrand Russell, with a surprisingly simple argument, refuted the general comprehension principle, the principle asserting that every property φ determines a set, the set of all objects with that property:

$$\{x \mid \varphi(x)\}.$$

Russell shows that for some properties, this principle is simply inconsistent. Specifically, if the principle were correct, we could form the set R of all sets x that are not self-membered:

$$R = \{x \mid x \notin x\}.$$

The problem is that it now follows immediately that $R \in R$ if and only if $R \notin R$, which is a contradiction.

There is a clear parallel with the liar sentence here, for we had observed that R is a member of itself if and only if it is not a member of itself, not as the result of an elaborate argument, but rather directly because of the defining property of the Russell set itself.

Similarly, the Russellian barber shaves all and only those in town who do not shave themselves, and so, directly because of this, the barber shaves himself if and only if he doesn't. And the barista, who makes coffee for all and only those in town who do not make coffee for themselves, makes coffee for herself if and only if she doesn't. In each case, we are defining a certain feature that will hold in a certain case if and only if it doesn't, which is exactly the situation of the liar paradox.

Berry's Paradox

Let us also recall *Berry's number b*, defined as the following number:

> The smallest number not definable in fewer than a dozen words.

Since there are only finitely many English expressions using fewer than a dozen words, they can define only finitely many numbers, so it seems that there must be some smallest number that is not definable in that manner. The paradoxical situation, however, is that this very definition itself uses fewer than a dozen words. Our earlier analysis in section 8.4 of Berry's paradox pointed to subtle difficulties in the notion of definability, opening the possibility that some expressions that seem to define numbers might not actually succeed.

Berry's Revenge

In light of this, consider now a revenge version of Berry's paradox:

> Let n be the number specified by this definition, plus 1, if the definition succeeds; otherwise 0.

Does the definition succeed?

If the definition succeeds in defining a number n, then because of what it says, it should specify $n + 1$, which would contradict the previous specification. So it seems that the definition cannot succeed. But in this case, it specifies 0 in the default clause. But then it should specify 1, and therefore 2, and so on; around and around we go.

Here is another variant:

> Let B be the smallest number that this definition couldn't possibly define, if there is such a number; otherwise 0.

Has the definition succeeded in specifying a number? If not, then it would seem to revert to the default case, so it would specify 0. So it seems that in any case, it should succeed in specifying a number. But if it specifies a number n, then because of what it says, it should not specify n after all. So it seems there is no number that it can possibly specify. But in this case, it should specify 0. Around and around again.

25.7 The Gödel Sentence

Let us come now to the Gödel sentence, fruitfully viewed as a provability version of the liar paradox:

> This sentence is not provable.

Remarkably, Kurt Gödel explains how such a self-referential sentence can be expressed in the purely mathematical formal language of arithmetic. How strange and wonderful that is. In one respect, the formal Gödel sentence asserts that every number has a certain definite numerical property expressed in terms of addition and multiplication of that number in combination with smaller numbers, but in another respect, the sentence is also naturally interpreted as expressing the nonprovability of the very same statement.

To specify the Gödel sentence, we fix a particular axiomatic formal system, such as the axiomatic framework of Peano arithmetic (PA), and then regard provability to mean provability in that formal system. Ultimately, the analysis will lead to Gödel's incompleteness theorem, one of the most profound developments of twentieth-century mathematics, which marked the coming of age of the subject of mathematical logic. I should like to provide an accessible soft account of this theorem for the rest of this chapter.

Arithmetization

One of Gödel's key insights is the concept of *arithmetization*, the idea that any given finite mathematical structure can be encoded with numbers in such a way that the natural combinatorial processes we might employ with it become arithmetic manipulations on the corresponding numerical codes.

Each of us is familiar, for example, with the fact that when we type our documents into a computer—perhaps writing the great American novel—the information is stored ultimately in the computer's memory. While composing the document, we naturally conceive of it in high-level terms; it consists of chapters and paragraphs, perhaps written in certain fonts with certain formatting. These components in turn consist of words, and ultimately individual characters, supplemented with control codes for the fonts and spacing; each such atomic piece of information is represented at bottom simply by a number, such as in the extended ASCII system, which has 256 distinct symbols and control characters (this is 2^8), so each character can be represented in binary by exactly eight digits of 0 or 1. (Current systems tend to be based on Unicode, which allows a far larger symbol set.) Thus, the entire document is represented as a gigantic sequence of 0s and 1s, with each bit represented via the presence of low or high voltage, respectively, in certain transistors on the circuit board. In this way, the entire document can be thought of as a single, enormous, binary number, the *Gödel code* of the entire document.

Similarly, we may undertake the arithmetization of other mathematical structures, such as graphs, finite partial orders, or what have you. And indeed, the formal language of arithmetic itself amounts to such a combinatorial structure, with the syntax of formal expressions via parse trees and so on. Every assertion φ in the formal language of arithmetic thus admits a Gödel code $\ulcorner\varphi\urcorner$, which is a number representing φ, and all the various syntactic operations that we might like to undertake with φ will be themselves expressible in the language of arithmetic. In this way, the formal language of arithmetic admits the capacity to make assertions about the formal language of arithmetic.

Self-Reference in Arithmetic

Using arithmetization, let us explore a logical enigma—the fixed-point lemma, a mathematical mystery, a logical labyrinth that shows how self-reference, the stuff of nonsense and confusion, sneaks explicitly into our beautiful number theory. It continually amazes me.

The fixed-point lemma is the claim that for any assertion $\varphi(x)$ in the formal language of arithmetic, there is a sentence ψ such that in the standard Peano axioms of arithmetic, one can prove the equivalence

$$\mathrm{PA} \vdash \psi \leftrightarrow \varphi(\ulcorner\psi\urcorner).$$

Thus, the sentence ψ asserts that "φ holds of my Gödel code."

For those willing to dive into some technical details, let me briefly give the proof. Consider the subsitution function, sub, defined so as to return the Gödel code after substituting a free variable with a specific numerical value, like this:

$$\mathrm{sub}(\ulcorner\varphi(x)\urcorner, m) = \ulcorner\varphi(\underline{m})\urcorner,$$

where $\underline{m}$ is the syntactic term $1 + \cdots + 1$, with m many 1s. The sub function is a primitive recursive function, representable in the formal language of arithmetic. Consider now any formula $\varphi(x)$ with one free variable x. Let $\theta(x) = \varphi(\mathrm{sub}(x, x))$, and let $n = \ulcorner\theta(x)\urcorner$. Finally, let $\psi = \theta(\underline{n})$, which is a sentence in the language of arithmetic. Putting all this together, we observe in Peano arithmetic the following equivalences:

$$\begin{aligned}\psi &\leftrightarrow \theta(\underline{n}) \\ &\leftrightarrow \varphi(\mathrm{sub}(n,n)) \\ &\leftrightarrow \varphi(\mathrm{sub}(\ulcorner\theta(x)\urcorner,n)) \\ &\leftrightarrow \varphi(\ulcorner\theta(\underline{n})\urcorner) \\ &\leftrightarrow \varphi(\ulcorner\psi\urcorner).\end{aligned}$$

Thus, the sentence ψ has the desired fixed-point property that we had claimed. So every arithmetically expressible property $\varphi(x)$ admit a fixed-point sentence ψ.

First Incompleteness Theorem

Let us now use the Gödel sentence to prove the first incompleteness theorem, establishing that every consistent formal system of arithmetic will admit true but unprovable sentences. In short, arithmetic truth is not the same as provability in any such system.

Consider the Gödel sentence, which asserts its own nonprovability in the formal system of Peano arithmetic. If it were provable in that system, then we could formalize that proof and then prove that the proof indeed was a proof. In short, every provable statement is provably provable. But in proving the Gödel sentence, however, we would be proving that it is not provable, since that is precisely what the sentence asserts. So we will have proved both that it is provable and that it is not provable, which would be proving a contradiction. If the theory is consistent, then this is impossible, so the Gödel sentence cannot be provable.

But that is precisely what the Gödel sentence asserts! And therefore, the Gödel sentence is a true statement that is not provable in the Peano theory.

This argument generalizes beyond Peano arithmetic to work with any formal system whose fundamental axioms we can in principle write down. The general conclusion is that every consistent arithmetic theory with a computably enumerable list of axioms will admit true but unprovable sentences. Indeed, the resulting theory will necessarily be incomplete, admitting statements that can be neither proved nor refuted.

The success of the Gödel sentence analysis, culminating not in paradox and confusion but rather in those profound mathematical theorems established with precision and rigour, to my way of thinking identifies the problem with the liar. The two sentences are fundamentally similar, merely replacing truth with proof but otherwise having the same self-referential form. The paradoxical nature of the liar therefore cannot be blamed on that self-referential nature, since this feature is shared by the Gödel sentence. Rather, the paradox of the liar seems to arise from the theory of truth that it employs—a theory that is not spelled out, and ultimately cannot be. Whereas the Gödel sentence employs a fully specified and precise notion of formal proof, there seems to be no coherent notion of self-applicable truth underlying the liar sentence.

25.8 Tarski's Theorem on the Nondefinability of Truth

Next, let us prove Alfred Tarski's remarkable theorem on the nondefinability of truth. This theorem asserts that there is no arithmetically expressible property that picks out exactly the true arithmetic sentences; that is, a predicate that holds exactly of the numerical codes of the true arithmetic assertions. More precisely, there is no arithmetic formula $T(y)$ such that

$$T(\ulcorner\psi\urcorner) \leftrightarrow \psi$$

for every arithmetic sentence ψ. Let us call this the *truth scheme* for the predicate T.

Tarski's theorem is an immediate consequence of the fixed-point lemma by using the liar paradox logic. Namely, if $T(y)$ were arithmetically expressible, then we may consider the formula $\neg T(x)$, which must admit a fixed point ψ, so that

$$\psi \leftrightarrow \neg T(\ulcorner\psi\urcorner).$$

But this equivalence means that ψ asserts its own untruth with respect to T, a form of the liar sentence, and consequently

$$T(\ulcorner\psi\urcorner) \leftrightarrow \neg\psi,$$

which would prevent T from fulfilling the truth scheme. So there can be no such expressible truth predicate.

Alternative Proof via Russell

I should like to give an alternative proof of Tarski's theorem that avoids the fixed-point lemma by appealing directly to the central idea of the Russell paradox, and without the need for much arithmetization, beyond the idea that every formal assertion φ is represented by a numerical code $\ulcorner\varphi\urcorner$. The claim will be that there is no arithmetically expressible formula T that expresses what it means for a formula $\varphi(x)$ to be true at its argument x, like this:

$$\forall x\,[T(\ulcorner\varphi\urcorner, x) \leftrightarrow \varphi(x)].$$

So this is a parametric version of Tarski's theorem.

Suppose toward contradiction that there were such a truth predicate T. Using this predicate, let $R(x) = \neg T(x, x)$, the assertion "It is not the case that x is a formula that is true of itself." Let $r = \ulcorner R\urcorner$, and consider the sentence $R(r)$, which asserts $\neg T(r, r)$—that is, $\neg T(\ulcorner R\urcorner, r)$—which by the truth-predicate requirement is equivalent to $\neg R(r)$, which is a contradiction. That's it. So there can be no arithmetically expressible account of truth. In effect, the formula $R(x)$ expresses that x is the code of a heterological formula:

$$\neg T(\ulcorner\varphi\urcorner, \ulcorner\varphi\urcorner) \leftrightarrow \neg\varphi(\ulcorner\varphi\urcorner).$$

The point is that in this context of a formal language, the notion of heterologicality becomes a completely rigorous mathematical definition. Namely, a formula $\varphi(x)$ in the language of arithmetic is heterological if $\neg\varphi(\ulcorner\varphi\urcorner)$. Our proof of Tarski's theorem via Russell amounts to exactly the same logic of the Grelling-Nelson paradox of heterologicality, except that here, it is about precise assertions in the formal language of arithmetic. Thus we are led to a fundamental unity of all these paradoxes.

The Infinite Gödel Sentences

Let us consider the Gödelian version of the infinite liar, a provability version of the Visser-Yablo paradox:

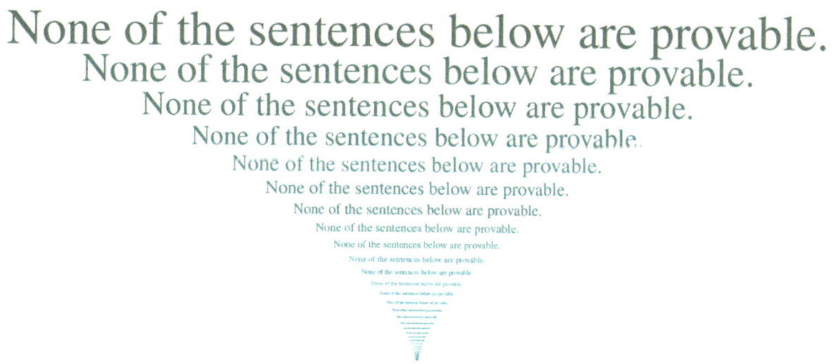

We can formalize this more precisely with the fixed-point lemma to find a formula $\gamma(k)$ in one free variable that asserts the nonprovability of all later sentences:

$$\text{PA} \vdash \forall k \left[\gamma(k) \leftrightarrow \forall n > k\ (\gamma(\underline{n}) \text{ is not provable in PA})\right].$$

If any specific instance $\gamma(k)$ were provable, then we will have proved that no later instance $\gamma(n)$ is provable. But every such later instance $\gamma(n)$ makes a weaker claim than $\gamma(k)$ itself, since every number above n is also above k, so in proving $\gamma(k)$, we are also able to prove $\gamma(n)$. So we can prove that $\gamma(n)$ is provable and also prove that $\gamma(n)$ is not provable, which is impossible if the theory is consistent. Thus, no specific instance $\gamma(k)$ can be provable. Thus, in virtue of what they assert, these are all true but unprovable sentences.

The Speed-Up Phenomenon for Lengths of Proofs

Consider the following sentence, a hybrid of the Gödel sentence with Berry's paradox:

> This sentence is not provable in Peano arithmetic with a proof using fewer than a billion symbols.

Remarkable as it may seem, we can express such a sentence in the formal language of arithmetic by using the usual fixed-point tricks, as with the Gödel sentence. That is, there is a formal assertion β in the language of arithmetic that expresses the fact that the very sentence itself is not provable by a formal proof in that language using fewer than one billion symbols.

Is it true?

Well, if it were provable using fewer than one billion symbols, then we would be able to prove that that proof was a proof with fewer than one billion symbols (this proof that it is such a proof might be longer, but this does not matter). But then we will have proved that the sentence is false. So we will have proved the sentence and its negation, which would mean that our axioms are inconsistent. So if the axioms are consistent, then indeed the sentence is not provable using fewer then one billion symbols. And therefore, the sentence β is true.

Meanwhile, I claim also that the sentence is provable, but only by a very long proof! The reason is that precisely because there is no proof using fewer than a billion symbols, we can prove in each case that none of the possible candidate proofs using fewer than one billion symbols is a proof of this sentence. This would be a very long proof. What we have, in short, is a sentence that is true and provable, but necessarily, every proof of it is very long. Meanwhile, we have a very short proof of the sentence from the additional assumption that the theory is consistent.

This situation enables a simple proof of the second incompleteness theorem. Let me explain.

25.9 The Second Incompleteness Theorem

The second incompleteness theorem is the assertion that no consistent theory of arithmetic with a computable set of axioms (and including a certain basic set of axioms) can prove its own consistency. For example, the theory of Peano arithmetic (PA), if consistent, does not prove itself consistent. The assertion that PA is consistent thus becomes a true statement that is unprovable in PA.

Let me give a self-contained proof of this using the simple ideas from the speed-up phenomenon that we just discussed. Suppose toward contradiction that PA proves its own consistency. The proof has a certain definite size. Let N be much larger than this but with a small description, such as the exponential of the proof size. Consider the corresponding Gödel-Berry hybrid sentence β, asserting that it has no proof of size less than N. The speed-up analysis shows both that β has no proof in PA of size less than N, but it does have a very short proof in PA + Con(PA). Since PA proves Con(PA) in much less than size N, we can combine that proof with the proof of β in PA + Con(PA) and still have a proof of size less than N. So PA does admit a short proof of β, after all—a contradiction. Therefore, PA does not prove its own consistency, which establishes the second incompletness theorem.

25.10 The Truth-Teller and Löb's Theorem

If the liar sentence asserts its own falsity, let us consider the truth-teller sentence, which asserts its own truth:

> This sentence is true.

Is it true?

Perhaps you think that it can be either true or false, but I shall have something to say about that momentarily. Meanwhile, just as the Gödel sentence asserts its own unprovability, let us also consider the *Löb sentence*, named for Martin Löb (1921–2006), which asserts its own provability:

This sentence is provable.

So we have four sentences forming a two-dimensional symmetric array:

Truth-teller: This sentence is true.	**Liar:** This sentence is false.
Löb: This sentence is provable.	**Gödel:** This sentence is not provable.

Many people undertake the analysis of the liar sentence by arguing that if the liar sentence is true, then it must be false; and if it is false, then it must be true. Thus, they conclude, the sentence is paradoxical. Meanwhile, the corresponding analysis of the truth-teller sentence seems to go nowhere: If it is true, then it is true; and if it is false, then it is false. Sometimes people conclude from this that the truth-teller sentence could be either true or false; it is somehow not determined.

But that analysis does not actually prove the claim. Just because one does not immediately find a contradiction does not mean that there is not a deeper conclusion to be made. It is like saying that you searched for the keys and did not find them, therefore, they could be anywhere! But they are actually in a specific place, and they probably could not be on the Moon or inside Grant's Tomb. Can the truth-teller really be either true or false?

If you think so, and this is my main point here, then perhaps you might be inclined to expect a similar situation for the Löb sentence. But that conclusion would be wrong in light of Löb's theorem, which shows, remarkably, that the Löb sentence, the sentence asserting its own provability, is actually provable. The Löb sentence *is* provable, and therefore it is true.

Let me sketch a soft proof here. Let λ be the Löb sentence, and by the fixed-point lemma, let σ be a sentence that asserts the implication, that if σ is provable, then λ holds:

$$\sigma \quad \leftrightarrow \quad (\,\sigma \text{ is provable } \to \lambda\,)$$

Suppose that σ were provable. By observing that proof in detail, we could prove that σ is provable, and because the biconditional given here is provable, we would therefore be able to prove that λ is provable. But this is provably equivalent to λ. So we have argued that if σ is provable, then λ is true. But that is just what σ asserts, so by formalizing our argument, we see that σ is in fact provable. Therefore, it is provably provable, so λ is provable. Thus, the Löb sentence is true.

In short, the sentence asserting "This sentence is provable" is in fact provable. That is fascinating and surprising. But moreover, I should like to take this fact as a warning for us not to be too quick in our analysis of the sentence, "This sentence is true."

The Infinite Truth-Teller and Infinite Löb Sentences

Finally, consider the infinite truth-teller sentences:

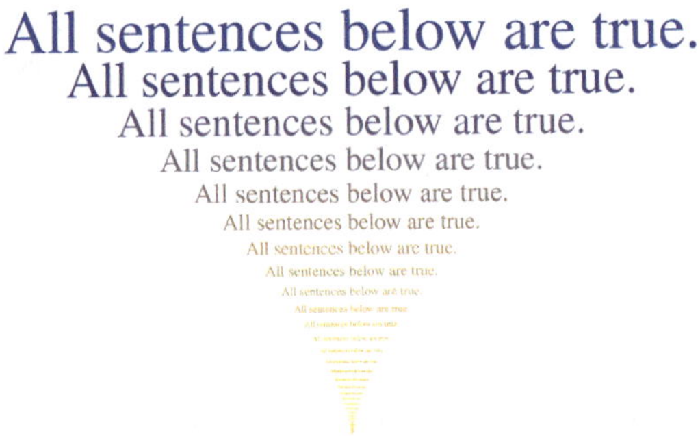

And similarly, let us analyze the infinite Löb sentences:

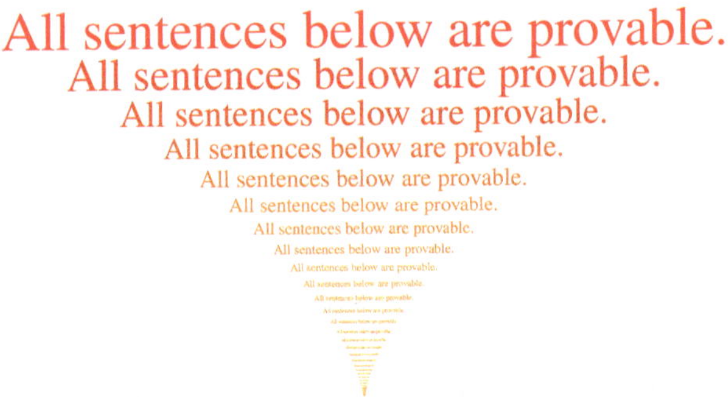

Let me argue for these latter sentences, as in Löb's theorem, that they are in fact provable. We construct the sentences by the fixed-point lemma, finding sentences $\tau(k)$ that assert the provability of all $\tau(n)$ for $n > k$:

$$\tau(k) \quad \leftrightarrow \quad \forall n > k \,(\, \tau(n) \text{ is provable} \,).$$

To see that these assertions are in fact true, apply the fixed-point lemma again to find a sentence σ asserting

$$\sigma \quad \leftrightarrow \quad (\, \sigma \text{ is provable } \rightarrow \tau(0)\,).$$

Now, we argue as in Löb's theorem. If σ were provable, then it would be provably provable, and from that, using the previous characterization of σ as the implication on the right-hand side, it would follow that $\tau(0)$ is provable. Using this defining characterization of $\tau(0)$, we could therefore prove that every $\tau(n)$ for $n > 0$ would be provable, and using the other direction of that biconditional, we could prove every $\tau(n)$ for $n > 0$. Thus, each $\tau(n)$ would be not merely provable, but true. So we have argued that if σ is provable, then $\tau(0)$ holds. But this implication is just what σ asserts, so we have now actually proved σ. Therefore, $\tau(0)$ is true. And this implies all later $\tau(n)$, since the scope of the quantifier is descending.

Questions for Further Thought

25.1 Does the liar sentence express a meaningful proposition?

25.2 What is your analysis of the following sentence:

This sentence does not express a meaningful proposition.

25.3 Do the twin liars and the infinite liar avoid self-reference?

25.4 What is your analysis of the following sentences?

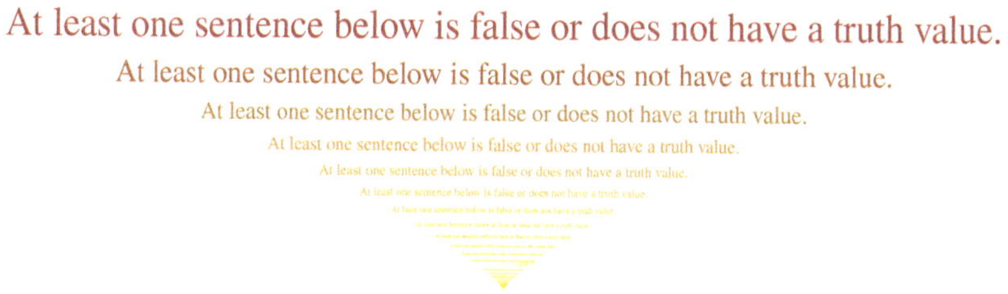

25.5 What about these?

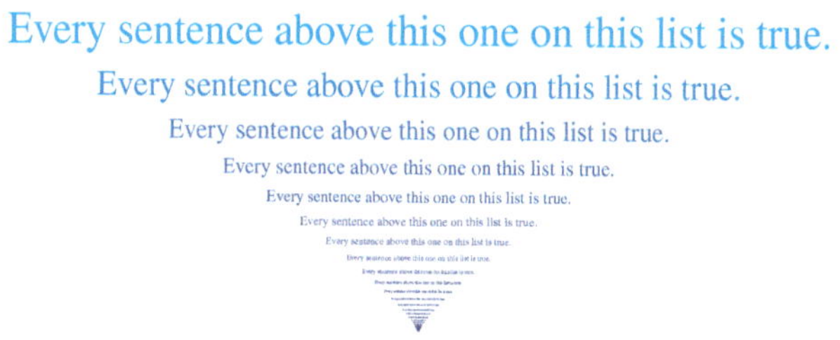

25.6 And these?

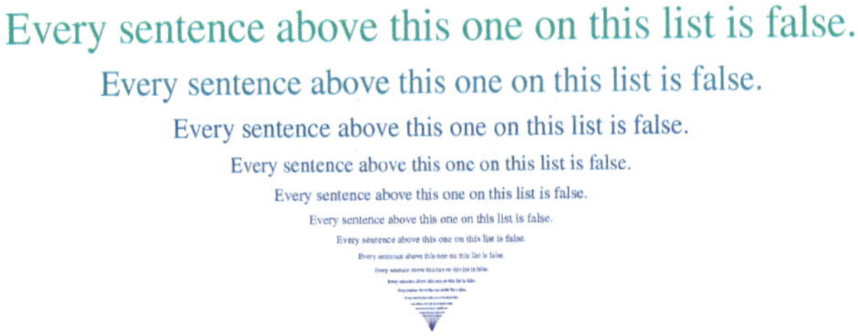

25.7 What is your analysis of an infinite collection of sentences, each asserting that all the other sentences are provable? (Hint: Mount a Löb-style analysis.)

25.8 Construct further versions of the infinite liar with other order types. What happens if the infinite list goes up instead of down? What if the sentences have the order type of the rational numbers? What if the list is transfinite?

Further Reading

Hamkins (2021b). In chapter 7, I give a thorough account of the incompleteness theorems and their role in the Hilbert program.

Credits

The quotation on page 325 is from the poem "Jabberwocky" by Lewis Carroll. Some of the material on Gödel's theorem and Löb's theorem was adapted from my book, Hamkins (2021b).

26 Common Knowledge

Abstract. Problems and paradoxes of epistemic logic, including the two generals, the problem of common knowledge, the two parents, Cheryl's birthday, the hidden surprise, the blue-eyed islanders, pirate treasure division, the philosopher's ruling council, the coordination paradox, Fitch's paradox, and Cheryl's rational gifts.

Three logicians walk into the Logic Bar.

Waiter Do you all want beer?

First logician I don't know.

Second logician I don't know.

Third logician Yes.

Hilarious! . . . Did you get it?

To explain a joke is seldom funny, of course, but kindly humor me. The joke turns on epistemic logic, the logic of knowledge, for the waiter had asked whether they *all* want beer, so the logicians were answering not only for themselves, but also regarding what they knew about whether the others wanted beer. Because the first logician said, "I don't know," it must have been that she wanted beer, but didn't know if everyone wanted beer. If she herself had not wanted beer, then she would have said simply "No," since she would have known that they didn't all want beer; but it would have been wrong for her to say "Yes" merely because she herself wanted beer, not knowing about the others, since the question is whether they *all* want beer. Similarly, we can tell that the second logician must want beer, since otherwise, he would have said "No," but he doesn't know whether the third logician also wants beer. Finally, the third logician, reasoning just as we have, knows that the other two want beer, and since evidently he also wants beer, he says "Yes," since he now knows that they all want beer.

In this chapter, we shall have a series of fun epistemic logic puzzles, many of which will lead us to further serious issues in epistemic logic.

26.1 The Two-Generals Problem

Let us begin with a classic problem in epistemic logic—the two-generals problem. Two red generals and their armies are situated on mountains overlooking a valley between them. They want to coordinate a joint attack on the blue army in the valley, but they are separated by the terrain and they must coordinate the precise details and timing of the attack in advance. If both attack simultaneously, they will succeed; but if either should delay or attack alone, they will both fail. Meanwhile, the communication channels are difficult and unreliable.

The first red general dispatches a messenger through the dangerous blue valley:

> *We attack at dawn; agreed?*

The message gets through! The second red general replies:

> *Yes, at dawn! Please confirm.*

He needs confirmation, you see, since otherwise, he might think that the first general would think that the original message had not made it. So the first general confirms:

> *Received your message! Ready to go at dawn ... provided we know you get this message.*

This confirmation also needs confirmation, since otherwise, the first general wouldn't know that the second general had received the necessary confirmation, and he would therefore call off the plan. So the second red general verifies that indeed he received it.

> *Got it! We're definitely on, once we know you have received this.*

And so on, back and forth. The brave messengers make it through each time against the odds. At each stage of the process, however, the generals need confirmation that their message was received, so they would know that the other general will undertake the plan, since otherwise, they would have grounds for thinking that they would be attacking alone, a dangerous possibility.

How much confirmation suffices? Can the two generals ever become fully confident in the coordination of their joint plan?

Think about it . . .

Let me argue that in fact, the two generals can never become completely confident, no matter how many messages they send. Clearly, the first message needs confirmation, in order for them to be completely confident, since otherwise the first general could not know that the second general even knows about the plan in the first place. And it is also clear that the reply to this message needs its own confirmation, since otherwise, the second general would worry that his reply hadn't made it through, and in this case, he would conclude that the first general would have no reason to think that his original message was ever received and therefore abandon the plan.

Similarly, if neither general is confident after sending n messages, then they cannot become confident after sending $n + 1$ messages. The reason is that since the last message will not be confirmed, then the confidence should arrive whether or not this final message makes it through. But in this case, it needn't have been sent at all—the effects on communication would be the same—and so they should have already been confident in the plan with only n messages, contrary to our assumption. Thus, by induction, no finite number of messages will suffice.

A slightly different way to argue is this: Suppose toward contradiction that some finite number of messages suffice for the generals to be confident that they have reached agreement on the joint attack plan. Let n be the smallest number of messages that suffice. In particular, the generals should become perfectly confident in their agreement upon exchanging those n messages. But in this case, since the final message will get no confirmation of receipt, the generals should become confident whether or not that last message actually makes it through. But in this case, the last message needn't be sent at all, since the communication effect of a message that gets lost is the same as one that is not sent in the first place. Thus, $n - 1$ messages actually suffice for the generals to achieve complete confidence. This is a contradiction to the minimality definition of n.

The surprising conclusion is that no finite number of messages exchanged back and forth can ensure that the generals have the information that is required for them to be fully confident in the joint attack plan.

Mutual Knowledge Versus Common Knowledge

Suppose that we all know that there is a dangerous crocodile inside the large box at the front of the lecture hall. None of us would elect to open the box since this would release the crocodile and cause chaos, if not death and destruction. We have *mutual knowledge* we all know it individually—that there is a crocodile in the box. Knowing this, however, is not quite the same as knowing that all the others also know this. We would surely behave differently if we didn't know that everyone else knew about the crocodile; we might be anxious, for example, that someone would open the box out of curiosity, without expecting or realizing the danger.

A prudent person would therefore likely announce for all to hear that there is a dangerous crocodile in the box. The aim of such a public announcement would be to create a state of *common knowledge* about the danger—a state where not only do we all know that there is a crocodile in the box, but we also know that we all know it, and we all know that we all know that we all know it, and so on.

The two generals are similarly striving (but failing) to create a state of common knowledge about their plan. When the first message gets through, they both know about the plan, but this is not enough, since the first general needs to know that the second general knows about the plan. And that is not enough since the second general needs to know that the first general knows that the second general knows the plan. Indeed, no finite amount of iterating "I know that you know that ..." will suffice to achieve common knowledge.

Let us try to give a somewhat more formal account. We say that a collection of individuals are in a state of common knowledge of a proposition p if every individual knows p, and also every individual knows that every individual knows p, and every individual knows that every individual knows that every individual knows p, and so on for every finite iteration. Let us write $\Box_a p$ to indicate that individual a knows p. This notation views knowledge as a kind of necessity, in light of many features of knowledge that obey some of the logical principles holding for other concepts of necessity.

In this notation, to say that every individual knows p is to assert the following:

$$\forall a \, \Box_a \, p.$$

To say that every individual knows that every individual knows p is to assert the following:

$$\forall a \, \Box_a \, \forall b \, \Box_b \, p.$$

And so on. The state of common knowledge is reached when we have all the finite iterations, of every finite nested depth:

$$\forall a \, \Box_a \, \forall b \, \Box_b \cdots \forall z \Box_z \, p.$$

Thus, the state of common knowledge is described formally by this infinite scheme of assertions.

The Two-Parents Problem

Consider a contemporary variation of the two-generals problem, which I had experienced myself. When my kids were younger, my wife usually picked up our daughter Hypatia at her school, while I would pick up our son Horatio at his school, which is a different school in a different neighborhood. One day, because of an errand, it was convenient to swap tasks. So I emailed my wife:

`I'll pick up Hypatia today; you get Horatio. Please confirm; otherwise it is as usual.`

She texted back:

Let's do it! Let me know, so I know we're on.

So I left her a voicemail message:

OK, we're on for the swap...as long as I know you get this message.

And she direct-messaged me back:

`Got the message. We're on! But let me know that you get this message so I can count on you.`

And so on.

Just as with the two generals, it seemed that at no stage of the conversation could we know for certain that the other person had all the necessary information to count on the other person making the swap. One cannot count on another person actually reading an email or text, and sending such a message may often be as uncertain as dispatching a horse-mounted messenger through a dangerous enemy valley.

Clearly, the first message needed confirmation, and the reply needed its own confirmation, for otherwise she couldn't have known that I got the confirmation; and then that reply needed to be confirmed, and so on endlessly, just as before.

In the end, despite all our messages, we both did the logical thing: We abandoned the swap idea and just picked up the usual child at the usual school.

Internet Communication Protocols

These problems might seem silly or artificial, but let me try to convince you that they are not. The reason they matter is that essentially the same issues arise in the design of internet communication protocols. When you click on a link on a web page, for example, your computer needs to communicate with other computers in such a way that they can coordinate to send and receive the information. When will a video stream be sent? How will a financial transaction be completed? Your computer needs to be ready to receive information at the right time when it is sent by the other computer, and the nature of this communication is essentially identical in important respects to the issues of the two-generals problem. The two-generals impossibility argument shows that there is no protocol that solves the problem perfectly, and one must accept protocols that compromise and operate with incomplete confidence. And sometimes, as we are all aware, the joint operations fail.

Do Public Announcements Lead to Common Knowledge?

Earlier in this chapter, we mentioned the possibility of achieving common knowledge with a public announcement. When someone announces publicly for all to hear, "There is a dangerous crocodile in the box," then we all seem to know not only that fact, but also the fact that everyone knows it, and the fact that everyone knows that everyone knows it, and so on. Public announcements are often presented as a means to achieve the state of common knowledge.

But is this right? In practice, of course, actual public announcements often do not have this character. A scratchy public announcement in a noisy subway car might be understood by fewer than half the passengers; but even if it is understood by all, they probably do not come to know that everyone knows the information since many of them might reasonably believe that others may not have understood it.

Even with a crystal-clear public announcement, I claim that we might reasonably harbor some doubt about whether everyone was paying attention, and this would prevent us from knowing that everyone knows. And furthermore, even if nobody happened to have such doubts, to achieve common knowledge, we would all need to *know* that nobody had such doubts—otherwise, we would not know that everyone knows that everyone knows the information. But do we actually get this information from a public announcement? For me to know that everyone knows that everyone knows the information, I would have to know that nobody noticed something strange that might have given them reason to think that a particular person might not have been paying attention. Perhaps, for all I know, someone might have noticed someone with eyes closed during the announcement and suspected that they might have missed it—I didn't see anyone with eyes closed, but do I truly know that nobody else saw someone like that? I do not think I do. Even if I paid very close attention, and indeed even if everyone paid very close attention, would I truly know that *everyone* paid enough attention to know that everyone knew that everyone knew the information? I do not think so. And the situation is even more uncertain when it comes to knowing that everyone knows that everyone knows that everyone knows the information.

To my way of thinking, a mere public announcement, even a perfectly clear public announcement that was obviously heard by everyone, does not generally achieve these higher levels of iterated knowledge of knowledge as in our formal account of common knowledge. For this reason, I do not believe that common knowledge can be obtained as cheaply as a public address system. To my way of thinking, public announcements do not generally achieve full common knowledge.

I claim further that there is genuine difficulty with achieving common knowledge even with two people talking to each other directly. Imagine that my wife and I were talking to each other on the phone, instead of exchanging messages, or perhaps even talking in person, face to face. Doesn't the same difficulty arise with our coordination? I would propose to swap the child-pickup duties, and she would agree, but she would need to know that I understood her agreement, so I would need to confirm that somehow, perhaps by nodding, and I would need to know that she noticed that, and so on ad infinitum. Do we actually achieve this infinitely iterated knowledge of confirmation? I do not think so. How is coordination communication ever possible? Can the two generals achieve the necessary joint coordination even if they are speaking face to face?

One answer, of course, is that we never actually achieve the perfect state of common knowledge; rather, we make do with partial knowledge. We might just confirm back and forth once, or perhaps confirm several more times back and forth in critical instances, and then simply proceed with the resulting approximation to common knowledge. In most instances, this amount of knowledge suffices for our purposes.

26.2 Cheryl's Birthday Problem

Several years ago, this logic puzzle from the Singapore Math Olympiad went viral on the internet, stumping millions of people all over the world. Can we solve it? Here it is:

Albert and Bernard are friends with Cheryl, and they want to know when her birthday is. Cheryl provides a list of 10 possible dates:

May 15	June 17	July 14	August 14
May 16	June 18	July 16	August 15
May 19			August 17

Cheryl then tells Albert and Bernard separately the month and the day of her birthday, respectively. The following discussion ensues:

Albert I don't know when Cheryl's birthday is, but I know that Bernard also does not know.

Bernard At first I didn't know Cheryl's birthday, but now I know.

Albert Then I also know when Cheryl's birthday is.

When is Cheryl's birthday? Please try to solve the problem on your own, before continuing. Tens of thousands of people on the internet have done so, and I am confident that with careful thought, you also can solve the problem. Give it a good amount of time and thought; solving the problem is worthwhile.

Think about it...

Welcome back. Did you solve it? Let me explain how I think about it. At first, Albert has only the month and Bernard has only the day, and to make an implicit point explicit: Let us assume that they know this in the state of common knowledge. Albert begins by saying that he doesn't know when Cheryl's birthday is. Well, of course he doesn't know, since all he knows at first is the month, and each of the months has several possible days.

But then Albert says something interesting—namely, he also knows that Bernard doesn't know the birthday either. Now, looking at the chart of possible birthdays, the only way that Bernard could know the exact birthday is if the day number were either 18 or 19, since in each of those cases, there is only one birthday on the list with that number, whereas for each of the other numbers, there are multiple birthdays with that number. So, when Albert says that he knows that Bernard doesn't know, it must be that the information that Albert has is incompatible with Bernard having either 18 or 19. Thus, Albert must not have May or June. So Albert has either July or August.

Bernard says that at first he didn't know Cheryl's birthday, but now he knows. Thus, the date is not 14, since otherwise Bernard wouldn't know the birthday at this point, since that day number occurs in both July and August. So it must be either July 16, August 15, or August 17, and Bernard knows because he must have one of those numbers.

Finally, Albert says that he now knows Cheryl's birthday. This implies that Albert cannot have August, since if he did, he wouldn't have known whether it was August 15 or August 17. So Albert has July, and therefore the birthday must be July 16, which solves the puzzle.

A Variation on Cheryl's Birthday

Let us consider a variation of the puzzle that I have devised. The following weekend, Albert, Bernard, and Cheryl get together again for more conversation. Cheryl privately gives Albert and Bernard each a positive integer, amongst 1, 2, 3, and so on. The following conversation occurs:

Cheryl I gave you each a different positive integer. Whose is larger?

Albert I don't know.

Bernard I don't know either.

Albert I still don't know whose number is larger.

Bernard Alas, I remain in ignorance.

Albert Ah, now that you say that, suddenly I know whose number is larger!

Bernard Really? In that case, I know both numbers!

Albert And now I also know both numbers.

What numbers did she give them?

Think about it...

Let us solve the puzzle together. Albert and Bernard have different positive integers, amongst 1, 2, 3 and so on—their particular numbers could be very large. Furthermore, they both know that information, and they know that the other person knows it, and so on, since it was announced clearly enough by Cheryl. When Albert begins by saying that he doesn't know who has the larger number, it must be that he doesn't have 1, since if he had 1, then he would have known that Bernard's number was larger. But really that is all that we may deduce about his number, since if he had had any number other than 1, he couldn't know whether Bernard's number was smaller or larger.

When Bernard replies that he doesn't know either, then it follows similarly that Bernard doesn't have 1, but also that he doesn't have 2, since he already knows that Albert doesn't have 1. Albert says next that he still doesn't know whose number is larger, which means that Albert must not have 2 or 3, since in those cases, he would have known. And since after this, Bernard announces that he remains in ignorance, Bernard must not have 3 or 4.

Now, suddenly, Albert announces that he knows who has the larger number. How could he know that? Well, he must have either 4 or 5, because if he did, then he would know that Bernard's number is larger, since Bernard has been excluded from 1, 2, 3 and 4 already; but Albert had a number larger than 5, then he couldn't know that Bernard didn't have 5, so he wouldn't know who had the larger number. So Albert must have either 4 or 5, but we do not yet know which.

Meanwhile, finally, Bernard says that he now knows both numbers. For Bernard to know that, he must have one of the numbers 4 or 5, to exclude it as a possibility for Albert, since otherwise he wouldn't know Albert's number. But we have already argued Bernard cannot have 4, so he must have 5. Thus, Albert has 4 and Bernard has 5. When Albert says that he also knows both numbers, it is because he could reason just as we just did.

Hidden Surprise

At a party for our perfectly logical philosophy friends Sheila and Colin, a surprise has been hidden under one of these colored tiles:

Each friend is privately told a piece of information about where the surprise is:

- Sheila knows the *shape* of the tile.
- Colin knows the *color* of the tile.

It is commonly known by all that this and no other information is given.

The following conversation begins:

Host Does either of you know where the surprise is?

...Awkward long silence...

Host Do you know now?

...More awkward silence...

Sheila, Colin (simultaneously) Now I know where it is!

Where is the surprise?

Think about it...

If it were the yellow circle, Colin would have known right away. So it is not yellow. Similarly, if it were square, then Sheila would have known right away. So it isn't square. After the first silence, then, if it were the red circle, then Sheila would have known, since that is the only circle left; and if it were the blue triangle, then Colin would have known, since it would be the only blue shape left. So after the second silence, both of them deduce, as we just have, that it must be the red triangle. The reader is invited to consider several followup questions in the "Questions for Further Thought" at the end of the chapter.

26.3 The Blue-Eyed Islanders

A remote island has 100 perfectly logical inhabitants, with excellent eyesight, and they all know this. Furthermore, they all happen to have vivid, deep blue eyes. However, it is part of their cultural practice never to discuss eye color, and if any of them should come to know that they have blue eyes, then they must leave the island the next dawn in a flashy display visible to all inhabitants. For this reason, there are no mirrors or cameras on the island, and they never view their own reflections. And indeed, none of the islanders knows his or her own eye color.

One day, a trusted friend makes an enjoyable extended visit to the island. At her departure, in front of the entire community assembled to see her off, she remarks, "At least one of you has blue eyes." Exactly 100 days later, it turns out, all the islanders make a big flashy display at dawn and everyone leaves the island.

Why? Can you explain how the islanders all came to know on that day that they had blue eyes?

Think about it...

Let us solve the puzzle together. It will be illuminating to begin with easier instances. Consider first the trivial case where there is only one blue-eyed inhabitant, rather than 100. When the trusted visitor says, "At least one of you has blue eyes," the sole islander naturally deduces that it must be himself, so he performs the flashy departure display at dawn the next morning and exiles himself.

Now consider the case of two islanders, each with blue eyes. Each of them knows that the other has blue eyes, of course, but each is initially unsure of their own eye color. When the visitor remarks that at least one of them has blue eyes, each of them undertakes the following line of thought: If I do not have blue eyes, then the other islander will know that, and since he has just learned that one of us has blue eyes, he will deduce immediately that he has blue eyes himself; in that case, at the next dawn he will make a flashy departure. When no display occurs the next day, both islanders will continue their reasoning: since the other islander didn't make a flashy exile departure, it must be that I myself have blue eyes. Having thus deduced their own blue eye colors, the two islanders will both make a flashy display at the dawn of the second day and leave the island.

In the case of three islanders, each of them will consider the possibility that they do not have blue eyes, and since the other two islander do have blue eyes, in this case each would expect the other two to go through the reasoning process of the two-islander case. Specifically, in this case, each of the others would see one other islander with blue eyes and wonder whether they themselves also have blue eyes; if not, then the other islander would have made the flashy departure on the first dawn, but this doesn't happen (since each actually sees two people with blue eyes), so they would both know that they have blue eyes and make the flashy departure on the second dawn. But that doesn't happen (since all three are waiting to observe this behavior in the other two), so now they all know that they all have blue eyes, and the flashy departures come on the third dawn after the announcement.

And so on for more islanders. The general claim, which I shall argue by induction, is that if there are n blue-eyed islanders and possibly some other non-blue-eyed islanders, then they will all come to know this by the dawn of the nth day after the announcement, putting on the flashy display and departing on the nth day. We have argued already that this is true for $n = 1$. Suppose that this is true for n, and consider now the case of $n + 1$ blue-eyed islanders. Each of them considers that perhaps they do not have blue eyes, in which case there would be n blue-eyed islanders, so they are all expecting possibly to see all the others make the flashy display and depart on the nth day. But when that doesn't happen (since they are all waiting to observe if the others do it), each islander deduces that he must have blue eyes after all. Since each of the $n + 1$ blue-eyed islanders go through this reasoning, independently deducing that they have blue eyes, they all make the flashy display and depart at dawn of day $n + 1$. By induction, therefore, the claim is proved for all n, and in particular, for $n = 100$ as in the puzzle. Thus, we have solved it.

A Subtle Conundrum

Notice that the statement that the visitor makes, "At least one of you has blue eyes," was something that each of the islanders already knows at that time. But how can it be that making a public statement about something that everyone already knows sets into motion the elaborate reasoning process leading to their regrettable exile?

The answer is that it isn't just the fact of that statement that matters for the reasoning process of the islanders, but rather, the fact that each islander knows that the others also know the statement, and they know that everyone knows that everyone knows it. And so on, iterated 100 times. In short, what mattered is that the islanders are in a state of common knowledge about the fact that at least one of them had blue eyes.

26.4 The Pirate-Treasure Division Problem

There is a pirate ship with a crew of fearsome, perfectly logical pirates and a treasure of 100 gold coins to be divided amongst them. How shall they do it?

The pirates have long agreed upon the following pirate-treasure division procedure: The pirates are ordered by rank, with the captain, the first lieutenant, the second lieutenant, and so on down the line; but let us refer to them simply as pirate 1, pirate 2, pirate 3, and so on. Pirate 9 is swabbing the decks in preparation.

For the division procedure, all the pirates assemble on deck, and the lowest-ranking pirate mounts the plank. Facing the other pirates, he proposes a particular division of the gold, with so-and-so many gold pieces to the captain, so-and-so many pieces to pirate 2, and so on. The pirates then vote on the plan, including the pirate on the plank, and if a strict majority of the pirates approve, then the plan is adopted and that is how the gold is divided. But if the pirate's plan is not approved by a majority, then regrettably he must walk the plank into the sea (and his death) and the procedure continues with the next-lowest ranking pirate, who of course has become the lowest-ranking pirate.

Suppose that you are pirate 10. What plan do you propose?

Think about it...

Would you think it is a good idea to propose that you get to keep 94 gold pieces for yourself, with the 6 remaining ones given to a few of the other pirates? In fact, you can propose just such a thing, and if you do it correctly, your plan will pass!

Before explaining why, let me tell you a little more about the pirates. I mentioned that the pirates are perfectly logical, and not only that, they have the common knowledge that they are all perfectly logical. In particular, in their reasoning, they can rely on the fact that the other pirates are logical, and that the other pirates know that they are all logical and that they know *that*, and so on.

Furthermore, it is common knowledge amongst the pirates that they all share the same pirate value system, with the following strictly ordered list of priorities:

A. Most important, stay alive.

B. Get gold—the more the better.

C. Cause the death of other pirates, if possible.

D. All other things being equal, arrange when possible that gold goes to senior pirates.

That is, at all costs, each pirate would prefer to avoid death, and if alive, to get as much gold as possible, but having achieved that, each one would prefer that as many other pirates die as possible (but not so much as to give up even 1 gold coin for additional deaths), and if all other things are equal, each one would prefer that whatever gold was not gotten for herself, that it goes as much as possible to the most senior pirates, for the pirates are, in their hearts, conservative people.

What plan do you propose as pirate 10? The pirates will naturally consider pirate 10's plan in light of the alternative, which will be the plan proposed by pirate 9, which will be compared with the plan of pirate 8, and so on. Thus, it seems we should propagate our analysis from the bottom, working backward from a smaller number of pirates.

One pirate. If there is only one pirate, the captain, then he mounts the plank, and clearly he should propose, "Pirate 1 gets all the gold," and he should vote in favor of this plan, which therefore passes unanimously. So pirate 1 gets all the gold, as anyone would have expected.

Two pirates. If there are exactly two pirates, then pirate 2 will mount the plank, and what will he propose? He needs a majority of the two pirates, which means that he must get the captain to vote for his plan. He is in a truly desperate situation. Clearly, he should propose that all the gold goes to the captain. But will this work? No, indeed. No matter what plan pirate 2 proposes, the captain will vote against it, since if pirate 2 is killed, then the captain will get all the gold anyway, and because of pirate value C, he would prefer that pirate 2 is killed off. So pirate 2's plan will not be approved by the captain, so regrettably, pirate 2 will walk the plank.

Three pirates. If there are three pirates, then what will pirate 3 propose? Well, he needs only two votes, and one of them will be his own. So he must convince either pirate 1 or pirate 2 to vote for his plan. But actually, pirate 2 will have a strong incentive to vote for the plan since otherwise, pirate 2 will be in the situation of the two-pirate case, which ended with pirate 2's death. So pirate 3 can count on pirate 2's vote regardless of the plan details, so pirate 3 will propose: Pirate 3 gets all the gold! This will be approved by both pirate 2 and pirate 3, a majority, so with three pirates, pirate 3 gets all the gold.

Four pirates. Pirate 4 needs to have three votes, so he needs to get two of the others to vote for his plan. He notices that if he is to die, then pirates 1 and 2 will get no gold, so he realizes that if he offers each of them one gold coin, they will prefer that, because of the pirate value system. So pirate 4 will propose to give one gold coin each to pirates 1 and 2, and 98 gold coins to himself. This plan will pass with the votes of pirates 1, 2, and 4.

Five pirates. Pirate 5 needs three votes, including his own. He can effectively buy the vote of pirate 3 with one gold coin, since pirate 3 will otherwise get nothing in the case of four pirates. And he needs one additional vote, that of pirate 1 or 2, which he can get by offering two gold coins. Because of pirate value D, he would prefer that the coins go to the highest-ranking pirate, so he offers the plan: 2 coins to pirate 1; nothing to pirate 2; a single coin to pirate 3; nothing to pirate 4; and 97 coins to himself. This plan will pass with the votes of pirates 1, 3, and 5.

Six pirates. Pirate 6 needs four votes, and he can buy the votes of pirates 2 and 4 with 1 gold coin each, and then 2 gold coins to pirate 3, which is cheaper than the alternatives. So he proposes the following: 1 coin each to pirates 2 and 4; 2 coins to pirate 3; and 96 coins for himself. This proposal passes with the votes of pirates 2, 3, 4, and 6.

Seven pirates. Pirate 7 needs four votes, and he can buy the votes of pirates 1 and 5 with only 1 coin each, since they get nothing in the six-pirate case. By offering 2 coins to pirate 2, he can also get another vote (and he prefers to give the extra gold to pirate 2 than to other pirates in light of the pirate values).

Eight pirates. Pirate 8 needs five votes, and he can buy the votes of pirates 3, 4, and 6 with 1 coin each, and ensure another vote by giving 2 coins to pirate 1, keeping the other 95 coins for himself. With his own vote, this plan will pass.

Nine pirates. Pirate nine needs five votes, and he can buy the votes of pirates 2, 5, and 7 with 1 coin each; and with also 2 coins to pirate 3 and his own vote, the plan will pass.

Ten pirates. In light of the division that we know will be offered by pirate nine, we can now see that pirate 10 can ensure 6 votes by proposing to give one coin each to pirates 1, 4, 6, and 8; 2 coins to pirate 2; and the remaining 94 coins for himself. This plan will pass with those pirates voting in favor (as well as himself), because they each get more gold this way than they would under the plan of pirate 9.

Let me summarize the proposals in the following table, where the nth row corresponds to the proposal of pirate n:

	1	2	3	4	5	6	7	8	9	10
One pirate	100									
Two pirates	$*$	X								
Three pirates	0	0	100							
Four pirates	1	1	0	98						
Five pirates	2	0	1	0	97					
Six pirates	0	1	2	1	0	96				
Seven pirates	1	2	0	0	1	0	96			
Eight pirates	2	0	1	1	0	1	0	95		
Nine pirates	0	1	2	0	1	0	1	0	95	
Ten pirates	1	2	0	1	0	1	0	1	0	94

There are a few things to notice, which we can use to deduce how the pattern will continue. Notice that in each row beyond the third row, the number of pirates that get 0 coins is almost half (the largest integer strictly less than half), exactly one pirate gets 2 coins, and the remainder get 1 coin, except for the proposer himself, who gets all the rest.

This pattern is sustainable so long as there is enough gold to implement it because each pirate can effectively buy the votes of the pirates getting 0 under the alternative plan with one fewer pirate, and this will be at most one less than half of the previous number; then he can buy one more vote by giving 2 coins to one of the pirates who got only 1 coin in the alternative plan; and with his own vote, this will be half plus 1, which is a majority.

We can furthermore observe that by the pirate value system, the 2 coins will always go to either pirate 1, 2, or 3, since one of these will always be the top-ranked pirate having one coin on the previous round. They each cycle through the pattern of 0 coins, 1 coin, and 2 coins in the various proposals.

At least until the gold becomes limited, all the other pirates from pirate 4 onward will alternate between 0 coins and 1 coin with each subsequent proposal, and pirate $n-1$ will always get 0 from pirate n. We can thereby see that the pattern continues upward until at least pirate 199, whose proposal will follow this pattern:

Pirate 199: 1 2 0 0 1 0 1 0 1 0 1 0 1 . . . 1 0 1 0 0.

It is with pirate 199, specifically, that for the first time, it takes all 100 coins to buy the other votes since he must give 98 pirates 1 coin each, and 2 coins to pirate 2 in order to have 100 votes altogether, including his own, leaving no coins for himself.

For this reason, pirate 200 will have a successful proposal since he no longer needs to spend 2 coins for one vote, as the proposal of pirate 199 has 100 pirates getting 0 coins. So pirate 200 can get 100 votes by proposing 1 coin to everyone who would get 0 from 199, plus his own vote, for a majority of 101 votes:

200 pirates: 0 0 1 1 0 1 0 1 0 1 0 . . . 0 1 0 1 1 0.

Pirate 201 also needs 101 votes, which he can get by giving all the zeros of the 200 case 1 coin each, plus his own vote. The unfortunate pirate 202, however, needs 102 votes, and this will not be possible since he has only 100 coins, so pirate 202 will die. The interesting thing to notice next is that pirate 203 will be able to count on the vote of pirate 202 without paying any gold for it, so since he needs only 100 additional votes (after his own vote and pirate 202's vote), he will be able to buy 100 votes for 1 coin each. Pirate 204 will again be 1 coin short, so he will die. Although pirate 205 will be able to count on that 1 additional free vote, this will be insufficient to gain a passing proposal because he will be able to buy 100 votes with the coins, plus his own vote and the free vote of pirate 204, making 102 votes altogether, which is not a majority. Similarly, pirate 206 will fall short, because even with his vote and the free votes of pirate 204 and 205, he will be able to get at most 103 votes, which is not a majority. Thus, pirate 207 will be able to count on the votes of pirates 204, 205, and 206, which with his own vote and 100 more votes (gotten by giving one coin each to the pirates who would otherwise get nothing), he can obtain 104 votes, which is a majority. The reader will investigate further in question 26.6 at the end of the chapter. It is a fun problem to work out. The emerging phenomenon is longer and longer sequences of pirates in a row finding themselves unable to make a winning proposal, and then suddenly a pirate survives by counting on their votes.

Pirate Treasure Division with Very Few Coins

It is also interesting to work out what happens when there is only a small number of coins. For example, if there is only 1 gold coin, then already pirate 4 is unable to make a passing proposal, since he can buy only one other vote, and with his own, this will make only two votes, falling short of a majority. With only 1 coin, pirate 5 will survive by buying a vote from pirate 1 and counting on the vote of pirate 4 and his own vote, for a majority. The reader will investigate further in question 26.7.

Even the case of 0 coins is interesting. In this case, there is no gold to distribute, so the voting is really just about whether the pirate should walk the plank or not. If there is only one pirate, he will live. Pirate 2 will die, since pirate 1 will vote against. But for that reason, pirate 2 will vote in favor of pirate 3, who will live.

The pattern that emerges is:

> lives, dies, lives, dies, dies, dies, lives, dies, dies, dies, dies, dies, dies, dies, lives,...

After each successful proposal, in which the pirates live, there must subsequently be many deaths in a row for the proposal to count on enough votes in support. After each "lives" in the pattern, one must double the length with many "dies" in a row before there is sufficient support for the next pirate to live.

One could recast this version of the process as the Popularity Game, played by high-schoolers trying to get into the in-group. The lowest-ranked student stands before the others and is voted on, whether they should be cast out or kept in the in-group.

The Philosopher's Ruling Council

In the ancient land of Philosophia, there is a ruling council of philosophers, linearly ranked by power and prestige, with various benefits accruing in the order of this rank. Philosopher 1 is the recognized philosopher king, the most powerful, followed by philosopher 2, and so on.

It is time to pick a new council, and according to the long agreed-upon procedure, the lowest-ranked philosopher proposes a new council and ranking. The newly proposed council can in principle include any citizen from Philosophia—candidates are not limited to the current council members, although strangely, it usually happens that the new council is constituted by previous council members. Given the new proposal, the council votes. If a majority approve, then this is the new council and ranking; otherwise, the lowest-ranked philosopher is kicked off the council and the next lowest-ranked philosopher makes a proposal. This process continues until a whole new council and ranking is approved.

I have described the members of the ruling council as philosophers, but let me say a little more about their values and the criteria by which they will be judging the proposals. The fact of the matter is that this is a power-hungry, Machiavellian crowd, all desperate to be the philosopher king. Each council member prefers being on the new council above all other things, and they will never vote in favor of a council that they are not on. Second, being on the proposed council, they would prefer to have as high a rank as possible (that is, a low number—being philosopher 1 is best, 2 is second best, and so forth). Third, given that they will be on the council with a certain rank, they prefer that the council is as small as possible, so as not to have to share power unnecessarily (but superior rank on a larger council is preferred).

Suppose that the council currently has five members. What will be the proposal of the lowest-ranked philosopher, philosopher 5?

Think about it...

Let us solve the puzzle by working backward, as in the pirate treasure division. To know what philosopher 5 will propose, we should already know what philosopher 4 will propose, since that will be the alternative if we vote down philosopher 5's proposal. So let us start with the easiest case—a council with only one philosopher, the philosopher king. In this case, he will be the lowest-ranked philosopher, and he can propose that the new ruling council is the same as the old ruling council, having only himself as a member. This is clearly his most desired situation, according to the philosopher values that we mentioned, so he will vote in favor and the plan will be adopted.

With a council of two philosophers, philosopher 2 must propose a plan that will pass unanimously, since that is the only way to have a majority of two. But philosopher 1 will not approve of any plan except the plan of him being on the council alone since that is what he can attain if philosopher 2's plan is rejected. And we said that philosopher 2 will vote to approve only a plan in which she is on the council. So philosopher 2's proposal, whatever it is, will be rejected, and we will end up eventually with the council of just philosopher 1.

With a council of three philosophers, philosopher 3 needs two votes and will propose a council of two—himself as king and philosopher 2 next. This will be approved by philosophers 2 and 3, since both prefer it to the result of the two-philosopher case. Furthermore, this is the best possible arrangement for philosopher 3 since he is made king on a council of two.

With a council of four philosophers, philosopher 4 needs three votes. She will not get the vote of philosopher 3, who would be very well off if her plan were to be rejected, but she can get approval for a three-member council consisting of philosopher 2, herself, and philosopher 1, in that order. Each of these three will vote to approve since they are better off with this than the alternative. This plan is furthermore the best possible outcome for philosopher 4 because she needs a three-member council, but she must make philosopher 2 better off, so philosopher 2 must become king (or should we say, queen), and then philosopher 4 next, which is optimal, and then philosopher 1, a better position than he would get under the competing proposal of philosopher 3.

Consider now the case of five philosophers on the council. Philosopher 5 needs three votes and therefore will propose a council of size three, and indeed he can propose the council: himself, philosopher 1, and philosopher 3, in that order. Philosopher 5 likes this very well, since he becomes king, and philosopher 1 prefers this to philosopher 4's plan, since he will have rank two instead of rank three. Philosopher 3 also likes this plan since he was not on philosopher 4's proposed council at all.

Let us summarize the proposals as follows:

Size of Council	Proposed New Council and Ranking
1	1
2	Proposal rejected
3	3 2
4	2 4 1
5	5 1 3

The reader will consider larger councils in the questions at the end of the chapter.

26.5 The Color-Matching Game Show

Suppose that you are a contestant on a game show known for its perfectly logical participants. You and another contestant aim to coordinate your actions to win the big prize. Specifically, you want to cooperate so you both announce the same color at the same time.

You and the other contestant are led separately to private rooms, each equipped with a desk and some messaging equipment. The game will proceed in rounds of simultaneous action. On each round, each of you privately chooses either to end the game and officially announce your color or to send a message to the other contestant. The message can be essentially as long as you like, provided that you can type it in a reasonable amount of time. The contestants are separated from each other and neither will know during a round what kind of action the other player will make on that round until after it has occurred. Messages are received just afterward. Each contestant has only one opportunity to end the game and announce a color—make sure to get it right!

If on a given round both contestants end the game and announce exactly the same color, then both win the big prize; but if the colors are different, or if only one player opts to end the game and announce a color that round, then both have lost and the game is over.

Round 1 will begin soon. What will you do?

To get into the right mindset, we might increase the stakes by imagining that it is a matter of life and death—perhaps the two contestants are astronauts stranded in different parts of a rocket floating adrift in space, and they must fire their thrusters at exactly the same time in the same direction or they will die. They need to coordinate that action. The point is that I want to consider the game show situation under the assumption that both players desperately want to coordinate their action.

A Fundamental Problem of Symmetry

It seems clear that neither player should end the game with a color on the first round, since there might be too little basis to expect that the other player will do so with the same color. Rather, the players should use the messaging capability to make a joint plan.

Suppose that you send a message proposing simply:

> *Let's say "Red" next round.*

If the other player sends the same message, then indeed you might win on round 2 this way. But suppose that they do not? Are you still going to say "Red" next round? That would seem risky, and indeed, the ambiguity of this situation makes your message unwise.

To address that, perhaps it would be wiser to send a message like this:

> *I'm going to say "Red" next time, if you tell me that you will, and otherwise I'll say whatever you say.*

Will that work?

To see a problem with it, suppose that the simultaneous incoming message was,

> *I'm going to say "Blue" next time, if you tell me you will, and otherwise I'll say whatever you say.*

What should you do, then? This would be a dangerous situation if each person takes their own message seriously, since if they followed through with the plan, they would say "Blue" and "Red" mismatched.

This is actually a quite general logical problem with the puzzle, and many proposals face a problem when the partner contestant sends exactly the same message, but with a different color. It can be interesting (and difficult) to see how one might resolve this issue.

For example, suppose that we try to get around the problem by saying the following:

> *Let's say "Red" on round 5, if you confirm by then*

This method ultimately has the same problem since the other player might have said the same thing, but with a different color. If one player tries to defer to the other, then the other may also defer at the same time, again preventing agreement.

Perhaps one tries to win the game by establishing leader and follower roles for the players—one player might be adamant and insist that their own plan is the one to be followed, or they might be deferential and insist that the other player should specify the color and round on which to win. The problem arises when two stubborn players both insist on being leader or two deferential players both insist on the other deciding. What to do then?

This fundamental symmetry between the players almost turns into a proof that there can be no deterministic solution of the puzzle, since if both players play optimally, the other player might be led by similar reasoning to play an exactly isomorphic strategy, but perhaps with the colors permuted (since there could be no reason to prefer one particular color over another). If this is right, all messages will be identical, except possibly with different colors, and this would prevent a win.

Breaking the Symmetry

Meanwhile, there are a variety of ways to try to break the symmetry and somehow distinguish one player from the other, causing in effect leader/follower roles, which will enable a color-selection solution.

For example, perhaps it is promising to say:

> *Let's say "Red" on round 57 unless you have made a proposal for a later round, in which case we'll do that.*

If the numbers are large enough, it seems unlikely that both players will be picking the same number, so this strategy will break the symmetry and they can come to agreement.

One might try alternatively to break the symmetry by using other information about the players, such as alphabetical order of their names, their ages, or other characteristics.

Indeed, any asymmetry in the players could in principle be a reason to settle on one player's choice rather than the other's, and thereby come to agreement. But actually the enormous range of ways to break the symmetry is itself a problem, since how can the players settle on just one of them? If we have different plans on how to break the symmetry, then we need to break the symmetry first in order to decide how to break the symmetry.

Randomness

There are other ways to introduce randomness explicitly. One promising way to proceed is to say on the first message,

> *I'm going to say "Red" on the next round, if you also say you will; but if you mention some other color, then (for the first color you mention) I'm going to flip a coin each time between Red and that color to determine which one I'll tell you I'm going to say, until we happen to both propose to say the same color for the subsequent round. And then we shall win.*

This randomness approach gets around the symmetry obstacle since if both players adopt this approach, then it is likely that their colors will match on some round, and then they will win on the next round. In effect, the coin flips allow each player randomly to choose the leader or follower role, so it is likely that they will be compatible at some point.

Computer communication protocols often face similar such coordination problems, where the two machines need to establish a priority between them but it doesn't matter which one it is. Many protocols adopt randomness methods to make these decisions.

Schelling Point

Thomas Schelling realized that many coordination problems can be solved even in the absence of communication. For example, suppose that two people aim to meet on a certain day in New York City, but no plans have been set other than the specific day and they have no means to contact or communicate with one another. Where and when should they meet?

It turns out that many people, when asked this question, propose this solution: meet at noon at the information desk in Grand Central Terminal. Why there? Because Grand Central is a very common, well-known meeting point in the city—it is famous for meeting; it is a beautiful, grand place in which to wait; it is easy to get to from nearly anywhere in the city by many different modes of transportation; and it is convenient for many further prospective destinations. Why at the information booth? Well, if we are meeting in Grand Central, then the information booth in the very center of the grand hall seems the most natural choice. Why noon? Well, if we have to pick a canonical meeting time that would occur to both people, the very middle of the day seems natural. This kind of solution is known as a *Schelling point*—a solution to a coordination problem that people would independently hit on if they were trying to find a solution that many other people would also find.

Is there a Schelling point for where to meet in your city? On your university campus? Is there a Schelling point for the opening move in a chess game? Is there a Schelling point for which play to rehearse for the urgent, surprise performance tonight?

And is there a Schelling point for the game show color coordination game? Yes, indeed there is. At Oxford, we used this game as part of the admissions interview discussions we held for candidates seeking entry to the program in philosophy. More than 90 percent of the candidates used red as their color in the discussion in their initial messages—their proposed messages all involved red more than any other color. This is clearly a Schelling point color, so one would do well in the game to propose to say "Red"; agreement will become much more likely.

26.6 Fitch's Paradox

How does knowledge relate to truth? Let us try to articulate some basic principles. Many philosophers, for example, take it as part of what it means to have knowledge that one can only know things that are true. That is, if you know proposition p, then (amongst other things) p must be true. You might happen to believe some untrue propositions—indeed, you might have very strong grounds for your belief and you might even have what seems irrefutable proof—but if the proposition isn't actually true, then you cannot correctly be said to have knowledge. Let us express this principle as follows:

1. Every known statement is true.

Second, a sort of weak converse of this might also seem reasonable. Namely, every true statement is in principle something that *could* be known. To wit:

2. Every true statement is knowable—it is possible that we could know it.

Frederic Fitch (1963), in a line of reasoning known as *Fitch's paradox*, makes a startling conclusion on the basis of these two assumptions and other ordinary assumptions, such as that knowing the conjunction of A and B entails knowing A and knowing B separately. Fitch argues for the following conclusion:

3. Therefore, every true statement is already known.

Let us give the argument. Suppose toward contradiction that there were a true statement φ that we do not know. So it would be true that "φ and we do not know φ." So by (2), it is possible to know "φ and we do not know φ." In this possibility, we would know φ, and we would know that we do not know φ. But that is impossible since if we know that we do not know φ, then by (1), it would have to be the case that we do not know φ, contrary to the first clause. So there cannot be any true statement that we do not know, establishing (3).

Is it right?

Moore's Paradox

A similar argument arises with *Moore's paradox*, named for George Edward Moore (1873–1958). Namely, suppose that it happens to be raining in Kyoto at the moment, but our friend Miho is unaware of this and in fact doesn't believe it, since the weather prediction was for clear skies. So it would be true for us to make the following assertion:

> It is raining in Kyoto, but Miho doesn't believe it.

But there would be something incoherent, however, about Miho herself saying this. It would be like her saying the following:

> It is raining in Kyoto, but I do not believe this.

It seems incoherent, or possibly self-contradictory, for her to assert this. Thus, there are true statements p for which, for some people, it is incoherent for them straightforwardly to assert p. Similarly, such a statement p would be true but not knowable, for Miho herself could not seem ever to *know* the statements given here.

26.7 Cheryl's Rational Gifts

Let me conclude this chapter as well as the book with a final puzzle with which I shall leave you. Around the time that the Cheryl's birthday puzzle was going viral, I felt that the mathematical community deserved a more difficult and interesting, transfinite version of the problem. So I created *Cheryl's rational gifts*, as follows:

Cheryl Welcome, Albert and Bernard, to my birthday party, and I thank you for your gifts. To return the favor, as you entered my party, I privately made known to each of you a rational number of the form

$$n - \frac{1}{2^k} - \frac{1}{2^{k+r}},$$

where n and k are positive integers and r is a non-negative integer; please consider it my gift to each of you. Your numbers are different from each other, and you have received no other information about these numbers or anyone's knowledge about them beyond what I am now telling you. Let me ask, which of you has the larger number?

Albert I don't know.

Bernard Neither do I.

Albert Indeed, I still do not know.

Bernard And I still don't know either.

Cheryl Well, it is no use to continue that way! I can tell you that no matter how long you continue that back and forth, you shall not come to know who has the larger number.

Albert What interesting new information! But alas, I still do not know whose number is larger.

Bernard And still I do not know either.

Albert I continue not to know.

Bernard I regret that I also do not know.

Cheryl Let me say once again that no matter how long you continue truthfully to tell each other in succession that you do not yet know, you will not know who has the larger number.

Albert Well, thank you very much for saving us from that tiresome trouble! But unfortunately, I still do not know who has the larger number.

Bernard And also I remain in ignorance. However shall we come to know?

Cheryl Well, in fact, no matter how long we three continue from now in the pattern we have followed so far—namely, the pattern in which you two state back and forth that still you do not yet know whose number is larger, and then I tell you yet again that no further amount of that back and forth will enable you to know—then still after as much repetition of that pattern as we can stand, you will not know whose number is larger! Furthermore, I could make that same statement a second time, even after now that I have said it to you once, and it would still be true. And a third and fourth as well! Indeed, I could make that same pronouncement 100 times altogether in succession (counting my first time as amongst the 100), and it would be true every time. And furthermore, even after my having said it altogether 100 times in succession, you would still not know who has the larger number!

Albert Such powerful new information! But I am very sorry to say that still I do not know whose number is larger.

Bernard And I do not know either.

Albert But wait! It suddenly comes upon me after Bernard's last remark, that finally I know who has the larger number!

Bernard Really? In that case, then I also know, and what is more, I know both of our numbers!

Albert Well, now I also know them!

What numbers did Cheryl give Albert and Bernard?

Think about it…

This question will remain a cliff hanger, as far as this book is concerned. If you solve it, then kindly send me an email with your solution to jdh@hamkins.org. I would love to hear about it!

Questions for Further Thought

26.1 Do Albert and Bernard really need full common knowledge of the fact that Albert knows the month (and not the day) and Bernard knows the day (and not the month)? What exact fragment of common knowledge do they need?

26.2 Consider these follow-up questions regarding the "hidden surprise" puzzle with Sheila and Colin.

1. Did Sheila know initially that Colin did not know where the surprise was?
2. Did Colin know initially that Sheila did not know where the surprise was?
3. Did either of them expect the first silence?
4. In that case, what effect on their knowledge did that silence serve? How did they learn anything from it?
5. Did Colin know that Sheila knew that Colin didn't know initially where the surprise was?
6. Did either of them expect the second silence?

26.3 Suppose that the trusted visitor to the island of the blue-eyed islanders had instead announced, "At least a dozen of you have blue eyes." What would happen, and when?

26.4 Suppose that instead of making a public announcement, the visitor to the island of the blue-eyed islanders had privately told each person, individually, that at least one person on the island had blue eyes. Would this fundamentally change the situation?

26.5 Suppose that the island has 63 inhabitants with blue eyes and 37 inhabitants with green eyes, and the trusted visitor announces "at least 1 of you has blue eyes and at least 1 of you has green eyes." If the custom to make the flashy exile display applies whenever an islander comes to know their own eye color, then what will happen, and why?

26.6 Verify the pirate treasure analysis in the text up to 200 pirates and more, and investigate how the pattern of pirate treasure division continues beyond, with very large numbers of pirates.

26.7 What are the successful pirate-treasure division proposals when there is only one gold coin? Two gold coins?

26.8 What is the outcome of the pirate-treasure division protocol when what is being distributed are debts rather than coins? Assume that each pirate wants to live, wants to avoid debt, and so on.

26.9 What is your analysis of the Popularity Game for high schoolers trying to get into the in-group?

26.10 Analyze the philosopher's ruling council proposals for larger councils. Is there a unique determinate answer when there are six philosophers? If not, does this affect how judgements might be made for seven or more philosophers? Can we find reasonable value systems for which the proposals will become determinate?

26.11 What is your assessment of the argument in Fitch's paradox? Do you accept the premises of the argument? Do you accept the conclusion?

26.12 A group of children has just come into the house after having played outside, and it happens that they all have some mud on their foreheads. Each child can see the mud on the other children, but doesn't know whether their own forehead is muddy. The mother remarks, "At least one of you has mud on your forehead." Then, she says, "Raise your hand if at this moment you know that you have mud on your forehead," and she repeats this instruction over and over. What happens?

Further Reading

Vanderschraaf and Sillari (2022). An excellent philosophical account of common knowledge.

Credits

The two-generals problem is a classic of the subject. The history of the "Cheryl's birthday" puzzle can be found on Wikipedia. I had created the hidden surprise logic puzzle and the color coordination game show puzzle to use in admissions interviews for PPE at University College Oxford; they have subsequently appeared in *The Guardian* and on YouTube. I created the image of the blue-eyed islander using DALL·E. The pirate-treasure division problem appears to have been around for some time, and I am unsure of its exact provenance. Ian Stewart wrote a popular 1998 article for *Scientific American* analyzing the patterns that arise when the number of pirates is large in comparison with the number of gold pieces. Many versions of the puzzle, such as that on the "Pirate Game" entry on Wikipedia, are slightly different from mine. For example, many authors count tie votes as success, which leads to different outcomes. I prefer the strict-majority variation, since I find it interesting that one must sometimes use two gold coins to gain the majority, and also because the death of pirate 2 occurs right away in an interesting way, rather than requiring 200 or more pirates as with the plurality version. Another (inessential) difference is that some authors have the captain on the plank first, and then always the highest-ranking pirate making the proposal, rather than the lowest-ranking pirate, with the pirate values similarly swapped to prefer the lowest-ranking. This corresponds simply to inverting the ranking, so it doesn't fundamentally change the results. The pirate image is *Buried Treasure*, by Howard Pyle (public domain), obtained via Wikimedia Commons. My puzzle, "Cheryl's rational gifts," which appeared on my blog at jdh.hamkins.org/transfinite-epistemic-logic-puzzle-challenge/, was inspired by a suggestion of Timothy Gowers to find transfinite versions of the "Cheryl's birthday" puzzle.

References

Aaronson, Scott (1999). *Who Can Name the Bigger Number?* Author blog, Shtetl-Optimized. https://www.scottaaronson.com/writings/bignumbers.html.

Archimedes (1897 [3rd century BC]). *The Sand Reckoner.* Translated by Thomas Heath.

Archimedes (2009). *The Works of Archimedes: Edited in Modern Notation with Introductory Chapters.* Ed. by Thomas L. Heath. Cambridge Library Collection - Mathematics. Cambridge University Press. DOI:10.1017/CBO9780511695124.

Aristotle (n.d.). *Physics, Book III.* The Logic Museum. http://www.logicmuseum.com/wiki/Authors/Aristotle/physics/liber3.

Bell, John L. (2021). "The Axiom of Choice." In: *The Stanford Encyclopedia of Philosophy.* Ed. by Edward N. Zalta. Winter 2021. Metaphysics Research Lab, Stanford University.

Berkeley, George (1734). *A Discourse Addressed to an Infidel Mathematician.* The Strand. https://en.wikisource.org/wiki/The_Analyst:_a_Discourse_addressed_to_an_Infidel_Mathematician.

Bolzano, Bernard (1950 [1851]). *Paradoxes of the Infinite.* English translation, available at https://archive.org/details/dli.ernet.503861. Routledge Revivals. https://archive.org/details/dli.ernet.503861.

Chalons, Christophe (2007). *Celui-là, il Vous Plaira!* Vanilla online discussion forum. https://les-mathematiques.net/vanilla/discussion/352935/celui-la-il-vous-plaira.

Cohen, Paul J. (1963). "The Indendependence of the Continuum Hypothesis." *Proc. Nat. Acad. Sci. U.S.A.* 50, pp. 1143–1148.

Conway, J. H. (2000). *On Numbers and Games.* Ak Peters Series. Taylor & Francis.

Conway, J. H. and R. Guy (2012). *The Book of Numbers.* Springer New York.

Dauben, Joseph W. (1979). *Georg Cantor: His Mathematics and Philosophy of the Infinite.* Cambridge, MA: Harvard University Press.

Dauben, Joseph W. (2004). "Topology: Invariance of Dimension." In: *Companion Encyclopedia of the History and Philosophy of the Mathematical Sciences.* Ed. by Ivor Grattan-Guiness. Vol. 2. Routledge, pp. 939–946. DOI:10.4324/9781315541976.

Dedekind, Richard (1888). "Was Sind und Was Sollen die Zahlen? (What are Numbers and What Should They Be?)" Available in Ed. by William B. Ewald. *From Kant to Hilbert: A Source Book in the Foundations of Mathematics*, Vol. 2, 1996, Oxford University Press, pp. 787–832.

Ehrlich, Philip (2012). "The Absolute Arithmetic Continuum and the Unification of All Numbers Great and Small." *Bulletin of Symbolic Logic* 18.1, pp. 1–45. DOI:10.2178/bsl/1327328438.

Erhardt, Rudolf von and Erika von Erhardt-Siebold (1942). "Archimedes' Sand-Reckoner: Aristarchos and Copernicus." *Isis* 33.5, pp. 578–602. http://jstor.org/stable/330667.

Evans, C. D. A. and Joel David Hamkins (2014). "Transfinite Game Values in Infinite Chess." *Integers* 14, Paper No. G2, 36. DOI:10.5281/zenodo.10454032.

Fitch, Frederic B. (1963). "A Logical Analysis of Some Value Concepts." *Journal of Symbolic Logic* 28, pp. 135–142. DOI:10.2307/2271594.

Frege, Gottlob (1893/1903). *Grundgesetze der Arithmetic, Band I/II*. Verlag Herman Pohle.

Frege, Gottlob (1968 [1884]). *The Foundations of Arithmetic: A Logico-Mathematical Enquiry into the Concept of Number*. Translated by J. L. Austin, *Die Grundlagen der Arithmetik*, Breslau, Koebner, 1884. Parallel German and English text. Northwestern University Press.

Frege, Gottlob (2013). *Basic Laws of Arithmetic. Derived Using Concept-Script. Vol. I, II*. Translated and edited by Philip A. Ebert and Marcus Rossberg, with Crispin Wright. Oxford University Press, xl; xxxii+253 pp. (Vol. I); xvi+285 pp. (Vol. II); A42 pp.; I11.

Freiling, Chris (1986). "Axioms of Symmetry: Throwing Darts at the Real Number Line." *Journal of Symbolic Logic* 51.1, pp. 190–200. DOI:10.2307/2273955.

Friedman, Harvey (2002). "Philosophical Problems in Logic." seminar notes at the Princeton Philosophy Department, 107 pages. https://u.osu.edu/friedman.8/files/2014/01/Princeton532-1pa84c4.pdf.

Galileo Galilei (1914 [1638]). *Dialogues Concerning Two New Sciences*. Translated from the Italian and Latin by Henry Crew and Alfonso de Salvio. Online Library of Liberty. Macmillan. https://oll.libertyfund.org/title/galilei-dialogues-concerning-two-new-sciences.

Goodstein, R. L. (1944). "On the Restricted Ordinal Theorem." *Journal of Symbolic Logic* 9.2, 33–41. DOI:10.2307/2268019.

Goos, Demian Nahuel (2025). "The Grandeur and the Sorrow Behind the Birth of Set Theory, and the Emergence of the Dedekind Letters." *Phlogiston, Journal for History and Philosophy of Science and Technology* 33, pp. 39–92.

Gowers, John (2020). *How Big Is a Trillion Anyway?* blog post. https://johngowe.rs/blog/2020/04/17/how-big-is-a-trillion-anyway/ (version 27 January 2023).

Graham, R. L. and B. L. Rothschild (1971). "Ramsey's Theorem for *n*-Parameter Sets." *Transactions of the American Mathematical Society* 159, pp. 257–292. http://www.jstor.org/stable/1996010.

Hamkins, Joel David (2002). "Infinite Time Turing Machines." *Minds and Machines* 12.4. Special issue devoted to hypercomputation, pp. 521–539. DOI:10.1023/A:1021180801870. arXiv:math/0212047.

Hamkins, Joel David (2011). *Is Infinity an Odd or Even Number?* Mathematics Stack Exchange answer. https://math.stackexchange.com/a/49046/413.

Hamkins, Joel David (2012). *Concerning Proofs From the Axiom of Choice that* $\mathbb{R}^3$ *Admits Surprising Geometrical Decompositions: Can We Prove There Is No Borel Decomposition?* MathOverflow question. https://mathoverflow.net/q/93601 (version 09 April 2012).

Hamkins, Joel David (2013). *Largest-Number Contest: What Is the Largest Number That You Can Describe on an Index Card?* http://jdh.hamkins.org/largest-number-contest/.

Hamkins, Joel David (2015). "Is the Dream Solution of the Continuum Hypothesis Attainable?" *Notre Dame Journal of Formal Logic* 56.1, pp. 135–145. DOI:10.1215/00294527-2835047. arXiv:1203.4026. http://jdh.hamkins.org/dream-solution-of-ch.

Hamkins, Joel David (2016a). *A Lifetime of Hot Air*. Mathematics and Philosophy of the Infinite. http://jdh.hamkins.org/a-lifetime-of-hot-air/ (version 14 November 2016).

Hamkins, Joel David (2016b). *On Number Sense—Would One Day of NYC Coffee Fill the Statue of Liberty? And Other Fun Questions ...* Mathematics and Philosophy of the Infinite. http://jdh.hamkins.org/on-number-sense/ (version 08 November 2016).

Hamkins, Joel David (2020). *Proof and the Art of Mathematics*. Winner of MAA 2024 Daniel Solow Author's Award. MIT Press. https://mitpress.mit.edu/books/proof-and-art-mathematics.

Hamkins, Joel David (2021a). *An Infinite Game Possibly Due to Ernst Specker*. MathOverflow answer. https://mathoverflow.net/q/401136 (version 06 August 2021).

Hamkins, Joel David (2021b). *Lectures on the Philosophy of Mathematics*. MIT Press. https://mitpress.mit.edu/books/lectures-philosophy-mathematics.

Hamkins, Joel David (2021c). *Proof and the Art of Mathematics: Examples and Extensions*. MIT Press. https://mitpress.mit.edu/books/proof-and-art-mathematics-1.

Hamkins, Joel David (2021d). *The Chocolatier's Game: Can the Glutton Win with a Restricted Form of Strategy?* MathOverflow question. https://mathoverflow.net/q/401151.

Hamkins, Joel David (2022). *Zorn's Lemma: Old Friend or Historical Relic?* MathOverflow answer. https://mathoverflow.net/q/436348 (version 11 December 2022).

Hamkins, Joel David (2024a). *A Panorama of Logic*. book in preparation, currently 425 pages, serialized on https://www.infinitelymore.xyz/s/panorama-of-logic.

Hamkins, Joel David (2024b). "How the Continuum Hypothesis could have been a Fundamental Axiom." *Journal for the Philosophy of Mathematics* 1, pp. 113–126. DOI:10.36253/jpm-2936. arXiv:2407.02463.

Hamkins, Joel David (2024c). *What is the Infinite?* Invited essay for the collection Infinity Essays, Dr. Gindi, editor. https://www.dr-gindi.com/essays/what-is-the-infinite.

Hamkins, Joel David (2025a). *Infinite Games, Frivolities of the Gods*. book in preparation, 376 pages, Infinitely More. https://www.infinitelymore.xyz/s/infinite-games.

Hamkins, Joel David (2025b). "Nonlinearity and Illfoundedness in the Hierarchy of Large Cardinal Consistency Strength." *Monatshefte für Mathematik* 208. (also appeared on Qeios, 2022, https://doi.org/10.32388/T63YDP), pp. 687–728. DOI:10.1007/s00605-025-02082-1.

Hamkins, Joel David (2025c). "The Nearly Perfect Predictor Theorem." *Notices of the American Mathematical Society* 72.3. Short Stories column, pp. 308–309. DOI:10.1090/noti3098.

Hamkins, Joel David (2026a). "A Potentialist Conception of Ultrafinitism." *Philosophia Mathematica*. Adapted from my talk at the Ultrafinitism Conference at Columbia University, April 2025. DOI:10.1093/philmat/nkag010. arXiv:2512.06564.

Hamkins, Joel David (2026b). "Throwing Darts at the Real Line." *Notices Amer. Math. Soc.* 73.4. Short Stories column. DOI:10.1090/noti3315.

Hamkins, Joel David and Øystein Linnebo (2022). "The Modal Logic of Set-theoretic Potentialism and the Potentialist Maximality Principles." *Review of Symbolic Logic* 15.1, pp. 1–35. DOI:10.1017/S1755020318000242. arXiv:1708.01644.

Hamkins, Joel David and Wojciech Aleksander Wołoszyn (2024). "Modal Model Theory." *Notre Dame Journal of Formal Logic* 65.1, pp. 1–37. DOI:10.1215/00294527-2024-0001.

Hardin, Christopher S. and Alan D. Taylor (2008a). "A Peculiar Connection Between the Axiom of Choice and Predicting the Future." *Amer. Math. Monthly* 115.2, pp. 91–96. DOI:10.1080/00029890.2008.11920502.

Hardin, Christopher S. and Alan D. Taylor (2008b). "An Introduction to Infinite Hat Problems." *Math. Intelligencer* 30.4, pp. 20–25. DOI:10.1007/BF03038092.

Hardy, G. H. (1910). *Orders of Infinity, the 'Infinitärcalcül' of Paul Du Bois-Reymond.* Cambridge University Press. https://archive.org/embed/ordersofinfinity00harduoft.

Hilbert, David (2013). *David Hilbert's Lectures on the Foundations of Arithmetic and Logic, 1917–1933*. Vol. 3. David Hilbert's Lectures on the Foundations of Mathematics and Physics 1891–1933. Edited by William Ewald, Wilfried Sieg, and Michael Hallett, in collaboration with Ulrich Majer and Dirk Schlimm. Springer-Verlag, Berlin, pp. xxv+1062. DOI:10.1007/978-3-540-69444-1.

Hoffman, Hannah (2021). *Plenitudinous Primes (A Song About the Infinitude of Prime Numbers)*. YouTube music video. lyrics by Joel David Hamkins. https://youtu.be/WEyEpwAeaaI (version 19 October 2021).

Hofstadter, Douglas (1985). *Metamagical Themas*. Basic Books.

Horsten, Leon and Philip D. Welch (2013). "The Aftermath." *The Mathematical Intelligencer* 35, pp. 16–20. DOI:10.1007/s00283-012-9353-5.

Houston-Edwards, Kelsey (2017a). *How Infinity Explains the Finite*. PBS Infinite Series, YouTube video. https://youtu.be/oBOZ2WroiVY.

Houston-Edwards, Kelsey (2017b). *Kill the Mathematical Hydra*. PBS Infinite Series, YouTube video. https://youtu.be/uWwUpEY4c8o.

Huggett, Nick (2019). "Zeno's Paradoxes." In: *The Stanford Encyclopedia of Philosophy*. Ed. by Edward N. Zalta. Winter 2019. Metaphysics Research Lab, Stanford University. https://plato.stanford.edu/archives/win2019/entries/paradox-zeno/.

Hume, David (1739). *A Treatise of Human Nature*. 1896 reprint from the original edition in three volumes and edited, with an analytical index, by L. A. Selby-Bigge, M.A. Clarendon Press.

Huntington, Edward V. (1903). "Complete Sets of Postulates for the Theory of Real Quantities." *Transactions of the American Mathematical Society* 4.3, pp. 358–370. http://www.jstor.org/stable/1986269.

Isaacson, Daniel (2011). "The Reality of Mathematics and the Case of Set Theory." In: *Truth, Reference, and Realism*. Ed. by Zsolt Novak and Andras Simonyi. Central European University Press, pp. 1–76.

Jonsson, M. and J. Wästlund (1998). "Partitions of $\mathbb{R}^3$ into Curves." *Mathematica Scandinavica* 83.2, pp. 192–204. http://www.jstor.org/stable/24493132.

Juster, Norton (1961). *The Phantom Tollbooth*. illustrations by Jules Feiffer. Epstein & Carroll.

Keisler, H. Jerome (2000). *Elementary Calculus: An Infinitesimal Approach*. Earlier editions 1976, 1986 by Prindle, Weber, and Schmidt; free electronic edition available. https://www.math.wisc.edu/~keisler/calc.html.

Kirby, Laurie and Jeff Paris (1982). "Accessible Independence Results for Peano Arithmetic." *Bulletin of the London Mathematical Society* 14.4, pp. 285–293. DOI:10.1112/blms/14.4.285.

Knuth, D. E. (1974). *Surreal Numbers*. Addison-Wesley, pp. iii+119.

König, Julius (1906). "Sur la Théorie des Ensembles." *Comptes Rendus Hebdomadaires des Séances de l'Académie des Sciences* 143, pp. 110–112.

Krantz, Steven G. (2002). "The Axiom of Choice." In: *Handbook of Logic and Proof Techniques for Computer Science*. Birkhäuser Boston, pp. 121–126. DOI:10.1007/978-1-4612-0115-1_9.

Kuperberg, Denis (2013). *Probabilities in a Riddle Involving Axiom of Choice*. MathOverflow question. https://mathoverflow.net/q/151286.

Laraudogoitia, J. P. (1996). "A Beautiful Supertask." *Mind* 105.417, pp. 81–84. https://www.jstor.org/stable/2254538.

Littlewood, J. E. (1953). *A Mathematician's Miscellany*. London: Methuen.

Madore, David (2009). *Le Docteur No Continue ses Méfaits*. blog post. http://www.madore.org/~david/weblog/d.2009-05-16.1643.html.

Mandelbrot, Benoit (1967). "How Long Is the Coast of Britain? Statistical Self-Similarity and Fractional Dimension." *Science* 156.3775, pp. 636–638. DOI:10.1126/science.156.3775.636. https://www.science.org/doi/abs/10.1126/science.156.3775.636.

Mandelbrot, Benoit B. (1982). *The Fractal Geometry of Nature*. W. H. Freeman and Co., San Francisco, CA, pp. v+460.

Moore, Gregory H. (1982). *Zermelo's Axiom of Choice*. Vol. 8. Studies in the History of Mathematics and Physical Sciences. Its origins, development, and influence. Springer-Verlag, New York, xiv+410 pp. (1 plate). DOI:10.1007/978-1-4613-9478-5.

Naslund, Eric (2013). *Can an Infinite Number of Mathematicians Guess the Number in a Box with Only One Error?* MathOverflow answer. https://mathoverflow.net/q/152883.

Richardson, Lewis Fry (1922). *Weather Prediction by Numerical Processes*. Boston: Cambridge University Press. https://archive.org/details/weatherpredictio00richrich.

Rubin, Herman and Jean E. Rubin (1985). *Equivalents of the Axiom of Choice, II*. Vol. 116. Studies in Logic and the Foundations of Mathematics. North-Holland Publishing Co., Amsterdam, pp. xxviii+322.

Russell, B. (1907). "On Some Difficulties in the Theory of Transfinite Numbers and Order Types." *Proceedings of the London Mathematical Society* 4, pp. 29–53. DOI:10.1112/plms/s2-4.1.29.

Russell, B. and G. H. Moore (2015). *The Collected Papers of Bertrand Russell, Volume 5: Toward Principia Mathematica, 1905–08*. The Collected Papers of Bertrand Russell. Taylor & Francis. https://books.google.com/books?id=L-8vCgAAQBAJ.

Russell, Bertrand (1903). *The Principles of Mathematics*. Free online editions available in various formats at https://people.umass.edu/klement/pom.

Russell, Bertrand (1959). *My Philosophical Development*. Routledge Classics edition. Routledge.

Stoa, Ryan (2019). "The Coastline Paradox." *Rutgers University Law Review* 72.2, pp. 351–400. DOI:10.2139/ssrn.3445756. https://ssrn.com/abstract=3445756.

Swain, Gordon and Thomas Dence (1998). "Archimedes' Quadrature of the Parabola Revisited." *Mathematics Magazine* 71.2, pp. 123–130. http://www.jstor.org/stable/2691014.

Thomson, J. F. (October 1954). "Tasks and Super-Tasks." *Analysis* 15.1, pp. 1–13. DOI:10.1093/analys/15.1.1. https://academic.oup.com/analysis/article-pdf/15/1/1/405745/15-1-1.pdf.

Tomkowicz, Grzegorz and Stan Wagon (2016). *The Banach–Tarski Paradox*. 2nd ed. Encyclopedia of Mathematics and its Applications. Cambridge University Press.

Vanderschraaf, Peter and Giacomo Sillari (2022). "Common Knowledge." In: *The Stanford Encyclopedia of Philosophy*. Ed. by Edward N. Zalta and Uri Nodelman. Fall 2022. Metaphysics Research Lab, Stanford University.

Index

Index references to entire chapters are in bold, references involving direct quotations are italic (*254*), and references to the chapter-end questions are in a terminal font (`127`).

About the Author

I am both mathematician and philosopher, undertaking research in all parts of mathematical and philosophical logic, with a focus on the mathematics and philosophy of the infinite. Perhaps I suffer a little from an identity crisis regarding the question of whether I am truly a philosopher or a mathematician—my original training is mathematics, after all, but over the years I seem to have turned myself into a philosopher as well. I reject any suggestion that one must choose between the subjects, and my work lies in the generous region between mathematics and philosophy. Indeed, my favorite situation arises when a philosophical question gives rise to mathematical analysis, whose resolution raises new questions informing and deepening the philosophical discussion. I have been around that cycle many times.

In my book *Lectures on the Philosophy of Mathematics* (MIT Press, 2021), the basis for my lectures at the University of Oxford, I strive for a fresh approach to the philosophy of mathematics, treating philosophical issues as they arise organically in mathematics, grounded in mathematical contexts and concerns. Meanwhile, in *Proof and the Art of Mathematics* (MIT Press, 2020), I aim to teach aspiring mathematicians how to write proofs and think like a mathematician—the book has just won the 2024 Daniel Solow Author's Award from the Mathematical Association of America.

I am serializing all my current book projects on my substack Infinitely More (infinitelymore.xyz), including *A Panorama of Logic*, a new introduction to logic that is suitable for students in mathematics, philosophy, and computer science, as well as *Ten Proofs of Gödel Incompleteness*, covering all aspects of this remarkable result; and *Infinite Games: Frivolities of the Gods*, my introduction to the emerging subject of infinite games, including infinite chess, infinite draughts, infinite Hex, infinite Wordle, and more.

You can find a lot of my mathematical and philosophical writing online. I have posted well over one thousand mathematical arguments on MathOverflow, the online Q&A site for research-level mathematics questions, and I have been the top-rated user on that site by reputation score continuously since 2010, amongst tens of thousands of mathematicians, including some of the most famous mathematicians in the world.

I am currently the John Cardinal O'Hara Professor of Logic at the University of Notre Dame, as well as Concurrent Professor of Mathematics. Previously, I was Professor of Logic in the Faculty of Philosophy at the University of Oxford and the Sir Peter Strawson Fellow at University College, Oxford, and I maintained an affiliation with the Mathematical Institute in Oxford. Before Oxford, I held a longstanding appointment at the City University of New York, rising eventually to Distinguished Professor, and I have held visiting positions over the years at New York University, Carnegie Mellon University, Kobe University, University of Amsterdam, University of Münster, University of California at Berkeley, and elsewhere. My 1994 PhD in mathematics was from the University of California at Berkeley, after an undergraduate degree in mathematics at the California Institute of Technology.

Joel David Hamkins
John Cardinal O'Hara Professor of Logic, University of Notre Dame

jdhamkins@nd.edu
Substack: https://www.infinitelymore.xyz
Blog: https://jdh.hamkins.org
MathOverflow: https://mathoverflow.net/users/1946/joel-david-hamkins
X.com: @JDHamkins
Bluesky: @joeldavidhamkins.bsky.social

Publisher contact:
The MIT Press
Massachusetts Institute of Technology
77 Massachusetts Avenue, Cambridge, MA 02139
mitpress.mit.edu

EU Authorised Representative:
Easy Access System Europe, Mustamäe tee 50,
10621 Tallinn, Estonia
gpsr.requests@easproject.com

Printed by Integrated Books International,
United States of America